Intelligent Systems, Control and Automation: Science and Engineering

Volume 74

More information about this series at http://www.springer.com/series/6259

Simon Peter van Rysewyk · Matthijs Pontier
Editors

Machine Medical Ethics

Springer

Editors
Simon Peter van Rysewyk
Graduate Institute of Humanities in Medicine
Taipei Medical University
Taipei
Taiwan

and

Department of Philosophy
School of Humanities
University of Tasmania
Hobart
Australia

Matthijs Pontier
The Centre for Advanced Media Research
VU University Amsterdam
Amsterdam
The Netherlands

ISSN 2213-8986 ISSN 2213-8994 (electronic)
ISBN 978-3-319-08107-6 ISBN 978-3-319-08108-3 (eBook)
DOI 10.1007/978-3-319-08108-3

Library of Congress Control Number: 2014947388

Springer Cham Heidelberg New York Dordrecht London

Printed on acid-free paper

Springer is part of Springer Science+Business Media (www.springer.com)

Preface

Machines are occupying increasingly visible roles in human medical care. In hospitals, private clinics, care residences, and private homes, machines are interacting in close proximity with many people, sometimes the most vulnerable members of the human population. Medical machines are undertaking tasks that require interactive and emotional sensitivity, practical knowledge of a range of rules of professional conduct, and general ethical insight, autonomy, and responsibility. They will be working with patients who are in fragile states of health, or who have physical or cognitive disabilities of various kinds, who are very young or very old. The medical profession has well-defined codes of conduct for interacting with patients, in relation to minimizing harm, responsible and safe action, privacy, informed consent, and regard for personal dignity.

Although there is general agreement in the field of machine ethics that medical machines ought to be ethical, many important questions remain. What ethical theory or theories should constrain medical machine conduct? Is theory even necessary? What implementation and design features are required in medical machines? In what specific situations will it be necessary for machines to share praise or blame with humans for the ethical consequences of their decisions and actions? Are there medical decisions for which machine support is necessary? These questions are truly twenty-first century challenges, and for the first time are addressed in detail in this edited collection.

The collection is logically organized in two parts. The essays in Part I address foundational questions concerning machine ethics and machine medical ethics ("An Overview of Machine Medical Ethics"–"Moral Ecology Approaches to Machine Ethics"). Part II focuses on contemporary challenges in machine medical ethics, and include three sections: Justice, Rights, and the Law ("Opportunity Costs: Scarcity and Complex Medical Machines"–"Machine Medical Ethics and Robot Law: Legal Necessity or Science Fiction?"), Decision-Making, Responsibility, and Care ("Having the Final Say: Machine Support of Ethical Decisions of Doctors"–"Machine Medical Ethics: When a Human Is Delusive but the Machine Has Its Wits About Him"), and Technologies and Models ("ELIZA Fifty Years Later: An Automatic Therapist Using Bottom-Up and Top-Down

Approaches"–"Ethical and Technical Aspects of Emotions to Create Empathy in Medical Machines"). The collection Epilogue is an ethical dialog between a researcher and a visual artist on machine esthetic understanding.

In "An Overview of Machine Medical Ethics", Tatjana Kochetkova suggests machine roles in medicine be limited to medical cases for which professional codes of medical conduct already exist. In such "consensus cases," machine algorithms in operant medical machines should be either top-down, bottom-up, or mixed (top-down-bottom-up). Kochetkova cautiously reasons that it is premature to accord medical machines full ethical status. Instead, prudence suggests they be designed as explicit, but not full, ethical agents by humans.

Oliver Bendel ("Surgical, Therapeutic, Nursing and Sex Robots in Machine and Information Ethics") attempts to shake loose the nature of machine medical ethics by classifying medical machines according to context (surgery, therapy, nursing, and sex), function, and stage of development. Bendel ponders the subfield of machine medical ethics in relation to its parent disciplines machine ethics and applied ethics, and asks whether machine medical ethics can function independently of these fields. Bendel argues that, in the best ethical case, a medical machine ought to interact with humans in order to respect and preserve their autonomy.

Mark Coeckelbergh ("Good Healthcare Is in the "How": The Quality of Care, The Role of Machines, and the Need for New Skills") investigates whether machines threaten or enhance good health care. He argues that "good health care" relies on expert know-how and skills that enable caregivers to carefully engage with patients. Evaluating the introduction of new technologies such as robots or expert medical machines then requires us to ask how the technologies impact on the "know-how" expertise of caregivers, and whether they encourage a less careful way of doing things. Ultimately, Coeckelbergh thinks machines require new skills to handle the technology but also new know-how to handle people: knowing how to be careful and caring with the technology.

In "Implementation Fundamentals for Ethical Medical Agents", Mark R. Waser identifies some broad guidelines for the implementation of ethics in medical machines while acknowledging current machine limitations. All ethical machines need top-down medical decision-making rules and bottom-up methods to collect medical data, information, and knowledge as input to those rules, codified methods to determine the source, quality and accuracy of that input, and methods to recognize medical situations beyond machine expertise and which require specialist human intervention. Waser thinks correct codification and documentation of the processes by which each medical decision is reached will prove to be more important than the individual decisions themselves.

In "Towards a Principle-Based Healthcare Agent", Susan Leigh Anderson and Michael Anderson present a top-down method for discovering the ethically relevant features of possible actions that could be used by a machine as *prima facie* duties to either maximize or minimize those features, as well as decision principles that should be used to influence its behavior. This deontic approach is challenged by Gary Comstock and Joshua Lucas in "Do Machines Have Prima Facie Duties?."

Among several arguments Comstock and Lucas present against the Andersons and their *prima facie* method, they argue that such duties do not uniquely simulate the complexities of ethical decision-making. To substantiate this claim, Comstock and Lucas propose an act-utilitarian alternative, they call Satisficing Hedonistic Act Utilitarianism (SHAU). They show that SHAU can engage in ethical decision-making just as sophisticated as *prima facie* based ethical deliberation, and can produce the same verdict as a *prima facie* duty-based ethic in the medical case investigated by the Andersons.

In contrast to the approach taken in the preceding chapters, the next three chapters argue against the idea of a single theoretical machine ethic and for the idea that hybrid top-down-bottom-up approaches offer a more promising ethical line ("A Hybrid Bottom-Up and Top-Down Approach to Machine Medical Ethics: Theory and Data" and "Moral Ecology Approaches to Machine Ethics"). Simon Peter van Rysewyk and Matthijs Pontier ("A Hybrid Bottom-Up and Top-Down Approach to Machine Medical Ethics: Theory and Data") describe an experiment in which a machine (Silicon Coppélia) run on a hybrid ethic combining utilitarianism, deontology, and case-based reasoning matches in its own actions, the respective acts of human medical professionals in six clinical simulations. Christopher Charles Santos-Lang ("Moral Ecology Approaches to Machine Ethics") makes an interesting point that the brains of human beings are "hybrids individually," by which he means that living brains can adapt our deliberations and judgments to present circumstances in contrast to ecosystem approaches to ethics, which promote hybridization across, rather than within, individuals. Santos-Lang urges, we design and build diverse teams of machines to simulate the best human teams, instead of mass-producing identical machines to simulate the best individual human.

Adam Henschke begins Part II Contemporary Issues in Machine Medical Ethics (Responsibility, Decision-Making and Care) ("Opportunity Costs: Scarcity and Complex Medical Machines"). Future medical machines that prioritize health care only for a minority of patients to the disadvantage of a majority is ethically unjustified, according to Henschke, especially when resources are scarce. Instead, in a depressed global economy, optimizing health care outcomes requires funding increases for existing health care resources, such as nurses, nursing homes, and family that provide care to their loved ones, rather than mass-producing expensive medical machines that may ultimately serve only the very rich.

In "The Rights of Machines: Caring for Robotic Care-Givers" entitled, "Rights for Robots?—Caring for Robotic Care-Givers," David J. Gunkel ponders the question of "machine rights" for health care robots. Gunkel identifies two "machine rights" options: health care robots are nothing more than instrumental tools and accordingly deserve no legal rights; health care robots are valued domestic companions and deserve at least some legal protections. Since each option turns out to have problems, Gunkel urges that the question of "machine rights" be taken more seriously by society.

Are medical machines liable for their actions and mistakes, as are "natural humans"? Addressing this question in "Machine Medical Ethics and Robot Law:

Legal Necessity or Science Fiction?", Rob van den Hoven van Genderen predicts that new legal amendments will enter existing law to represent intelligent machines but only on behalf of a real legal actor, a natural human being. Since machines are best viewed as our assistants, workers or servants, they do not qualify as natural persons, and ought never to have full human rights and obligations. According to van den Hoven van Genderen, the legal system is under human control, and cannot ever be shared with machines.

Beginning the next section in Part III, Decision-Making, Responsibility, and Care, Julia Inthorn, Rudolf Seising, and Marco E. Tabacchi propose that Serious Games machines can share ethical responsibility with human health care professionals in solving medical dilemmas ("Having the Final Say: Machine Support of Ethical Decisions of Doctors"). The authors show that Serious Games improve upon current machines in clinical decision-making because they can integrate both a short and long perspective and enable learning with regard to bottom-up decision processes as well as top-down rules and maxims. Though there is a reluctance to use machine support in medicine, the possibilities of experiential learning ought to be considered an important aspect of behavioral change that could be used to improve ethical decision-making in medicine. The authors also provide an informative historical overview of decision support systems in medicine.

What are the prospects of "robotic-assisted dying" in medical contexts? Ryan Tonkens ("Ethics of Robotic Assisted Dying") proposes that if we develop robots to serve as human caregivers in medical contexts, and given that assistance in dying is sometimes an important aspect of geriatric care, it is ethical for such robots to facilitate and assist in the dying of those patients at the eligible patient's sound request. A major benefit of robotic-assisted dying is that the robot would always assist those consenting patients that are genuinely eligible, and thus such patients would not be at the mercy of a willing physician clause in order to have some control over the timing and manner of their death. At the same time, specialist humans must remain involved in establishing strict regulations and safety protocols concerning end-of-life situations and be present in the event of machine malfunction.

According to Blay Whitby ("Automating Medicine the Ethical Way"), unreliable technology and human errors in Information Technology (IT) resulting from poor user interfaces are two outstanding ethical problems. Whitby calls for improved ethical awareness and professionalism in IT workers in order to achieve ethically acceptable medical machines. Lessons from the aviation industry suggest that issues of acceptance and resistance by professionals can be successfully managed only if they are fully engaged in the operational and procedural changes at all stages. Negotiation over procedures and responsibility for errors in aviation is complex and informative for other fields, including machine ethics.

In "Machine Medical Ethics: When a Human Is Delusive but the Machine Has Its Wits About Him", Johan F. Hoorn imagines an advanced dementia patient under the care of a health care robot and asks: "Should the robot comply with the demand of human autonomy and obey every patient command?" To help answer this question, Hoorn offers a responsibility self-test for machine or human that differently prioritizes top-down maxims of autonomy, nonmaleficence, beneficence, and justice. The self-test comes in seven steps, ranging from "I do something" (to act, with or without self-agency), to "My "higher" cognitive functions are supposed to control my "lower" functions but failed or succeeded" (to act, with or without self-control).

In "ELIZA Fifty Years Later: An Automatic Therapist Using Bottom-Up and Top-Down Approaches", Rafal Rzepka and Kenji Araki present a machine therapist capable of analyzing thousands of patients' cases implemented in an algorithm for generating empathic machine reactions based on emotional and social consequences. Modules and lexicons of phrases based on these theories enable a medical machine to empathically sense how patients typically feel when certain events happen, and what could happen before and after actions. The authors suggest that this bottom-up method be complemented by a top-down utility calculation to ensure the best outcome for a particular human user.

Neuromachines capable of measuring brain function and to iteratively guide output will be a major development in neuromodulation technology. According to Eran Klein, the use of closed-loop technologies in particular will entail ethical changes in clinical practices ("Models of the Patient-Machine-Clinician Relationship in Closed-Loop Machine Neuromodulation"). Klein thinks current ethical models of the clinical relationship are only suited to certain forms of neuromodulation, but new models ought to be more comprehensive as new neuromodulatory technologies emerge. Klein assesses design, customer service, and quality monitoring models as candidates for a new ethic and urges that any successful theoretical approach ought to incorporate Aristotelian concepts of friendship.

Steve Torrance and Ron Chrisley ("Modelling Consciousness-Dependent Expertise in Machine Medical Moral Agents") suggest that a reasonable design constraint for an ethical medical machine is for it to at least model, if not reproduce, relevant aspects of consciousness. Consciousness has a key role in the expertise of human medical agents, including autonomous judging of options in diagnosis, planning treatment, use of imaginative creativity to generate courses of action, sensorimotor flexibility and sensitivity, and empathetic and ethically appropriate responsiveness.

An emerging application of affective systems is in support of psychiatric diagnosis and therapy. As affective systems in this application, medical machines must be able to control persuasive dialogs in order to obtain relevant patient data, despite less than optimal circumstances. Kim Hartman, Ingo Siegert, and Dmytro Prylipko address this challenge by examining the validity, reliability, and impacts of current techniques (e.g., word lists) used to determine the emotional states of speakers from speech ("Emotion and Disposition Detection in Medical Machines: Chances and Challenges"). They discuss underlying technical and psychological models and examine results of recent machine assessment of emotional states obtained through dialogs.

Medical machines are affective systems because they can detect, assess, and adapt to emotional state changes in humans. David Casacuberta and Jordi Vallverdú ("Ethical and Technical Aspects of Emotions to Create Empathy in Medical Machines") argue that empathy is the key emotion in health care and that machines need to be able to detect and mimic it in humans. They reinforce modeling of cultural, cognitive, and technical aspects in health care robots in order to approximate empathic bonds between machine and human. The emotional bonds between human and machines are not only the result of human-like communication protocols but also the outcome of a global trust process in which emotions are cocreated between machine and human.

In Epilogue, Dutch visual artist Janneke van Leeuvan and Simon van Rysewyk discuss whether intelligent machines can appreciate esthetic representations as a simulacrum of human esthetic understanding. The dialog is illustrated by selections from van Leeuwen's thoughtful photographic work, "Mind Models."

The book editors Simon Peter van Rysewyk and Matthijs Pontier wish to warmly thank Springer for the opportunity to publish this book, and in particular, to acknowledge Cynthia Feenstra and Nathalie Jacobs at Springer for their assistance and patience. We wish to thank Jessica Birkett (Faculty of Medicine, University of Melbourne) for reviewing author chapters, and all authors that feature in this book for their excellent and novel contributions. Simon Peter van Rysewyk acknowledges support from Taiwan National Science Council grant NSC102-2811-H-038-00. Thank you all.

Taiwan — Simon Peter van Rysewyk
The Netherlands — Matthijs Pontier

Contents

Part II Contemporary Challenges in Machine Medical Ethics: Justice, Rights and the Law

Part III Contemporary Challenges in Machine Medical Ethics: Decision-Making, Responsibility and Care

Part IV Contemporary Challenges in Machine Medical Ethics: Medical Machine Technologies and Models

Part I
Theoretical Foundations of Machine Medical Ethics

An Overview of Machine Medical Ethics

Tatjana Kochetkova

Abstract This chapter defines the field of medical ethics and gives a brief overview of the history of medical ethics, its main principles and key figures. It discusses the exponential growth of medical ethics along with its differentiation into various subfields since 1960. The major problems and disputes of medical ethics are outlined, with emphasis on the relation between physicians and patients, institutions, and society, as well as on meta-ethical and pedagogic issues. Next, the specific problems of machine ethics as a part of the ethics of artificial intelligence are introduced. Machine ethics is described as a reflection about how machines should behave with respect to humans, unlike roboethics, which considers how humans should behave with respect to robots. A key question is to what extent medical robots might be able to become autonomous, and what degree of hazard their abilities might cause. If there is risk, what can be done to avoid it while still allowing robots in medical care?

1 The Reality of Machine Medical Ethics

> A hospital patient needs to take her regular medication, but is watching his/her favorite TV program and reacts angrily to the reminder about taking medicine. If you were a nurse, how would you react? *How must a robot nurse react?*

This case study originates from a project conducted by robotic researchers Susan and Michael Anderson (Anderson and Anderson 2005). They programmed a NAO robot[1] to perform simple functions like reminding a "patient" that it is time to take

[1] A NAO robot is a programmable autonomous humanoid robot developed by French company Aldedaran Robotics. It was first produced at the beginning of the 21st century.

T. Kochetkova (✉)
Institute for Philosophy, University of Amsterdam, Amsterdam, The Netherlands
e-mail: tania.j.meira@gmail.com

S.P. van Rysewyk and M. Pontier (eds.), *Machine Medical Ethics*,
Intelligent Systems, Control and Automation: Science and Engineering 74,
DOI 10.1007/978-3-319-08108-3_1

prescribed medicine [29]. NAO brings a patient tablets and declares that it is time to take them. If the treatment is not observed (i.e., if the patient does not take the tablets), the robot should report this fact to the doctor in charge.

But suppose we program the robot to react also to a patient's mental and emotional states. This makes the situation much more complicated: a frustrated patient can yell at the robot, refuse to take pills or refuse to react at all, or do something else not included in the narrow algorithm that guides the robot. In order to react accordingly, the robot now needs to be more flexible: it has to balance the benefits that a patient receives from the medicine (or treatment) against the need to respect the patient's autonomy. In addition, the robot has to respect the independence and freedom of the patient. If, for instance, the disease is not too dangerous and the patient forgets to take a pill while watching his or her favorite television program, another reminder of the robot could bring him more displeasure (i.e., harm) than good. If skipping the medication had more serious consequences, then the robot would have to remind the patient, and if necessary, even notify the patient's doctor. The robot thus needs to make decisions based both on the situation at hand, and also on its built-in "value hierarchy": different principles might lead to different decisions for the same situation [4, 6].

In the near future, robots like the one in the Andersons study may become widespread. The question of how to give them a complex hierarchy of values therefore becomes increasingly important. In addition to having to think about their own ability to carry out their responsibilities (e.g., they must know when it is time to recharge their batteries, or else they might leave patients unattended and in potential risk), they will also need to make appropriate choices for their patients. This implies an in-built sense of justice when tackling even mundane tasks: if, for instance, they are supposed to change the channel of a TV set that several patients are watching together, they will have to take into account variables such as the patients conflicting desires, and how often each patient's TV wishes have been fulfilled in the past.

Reasons such as these explain the relevance of machine medical ethics. Machine medical ethics faces at least three challenges. Foremost, there is the need to ensure the safe use of medical robots, whose presence in the health sector is increasing. By the middle of the 21st century, about 25 % of West Europeans will be over 65 years old; there will be an increasing demand on the healthcare system that will only be met by using advanced technology, including robotics. Some remotely operated robots are now already routinely being used in surgery. Other expected applications of medical robots in the near future are [19]:

- Assisting patients with cognitive deficits in their daily life (reminding them to take medicine, drink or attend appointments).
- Mediating the interaction of patients and human caregivers, thus allowing caregivers to be more efficient and reducing the number of their physical visits.
- Collecting data and monitoring patients, preventing emergencies like heart failure and high blood sugar levels.
- Assisting the elderly or the disabled with domestic tasks such as cooking or cleaning, thus making it possible for them to continue living independently.

The demand for robots in the healthcare sector is already quite palpable, at least in the West, and I suspect this demand will only increase. This will include robots that can perform some human tasks but are quicker to train, cheaper to maintain, and are less bored by repetitive tasks, with the ultimate purpose being to take over tasks done by human caretakers and to reduce the demand for care homes [29]. It is clear that the behavior of such robots must be controlled by humans and within the ambit of human ethical values, otherwise the robots would be useless and possibly even dangerous: if their behavior is not always predictable, they could potentially cause harm to people. A robot with no programming on how to behave in an emergency situation could make it worse. To avoid such problems, it is necessary to build into the robots basic ethics that apply in all situations.

Second, there is a certain fear of autonomously thinking machines, especially in the West, probably due to uncertainty about whether they will always behave appropriately. Science fiction is full of such fears. The creation of ethical constraints for robots can make society more receptive to research in the field of artificial intelligence by allowing it to deal better with robots in ethical situations. In fact, in such situations, robots *without* ethical constraints would appear to be too risky to be allowed in society.

A third reason for the increasing interest in machine ethics is the question of who can ultimately make better ethical decisions: humans or robots? Humans use their intuition for moral decisions, which can be a very powerful heuristic [14, 25]. Yet humans can be bad at making impartial or unbiased decisions and are not always fully aware of their own biases. As for robots, Anderson claims that they may be able to make better decisions than people, because robots would methodically calculate the best course of action based on the moral system and principles programmed into them [4]. In other words, robots may behave better than humans simply because they are more accurate [4]. Yet it is not entirely clear whether such methodic consideration of all possible options by a robot will always be an advantage for decision making. As Damasio's research shows, people with brain damage actually do methodically consider all options, yet this does not guarantee that their decisions will be better, since their mental impairment forces them to consider, and perhaps take, many options that healthy people would immediately see as bad [10, 14].

A final reason for the growing relevance of machine ethics is the lack of consensus among experts on the ways to handle major ethical dilemmas, which makes it more difficult to transfer decision making to machines. Answers to ethical dilemmas are rarely clear. For instance, a classical problem like the train accident dilemma, discussed below, would be solved by different theories in different ways: utilitarians believe that the sacrifice of a single life in order to save more lives is right, while deontologists believe such a sacrifice is wrong since the ends cannot justify the means. There is no consensus on other vital issues such as whether abortion is permissible, and if so, under what circumstances. A medical robot, performing the role of adviser to a patient, for example, may have to take such facts into account and realize that it needs to shift the burden of making the right decision to a human being. But this may not always be possible: in another dilemma involving a possible car accident where one or several people would inevitably die, an autonomous car with an automatic navigation system would either lose the ability to act independently or have to take a random decision. Neither solution seems really acceptable [15].

2 The Development of Machine Medical Ethics: A Historical Overview

Machine medical ethics recently emerged as a branch of ethics. To fully understand *ethics*, it is important to see it as the critical and reflexive study of morality, i.e., as the rational scrutiny of values, norms, and customs. This critical stance differentiates ethics from morality: ethics is the *philosophical* study and questioning of moral attitudes. Even though machine medical ethics includes both normative and applied components, it is the latter which is recently gaining research prominence.

Since the 1950s, medical ethics has experienced exponential growth and differentiation in various subfields, hand in hand with the technological, political, and cultural changes of this time. Previously, the relation between medical professionals and patients had been paternalistic: all decisions were supposed to be taken by a professional (or a group of professionals) in the best interests of a patient and were then communicated to the patient. This paternalism was based on a knowledge gap between the medical professional and the patient and between the professional and the public, as well as on the relative clarity of medical decisions and the limited number of choices.

Paternalism in doctor-patient relationships has been undermined by public knowledge about atrocities in medical experiments conducted during the Second World War. The post-war Declaration of Geneva (1948) initiated the shift towards a more liberal model of doctor-patient relationship, promoting informed consent as an essential ethical value.

Since 1960 and up to the beginning of the 21st century, due to the growth of public education, empowerment of the general public, accessibility of medical knowledge, as well as new developments in medical technology and science, the general public has become more informed about available medical information and patients are participating in clinical decision-making. This has changed the relation between healthcare professionals and patients quite significantly: the paternalistic model is anachronistic, and in most cases shared decision-making model is the norm [8, 18, 30].

Concomitantly with the shift away from paternalism in medicine, ethics itself underwent a change in focus towards application. Scientific and technological development has given rise to various new choices and specific ethical problems. This led to the origin of bioethics (term introduced already in 1927 by Fritz Jahr). In the narrow sense, bioethics embraces the entire body of ethical problems found in the interaction between patient and physician, thus coinciding with medical ethics. In the broad sense, bioethics refers to the study of social, ecological, and medical problems of all living organisms, including, for instance, genetically modified organisms, stem cells, patents on life, creation of artificial living organisms, and so on.

Along with the appearance of bioethics and the shift away from paternalism and the consequent decrease of the role of the doctor as sole decision-maker, the idea that machines could also be a part of the process of care became more acceptable. Together with great progress in medical technology, this resulted in the emergence of the field of machine medical ethics. The main aim of machine medical ethics is to guarantee that medical machines will behave ethically. Machine medical ethics is an application of ethics and a topic of heated debate and acute public interest.

The reasons for this change in focus towards application have been widely discussed in bioethics. Among its causes is the growth of human knowledge and technological possibilities, which brought along a number of new ethical problems, some of which had never been encountered before. For example, should we switch to artificial means of reproduction? Is it acceptable to deliberately make human embryos for research or therapeutic purposes? Is it worthwhile to enhance humans and animals by means of genetic engineering or through digital technologies? In addition, there are also new problems concerning the usage of robots, brought about by rapid progress in the development of computer science. For example, is it acceptable to use robots as work force if their consciousness evolves, as they become AMAs (artificial moral agents)? Suddenly, the area of human-robot interactions is saturated with ethical dilemmas.

Given the increasing complexity and applicability of robots, it is quickly becoming possible for machines to perform at least some autonomous actions which may in turn cause either benefit or harm to humans. The possible consequences of robot errors and, accordingly, the need to regulate their actions is a pressing ethical concern. It is not simply a question of technical mistakes, like autopilot crashes, and their consequences, but also of cases in which robots have to make decisions that affect human interests. An obvious example in the field of medicine is the activity of robot nurses i.e., mobile robotic assistants [3]. Robot nurses have been developed to assist older adults with cognitive and physical impairments, as well as support nurses. Mobile robotic assistants are capable of successful human-robot interaction, they have a human tracking system and they can plan under uncertainty and select appropriate courses of actions. In one study, the robot successfully demonstrated that it can autonomously provide reminders and guidance for elderly residents in experimental settings [26].

Presently, medical robots are already in use in various areas. In surgery, operations involving robotic hands tend to have higher quality and involve fewer risks than traditional operations managed by humans. Robots are also being used in managing large information files ("Big Data"). For instance, the market share of "exchange robots", computer algorithms for earning their owners money in the stock market, is set to become more widespread, since their results are better than those of human traders. The relation between the quality of electronic and live traders is now the same as it was for chess players and chess programs on the eve of the match between the human player Kasparov and the program Deep Blue. As we all know, the program won.

This particular case does not seem very dangerous, but is there an element of risk involved in the success of intelligent machines in other areas? These questions increasingly concern not only the broad public, but also designers and theorists of Artificial Intelligence (AI) systems. The main challenge to solve is how to ensure safety with AI systems. Devices found only in fiction, like Isaac Asimov's famous Three Laws of Robotics, seem increasingly necessary [5, 12, 16]. In recent decades, such issues have been debated in a broad range of publications in computer science and machine ethics. The increasing success of various robot-related projects has stimulated research on the possibility of built-in mechanisms to protect people from unethical behavior by computer-controlled systems.

Currently, the demand for producing ethical robots for the aging population in developed countries exceeds medical services: the demand for service robots in restaurants, hotels, nurseries, and even at home has been growing. The entire service sector, it seems, is impatiently waiting for robots with reasonably good quality and affordable prices to appear. It would seem that all mechanical labor in today's increasingly educated society is regarded as something best shifted to the hands of robots.

But, problems in the production of such robots go beyond technological difficulties. Separating the mechanical and communicative component of specialized work (e.g., for nurses) is sometimes very difficult, or even impossible. They are subtly intertwined, which makes ethical programming of robots necessary for nearly all tasks involving interactions with humans. For instance, in the situation described in the introduction—a patient reacts negatively to the reminder to take medicine—communicative and ethical capacities must be intrinsic to a robot for it to be able to react ethically.

These difficulties might lead to the question whether machine ethics is possible at all, i.e., whether problems with ensuring AMAs behave ethically is a theoretically tractable problem. I think these difficulties are tractable from an engineering perspective. The real difficulty to be solved lies in improving robotic software, perfecting the sets of rules, and ensuring the correct elaboration of in-coming data. With robots as explicit AMAs, the challenge is to make their complex behavior predictable. However complicated this challenge proves to be, it basically is a matter of improving the degree of complexity of already existing robots, not the theoretical question of building entirely new AI systems. This supports my optimism about the prospects of robotics to ensure safe robot use in the real world.

3 Key Issues in Machine Medical Ethics

The key issues of machine medical ethics are linked to problems of AI. In current AI discussions, three major issues dominate: computability, robots as autonomous moral agents, and the relation between top-down, bottom-up and hybrid theoretical approaches. Each will be briefly considered in turn.

The computability of ethical reasoning concerns the conditions for the very existence of machine medical ethics. Indeed, ethics, as seen above, can be defined as a reflection on the normative aspects of human behavior. Presently, machine ethics, as the study of how machines should behave with respect to humans, attempts to create computer analogs for the object of ethical study—values and norms—so as to make ethics computable and ultimately permit its implementation in robots and machines [11, 29]. The hope is that ethics can be made translatable into computer language, i.e., ultimately presentable as a complex set of rules that can be made into a computer algorithm [24]. There already are programs that allow machines to imitate aspects of human decision-making, a first step towards the creation of robots that will be able to make such decisions by themselves. Some of these programs are discussed by Bringsjord and Taylor [7].

One approach to achieving a computable machine ethics is a complete and consistent theory that can guide and determine the actions of an intelligent moral agent in the enormous variety of life situations. With such a theory, a machine with AI could in principle be programmed to deal appropriately with all real-life situations. The basic question is then, of course, whether such a universally applicable theory actually exists: if it does, then machine ethics would be basically busy with programming it into computers. It may be, however, that no single ethical theory is or can truly be complete: completeness, albeit attractive, may ultimately turn out to be an unattainable ideal. The absence of unconditionally correct answers to ethical dilemmas, and changes in ethical standards through time, suggest that it is not prudent to hope for a "perfect" theory of ethics in attempting to build ethical machines. Rather, work on ethically programmed robots should start with the idea of ethical gray areas in mind, areas in which one cannot determine which behavior is unconditionally right on a case by case basis [11].

Rather than concentrating on one single system or theory of ethics (for which often intractable dilemmas can be found), it seems more productive to strive towards a hierarchic system including a plurality of ethical values, some of which are subordinate to others. For example, John Rawl's theory of "equilibrium", which is similar to, but more complicated than, the one used by the hospital robot in Anderson's experiment, is a candidate.

The study of machine ethics might thus advance the study of ethics in general. Although ethics ought to be a branch of philosophy with real-life impact, in practice theoretical work and discussions between philosophers often drift toward the consideration of unrealistic "academic" situations [11]. The application of AI to ethical problems will help understand them better, make them more realistic, and lead to better solutions, while perhaps even opening the ways for new problems to be considered.

Secondly, there is a widespread agreement in AI that a robot capable of making decisions and acting autonomously is an AMA [22]. This does not require any intrinsic consciousness or personal reflection per se: if an autonomous decision or action is performed by a machine, the system is an AMA regardless of its endogenous higher-order states. The possibility of robot AMAs led to an ongoing philosophical debate on the foundations of AI, about whether strong or weak AI is logically possible [27]. However, this question is much older than machine ethics. It was introduced by Alan Turing in the 1940s, and before that, science fiction book published in the early twentieth century featured "thinking machines" and "autonomous robots".

Third, there is the problem of how to program AMAs so that they behave in a way beneficial to humans. What ethical principles or rules should be adopted? And how can they be programmable in robot algorithms? Three major solutions to this problem are currently available: top-down, bottom-up and hybrid. According to top-down approaches, moral principles are used as rules for the selection of actions (for instance, utilitarian, or deontological theories). The problem with top-down approaches is that they cannot provide a general theory of intelligent action and machine learning for real-world tasks [1]. Their opposite, bottom-up approaches, seek to provide educational environments that select appropriate ethical behavior via experiential learning (e.g., trial and error, approval, disapproval). A problem with bottom-up approaches is that it is unclear how ethical rules and maxims fit into machine learning [1]. Because both top-down and bottom-up approaches each have problems as well as advantages, some authors reasonably suggest a hybrid approach which combines aspects of both [1, 3]. Whatever approach turns out to work in AMAs, medical machine ethics will certainly have a significant impact on the socialization of humanoid robots. This is because AMAs with functions overlapping with human activities will occur in a large variety of scenarios, unavoidably including situations of inherent moral ambiguity.

Finally, there is the important issue of dealing with ethical dilemmas for which no expert consensus currently exists. For instance, in philosophical ethics, deontologists, virtue ethicists and consequentionalist theorists fundamentally disagree about the solution to ethical problems. A reasonable response to this situation is that decisions in such dilemmas ought not to be dealt with by robots or machines at all, but *only* by humans, at least until expert consensus exists, thus forming a provisional limit on the freedom of AMAs. Thus, I suggest that the use of medical machines be provisionally pegged only to medical cases for which robust medical consensus exists, and to no other type of case.

4 What Kind of Moral Agents Do We Want Robots to Become?

For machine ethics, the major question is how machines should ethically behave. From this perspective, there are two general possibilities for building a robot guided by ethical principles. The first is to simply program the machine to behave ethically (according to some agreed-upon description of ethical behavior), or at least avoid bad deeds [2]. Such a robot carefully performs correct actions, in accordance to its programming, but cannot explain their meaning or the reasons for their being correct. It is ultimately limited in its behavior, which was conceived and programmed by its designers, who in this case are the real responsible moral agents. The problem with this alternative is that, in a situation that does not fit the programming (i.e., which is outside of the description of ethical behavior used its programming), the robot's behavior might be random [2].

There is, however, another way. It involves the creation of a machine truly capable of autonomous action: an "explicit moral agent," in J.H. Moor's classification [22]. Even though every system that behaves in a way that affects humans is in principle an AMA, only a system that is capable of making autonomous moral decisions is an explicit moral agent.

It is worthwhile to consider Moor's [22] classification in some detail. As he defined it, "ethical-impact agents are machines that have straightforward moral impact", like video cameras on the streets, used to detect or prevent crime. "Can a computer operate ethically because it's internally ethical in some way?" [22] To make things clear, Moor introduced a threefold differentiation between implicit, explicit, and full moral agents [22]:

- *Implicit moral agents* are robots constrained "to avoid unethical outcomes." They do not work out solutions to make ethical decision themselves, but are designed in such a way as to behave ethically. For instance, automated teller machines or automatic pilots on airplanes are implicit moral agents.
- *Explicit moral agents* are robots that can "do ethics like a computer can play chess" [17, 19–20, 22]. Explicit moral agents apply ethical principles such as Kant's categorical imperative, or Rawl's reflective equilibrium to concrete situations in order to choose their course of action. The crucial point is that explicit moral agents autonomously reach ethical decisions with the help of some procedure. Such machines could also, if need be, justify their decisions.
- *Full moral agents* are beings like humans, who possess "consciousness, intentionality and free will" [20, 22]. Full moral agents are accountable for their actions. The robots which are currently being worked on are not full moral agents in any sense, and it is not clear whether the creation of computational full moral agents is morally acceptable or rationally justifiable.

There are also important problems concerning the feasibility and desirability of making robots that are explicit moral agents. First, there is no single ethical theory enjoying general consensus that could be programmed into robots: machine ethicists typically do not agree with each other. In other words, there is no agreement on how autonomous robots should be programmed, on what should be programmed into them. One solution was proposed by Anderson et al. [4], who suggests that robots should be programmed to act ethically only in situations about which there is ethical consensus among theorists on what the best behavior should be; in other, more polemic cases, the robot should simply not act but yield to humans for decisions. The boundaries of moral consensus are thus, according to Anderson et al. [4], also the boundaries of robotic action.

A second problem is the possibility that AMAs may cross the symbolic boundaries between humans and machines [21, 23, 28]. This is a general problem with maintaining ethical boundaries between humans and machines, especially when such machines can have different levels of ethical development. A robot created especially to perform hazardous jobs is perfect as an explicit, but not a full, moral agent. If the same robot becomes a full moral agent, it is immediately far less suited for hazardous, or routine, repetitive jobs, since human moral obligations to full moral agents are completely different and far more demanding. Moral obligations to merely explicit moral agents, on the other hand, are much simpler: to humans, such moral agents may still be seen as equipment, which is precisely what makes them more suited for hazardous or routine works: their consent, rights and interests are non-existent and thus irrelevant. From full moral agents, we need consent and agreement; from explicit moral agents, it is acceptable to expect simply obedience to human commands.

The question of the moral agency of robots has been widely discussed by various science fiction writes, like Asimov, Spielberg, Kubrick and others. Asimov made an attempt at formulating machine ethics in 1942, in his guidelines for the behavior of robots. So machine ethics is older than its name, since Asimov was talking about it in 1942! Although science fiction is a form of literature, the problems raised by Asimov's stories are not in essence fictional: they represent problems involved with AMAs crossing the symbolic boundary between humans and machines, a particular case of the more general problem with maintaining ethical boundaries between humans and machines.

Let us look briefly at Asimov's guidelines, formulated as his famous "Three Laws of Robotics" (2004):

1. A robot may not injure a human being or, through inaction, allow a human being to come to harm.
2. A robot must obey all orders of the person, except when such orders conflict with the First Law.
3. A robot must protect its own well-being to the extent that it does not conflict with the First or Second Law.

Leaving aside details in their formulation, these three laws are inapplicable to full moral agents, which Asimov's robots apply to in his stories. One clear example is found in the story "The Bicentennial Man", in which the protagonist robot Andrew gradually fights for full freedom, and for the recognition of itself as a human being, finally obtaining this status in court. Andrew is clearly an exceptional robot. Asimov first demonstrates how it can engage in creative activities and have fun. His owner even sought permission for it to be paid for its work. Then we see that the subordinate position of the robot brings about suffering: in one episode, Andrew, who wore by that time human clothes, was attacked by vandals who forced it to undress by using the second law of robotics. To obtain the right to personhood, Andrew is even ready to accept suicide: it lets its positronic brain be reprogrammed, so that it will die in a 100 years. The robot surgeon who performs the operation initially refuses to do so: it cannot cause harm to a human being. But Andrew corrects him: the robot cannot harm a person, but Andrew is a robot. Asimov made his robot character behave in a way that is worthy of freedom: robot Andrew's moral sense is often superior to that of the people around him. This raises important questions about robots ever becoming autonomous agents.

The shift of the symbolic boundaries between human and machine is the result of the ongoing mergence of nanotechnology, biotechnology, information technology and cognitive science, otherwise called NBIC-convergence [9, 20, 28]. While some shifting and re-arranging of symbolic boundaries is inevitable, certain boundaries should still arguably be preserved.

The ethical boundary between human and machines is as follows: humans are both moral agents and moral subjects (recipients of ethical behavior), while machines have thus far not been moral subjects, even though they can be moral agents. A robot programmed to behave ethically can be an AMA without becoming a moral subject.

One can now argue that it is important to ensure that robots will not become moral subjects (which is, do not become truly sentient or self-reflexive) because this would lead to the deontological imperative to not utilize them purely as means to external goals. If, for instance, a household robot becomes capable of self-reflection and feeling pain, it might be no longer be suitable for use, at least not from an ethical point of view. Therefore, it is also advisable, in my view, to produce only robots that, while capable of highly complicated technical behavior and led by programs that incorporate human ethics, still are not capable of genuine emotions or self-reflection, i.e., who are not full moral agents. Otherwise, they would stop being pieces of equipment and become, for all intents and purposes, artificial subjects, which would defeat the purpose of using them as a mere means to save time, money and energy [13].

5 Summary

I have completed a brief overview of the recent history of medical robot ethics, a new field at the crossroads between biomedical ethics and machine ethics. The growing need to provide affordable care for an aging global population prompts the question of medical machines and machine medical ethics. In this chapter, I identified the major challenges of the field. These are summarized below together with my response:

1. *Is ethics translatable into computer languages?* It is clear that natural language is richer and more complicated than computer languages. The adequate translation between the two remains a vexing problem. This relates to the question about the limits of machine intelligence. There is no final answer to this question, yet it is clear that for typical, unproblematic situations, such as "consensus cases", one can build an algorithm that would guide machine behavior. Since the majority of care situations are typical in this sense, machines can be of medical use.
2. *What is the correct approach to program AMAs so that they are beneficial to humans?* Which ethical principles should be adopted? The choice falls between the three major approaches to the design of AMA's: top-down, bottom-up and hybrid.
3. *How should mechanical and communicative components necessary for specialized medical tasks be separated in AMAs?*

4. *How should ethical dilemmas be dealt with in which there is no consensus among experts?* A reasonable suggestion is to limit the use of medical robots to consensus medical cases only. Such cases are frequent enough to justify the use of robots in medical and other fields (e.g., physiotherapy).
5. *What kind of moral agents do we want robots to become? Shall robots become explicit or full moral agents?* Robots as explicit moral agents seem to me preferable, because being full moral agents would create an ethical conflict with their functionality. Their use would imply also their consent, respect of their rights and freedoms. Why is this problematic? Because the basic demand from healthcare for robots is to save human efforts to provide for an aging population. Such use of artificial moral agents might be associated with ethical and legal difficulties. Indeed, if we extend the logic of contemporary western legislation to strong AI, the rights and freedoms would be the same as human rights and freedoms and therefore this artificial labor might have the same costs as the human workforce. Thus, I argue that medical robots that are explicit, but not full moral agents would be best suited for the societal demand for them: to enable healthcare systems to provide high-quality and affordable healthcare for the growing elderly population.

References

1. Allen C, Smit I, Wallach W (2005) Artificial morality: top-down, bottom-up, and hybrid approaches. Ethics Inf Technol 7:149–155
2. Allen C, Varner G, Zinser J (2000) Prolegomena to any future artificial moral agent. J Exp Theor Artif Intell 12(3):251–261
3. Allen C, Wallach W, Smit I (2006) Why machine ethics?. Intell Life Syst 21(4):12–17. www.computer.org
4. Anderson M, Anderson SL, Armen C (eds) (2005) Machine ethics.In: AAAI Fall symposium, tech report FS-05-06. AAAI Press
5. Asimov I (2004) I-robot. Bantam Books, New York
6. Beavers AF (2011) Moral machines and the threat of ethical nihilism. In: Lin P, Bekey G, Abney K (eds) Robot ethics: the ethical and social implication on robotics. MIT Press, Cambridge, Mass
7. Bringsjord S, Taylor J (2011) The divine-command approach to robot ethics. In: Lin P, Bekey G, Abney K (eds) Robot ethics: the ethical and social implication on robotics. MIT Press, Cambridge, Mass
8. Campbell A, Gillet G, Jones G (2005) Medical ethics. OUP, Oxford
9. Canters P, Kochetkova T (2013) Ethiek. Boom Lemma, Damon
10. Christensen B (2009) Can robots make ethical decisions?. http://www.livescience.com/5729-robots-ethical-decisions.html
11. Churchland P (2011) Braintrust: what neuroscience tells us about morality. Princeton University Press, Princeton, pp 23–26
12. Clarke AC (2000) 2001: a space odyssey. ROC, New York
13. Coeckelbergh M (2010) Robot rights? Toward a social-relational justification of moral consideration. Ethics Inf Technol 12(3):209–221
14. Damasio A (2010) Self comes to mind: constructing the conscious brain. Pantheon, New York
15. Floridi L, Sanders JW (2004) On the morality of artificial agents. Mind Mach 14(3):349–379

16. Gibson W (2004) Neuromancer. ACE, New York
17. Gips J (1995) Towards the ethical robot. In: Ford K, Glymour C, Hayes P (eds) Android epistemology. MIT Press, Cambridge, pp. 243–252
18. Jackson E (2009) Medical law: text, cases, and materials. OUP, Oxford
19. Jervis C (2005) Carebots in the community. Br J Healthc Comput Inf Manage 22(8):1–3
20. Konovalova LV (1998) Prikladnala Etika. Instituut Filosofii, Moscow
21. Kurzweil R (2005) The singularity is near: when humans transcend biology. Viking Adult, New York
22. Moor JH (2006) The nature, importance, and difficulty of machine ethics. Intell Life Syst 8:18–21
23. Moravec H (2000) Robot: mere machine to transcendent mind. Oxford University Press, Oxford
24. Nissenbaum H (2001) How computer systems embody values. Computer 34(3):120, 118–119
25. Picard RW (2003) Affective computing: challenges. Int J Human Comput Stud 59(1–2):55–64
26. Pineau J, Montemerloa M, Pollackb M, Roya N, Thruna S (2003) Towards robotic assistants in nursing homes: challenges and results. Robot Auton Syst 42:271–281
27. Searle JR (1980) Minds, brains, and programs. Behav Brain Sci 3(3):417–457
28. Swierstra T, Boenink M, Walhout B, Van Est R (2009) Leven als bouwpakket. Ratenau Instituut
29. Wallach W, Allen C (2010) Moral machines: teaching robots right from wrong. The MIT Press, Cambridge
30. Warren R (2011) Paternalism in medical ethics: a critique. In: Journal of The University of York Philosophy Society, 10

Surgical, Therapeutic, Nursing and Sex Robots in Machine and Information Ethics

Oliver Bendel

Abstract Machine medical ethics is a novel field of research for ethicists, philosophers, artificial intelligence experts, information scientists, and medical specialists. I identify surgical, therapeutic, nursing and sex robots as the primary types of medical machines in this context. I raise general questions about machine ethics with a view to its development and application, specific questions about medical machine ethics (the term and concept which I prefer), and broad questions spanning multiple non-machine ethics, including information, technology, business and legal ethics, and interrelationships between these diverse ethics and machine medical ethics. Samples of each type of question are provided in my descriptions of surgical, therapeutic, nursing and sex robots. In particular, progress in information and technology ethics is needed in order to solve moral problems involving medical machines, and progress in machine ethics to prevent some of the problems.

1 Introduction

Robots have reached several peaks of attention. They reached the first peak in fiction, in the poems and stories by Homer and Ovid, by Stanislaw Lem and Isaac Asimov, and in movies as different as "Metropolis", "Star Wars" and "WALL-E". They reached the next peak in industry, with production machines that assemble other machines. Finally, they reach a peak of attention in everyday life where robots can take all kinds of forms. Some are inconspicuous, as for instance vacuum or mowing robots. Others such as ASIMO and Qrio are conspicuous because they are similar to man (humanoid)—so were Talos, the legendary Cretan guardian, and Pandora, the artificial woman with her box.

O. Bendel (✉)
School of Business, Institute for Information Systems,
University of Applied Sciences and Arts Northwestern Switzerland
FHNW, Basel, Switzerland
e-mail: oliver.bendel@fhnw.ch

S.P. van Rysewyk and M. Pontier (eds.), *Machine Medical Ethics*,
Intelligent Systems, Control and Automation: Science and Engineering 74,
DOI 10.1007/978-3-319-08108-3_2

Robots are present in everyday life, and living with them can be different for different people. It can tip to extremes, and the extremes can become everyday routines, for instance for patients. Surgical, therapeutic, nursing and sex Robots are conquering the healthcare systems in households, doctors' practices, and clinical centers. They supplement and complement human workforces and fellow human beings. They can do some things better, some things just as well, and some things worse. Usually they are very expensive, on the other hand they can generate cost savings. Then suddenly they multiply to an extent that makes them a popular topic of mass media.

This article takes the perspective of ethics without prioritizing on robot ethics as one might expect in this context. The issue is scientific ethics, and it is very important that it gains a new self-confidence and makes it to the headlines of newspapers and TV programs. With the combination of robots and ethics, this might well succeed.

In the beginning, the article explains human and machine ethics as well as applied ethics, with special consideration of information ethics, technology ethics, and medical ethics. Then follow questions related to machines as subjects and humans as objects. Not lastly, this shows how to distinguish machine and human ethics (including their specific ethics) from each other. At the end, it evaluates the questions, and as if the future had not been dealt sufficiently before, it takes an outlook into the future.

2 The Machine in Morality

Different ethics are dealing with the machine in morality. For thousands of years, ethics has been human ethics—leaving out the morality of the gods, for instance of the Olympus for once. This means ethics related to humans as subjects and often also humans as objects. Other ethical concepts seem to have occurred only in science fiction and in think tanks. The situation has become more complicated ever since the occurrence of (partly) autonomous systems, since machines—which is exactly what I mean here—decide and act independently of humans. Machine ethics can be seen on the same level as human ethics, therefore it is important to clarify both terms. Within human ethics, several disciplines of applied ethics may feel competent, especially information ethics and technology ethics. Further, the gap between the world of machines and the world of medicine has to be bridged, therefore I have to introduce medical ethics.

2.1 Human Ethics

Ethics as science is a discipline of philosophy and morality is its object. Empirical or descriptive ethics describes morality ("Moral" in German [33]) and the will for moral behavior ("Moralität" in German [33]), normative ethics rates it, criticizes it, and where necessary, gives reasons for making changes in behavior. Normative ethics does not finally refer to religious or political authorities or to what is natural, customary, or well-proven [27]. One can also refer to the will for moral behavior using normative ethics, and point out discrepancies between attitudes and behaviors. Metaethics analyses moral terms and propositions from a semantic perspective. Applied ethics is segmented into specific disciplines such as political ethics, medical ethics, business ethics, media ethics and information ethics. Theonomous ethics, referring to the Divine or God, is not part of ethics as a science, nor therefore is theological ethics. The proper subject of morality in human ethics is the human being, and the object is the human being, but in a specific ethics, for instance, an animal ethics, the object might be an animal.

2.2 *Machine Ethics*

According to that, ethics normally relates to the morality of humans, of individuals, and groups, and in a certain way also to the morality of organizations, for instance to the perspective of business ethics or more precisely corporate ethics. Differing from that principle, the issue can also be the morality of machines such as agents, chatbots, robots, UAVs (unmanned aerial vehicles), and unmanned cars, or in other words more or less autonomous programs and systems. Here, one can speak of machine ethics and classify it as information ethics (under computer ethics and network ethics, respectively) or as technology ethics or one can even place it on a par with human ethics [8]. The term "algorithm ethics" is used partly as a rough synonym, and partly in discussion about search engines, dropbox lists and big data; in what follows, I will not refer to this term.

Classification of machine ethics under the umbrella concept "information ethics/technology ethics" means it remains within the ambit of human ethics. This could imply one does not see machines as real subjects of morality, but assigns this privilege to humans alone. In this case, machine morality would be "on loan" from humans and given to autonomous machines with the understanding that it could be withdrawn at any time. Classifying machine ethics as equivalent to human ethics means it is an ethics in its own right, with systems and machines occurring as moral subjects with justification. This is the case when machines assume a life of their own and in certain extremes, independently move away from the ideas and wishes of their human inventors. Maybe this possibility is the source of the idea that morality is not something that lifts humans above other creatures, mechanical or organic.

If one understands machine ethics as an ethics in its own right, one can try to develop new normative models that match machines, models that they can process easily. Just as well one can try to refer to more classic normative concepts from philosophical ethics, and in this vein I think deontological ethics is a suitable candidate for machine morality [31]. A machine can easily process a duty or a top-down rule. An example: One can teach a machine to tell the truth (which will always be the truth in the situation), or to by-pass animal or human obstacles, or to brake to avoid them. Is a machine able to do more than follow such instructions? Is it able to consider the consequences of its actions, and act responsibly? Is it hence able to be obliged by consequentialist ethics or responsibility ethics? Thus, it seems that classic normative models, such as can be traced back to Aristotle or Immanuel Kant, are generally amenable for machine processing [12].

Robot ethics is a germ cell as well as a specialty of machine ethics [14, 30]. The central ethical issues are whether a robot can be a subject of morality, and how to implement machine morality. This discipline also focuses on mimicry, gesture and natural-language skills, as far as these are embedded in a moral context. If a human being approaches a robot, and the robot makes a face and says, "You are so ugly!", this takes us right to the heart of design, with one foot already on the field of morality. Thus, one can ask not only about duties, but also about the "rights" of robots. However, machines, being different to animals, are not normally granted rights. Not lastly, one can understand robot ethics in a very different sense; namely, in relation to the development and production as well as the consequences of robot applications. With this view, it can be located under technology ethics and information ethics. It follows that the term "robot ethics" is complex as well as easy to misunderstand. Although this article addresses ethics with a focus on robots, I prefer the terms "machine ethics" and "information ethics" or "technology ethics", as it concerns machines as subjects on the one hand, and humans as objects on the other, and the mentioned terms are clearer and less amenable to misunderstanding.

2.3 Disciplines of Applied Ethics

2.3.1 Information Ethics and Technology Ethics

The object of information ethics is the morality of those who offer and use information and communication technologies (ICT), application systems and new media [9, 29]. It inquires how these persons, groups and organizations behave in aspects of morality and ethics (empirical or descriptive information ethics) and how they should behave (normative information ethics). Those who do not offer and use ICT and new media but are involved in their production or are affected by their effects are also relevant. So, information ethics focuses on morality in the information society and analyses how its members behave, or should behave, in moral terms. It also analyses the relationship of the information society to itself, to non-technology affine members, and to low tech cultures under ethical aspects. Meta-information ethics analyses moral propositions or statements, starting for instance with the information technological terms, and it locates and compares concepts of information ethics. Computer, network and new media ethics classify under information ethics. This makes internet ethics a part of it, and machine ethics—as far as the machines are enriched with information technology (IT)—also seems to be close to it. The relationship between machine and information ethics was addressed above.

Technology ethics relates to moral issues of the use of technique and technology [13]. It can focus on the technology of vehicles or arms as well as on nanotechnology. There are manifold relations to science ethics. In the information society, technology ethics is also closely connected to information ethics. Not lastly, it has to cooperate with business ethics, in as far as companies are involved in the development and marketing of technological products, and these are demanded and used by customers. As robots and machines in general are technologies, technology ethics also seems to be very close to robot ethics and machine ethics. What was said in the previous paragraph also applies to the relation to machine ethics, and for robot ethics I refer to the above discussion. The section of robot ethics that analyses the effects of robots on humans is closely connected to technology ethics.

2.3.2 Medical Ethics

The object of medical ethics is the morality in medicine. Empirical medical ethics analyses moral thinking and behavior as related to the treatment of human illness, and the promotion of human health. Normative medical ethics deals "mit Fragen nach dem moralisch Gesollten, Erlaubten und Zulässigen speziell im Umgang mit menschlicher Krankheit und Gesundheit [English translation: "with questions of what is wanted, allowed and permitted morally, especially in terms of human illness and health"]" [34, p. 10]. Furthermore, the handling of animal illness and health can be reflected and transferred to human conditions.

Medical ethics shows increasing overlap with information and technology ethics. Medicine has long used instruments for diagnosis, treatment and surgery, and more recently some of these instruments have become machines, or parts of machines. Some machines are more than instruments, however, as for instance (partly) autonomous robots and information and consulting systems. In machine medical ethics or medical machine ethics, medical ethics now also meets with machine ethics.

3 Machine Medical Ethics Versus Medical Machine Ethics

Machine medical ethics or medical machine ethics defines a special range of medical ethical or machine ethical applications. Machine medical ethics opens a new field to medical ethics. Medical ethics deals in the present context with specific moral questions, related to machines as moral subjects. The question is how medical ethics copes with this requirement: on the one hand, it has to acquire knowledge about (partly) autonomous machines and their moral skills; on the other, it needs to acknowledge the conventional view of humans as privileged subjects in applied ethics.

Medical machine ethics develops from a machine ethics that structures its object, and finds the health sector as its field of primary application. As moral machines, it mainly identifies robots, but possibly also special information and consulting systems, intelligent houses and other "housing and living machines". This makes medical machine ethics close to robot ethics in a narrower sense and with relation to medicine, and it searches for other allies. Medical machine ethics is acquainted with non-human subjects of morality.

The situation reminds us of the discussion and development in business ethics. Göbel [22] distinguishes between (1) the application of ethics to economy, (2) the application of economics to morality, and (3) the integration of ethics and economics. Some aspects seem to support letting professional ethicists such as philosophers work on business ethics in the sense of conventional applied ethics. But, economists might be just as prejudiced as theologians who strongly intervene in business ethics. However, the question is whether machine ethics, understood as equivalent to human ethics, or medical ethics as specific ethics, should dominate or initiate. I propose that machine ethics, in close cooperation with medical ethics and medicine, develops a medical machine ethics, and one important argument supporting this was presented in the last paragraph.

At this point, one might ask who machine ethicists really are. Are they professional ethicists, as the name suggests? Or are they a motley crew of ethicists, other philosophers, experts for artificial intelligence, robotic experts, computer scientists, and more recently also business information scientists because of the economic implications? This might come closer to the truth, I suspect, and is supported by a rough analysis of the pertinent literature (e.g., [3, 35]). Another proposal would merit machine ethics to be the object of ethicists, in close cooperation with other disciplines. I presume no one objects to other science disciplines giving essential impulses. All in all, machine ethics is too cross-disciplinary for a field to make a fight for competencies reasonable. There is another aspect: in the present further training market it is easy to certify as an ethicist. Nothing objects to that, and in the end it is the results achieved with correct methodology that count.

Although the discipline of machine medical ethics or medical machine ethics is still very young it already has yielded promising results in theory and practice. The MedEthEx by Susan and Michael Anderson is but one salient example. Wallach and Allen [35] summarize the sense and purpose of the machine: "MedEthEx learns how to weigh duties against each other from the decisions made about specific cases by medical ethics experts when duties conflict." [35, p. 127] They point out: "The Andersons would not claim that MedEthEx is suitable for autonomous decision making in the clinic, although they do see this kind of software being useful in an advisory role." [35, p. 128] This role could be used not only by humans, but also by robots, meaning other machines which however would have to face the same ethical challenges as MedEthEx.

4 Robots in Healthcare

Robots have become inevitable in healthcare. A TA-SWISS study published in 2013 under the title "Robotik in Betreuung und Gesundheitsversorgung [Robotics in nursing and healthcare]" represents the opportunities as well as the risks in applied ethics [5]. Indeed, Switzerland is a land of robots: research of universities such as the University of Zurich, ETH Zurich and ETH Lausanne is widely acknowledged, and some Swiss researchers have an international reputation [32]. However, almost no research is devoted to the moral behavior of machines. The US-American study "Healthcare and Medical Robotics" [1] analyses the market for medical robots from 2010 to 2016 and gives an optimistic outlook.[1] By now, the reality of artificial assistants has become so complex it needs systematic analysis.

I propose to classify medical robots in this context in categories of surgical, therapeutic, nursing and sex robots [11]. Obviously, there is overlap between the different fields of operation. Surgery can be a form of therapy, and it is not always possible to segregate therapy and nursing precisely. Not everyone will be convinced that sex robots should be added to this category, but a sex life that fulfills the individual needs surely contributes to health and wellbeing. Probably one can identify additional types now and in future, for instance, tiny diagnostics robots that move through bodies, or medically trained counseling robots, or robots as "avatars" and tools of non-mobile people while controlled by way of brain-computer interfaces (BCIs). Furthermore, certain artificial limbs could be considered robots [6, p. 23].

The term "service robot" requires consideration. One could call all of the aforementioned robots "service robots" to segregate them from industrial robots [20, 21]. Depending on the language, the term "service" might sound strange in certain fields of application. Robots of this type are often described and discussed by the general public, which is not prima facie an objection to the use and dissemination in academic language, but against the use and dissemination in everyday language. Surely, it would be desirable for the general public and academic science to agree on a shared terminology.

Decker [20] differentiates between "personal/domestic robots" and "professional service robots". In the first category, he includes robots for "handicap assistance" (e.g., "robotized wheelchairs") next to personal assistants, vacuum and pool robots, and in the second category, he lists machines in the field of "medical robotics" (e.g., "diagnostic systems" and "rehabilitation systems"). Of course, there may be robots that can be used in the personal/domestic as well as the professional contexts, and therapy and nursing robots are of that kind.

---

[1] The company informs on its website: "A new study by ABI Research, 'Healthcare and Medical Robots' foresees the global market for medical robotics growing from just under $790 million in 2011 to nearly $1.3 billion in 2016, driven largely by sales of advanced surgical robots and related automated radiosurgical systems" [2].

Healthcare robots can be distinguished further by other criteria. Some appear as instruments used in production halls or fitness studios; others look like humans, being more or less humanoid or anthropomorphous. A few of them are in between. A robot like JACO that consists of an arm with a hand with three fingers is like an industrial robot and a living being at the same time. Robot gender is a category based on appearance. Female chatbots, for example, are “fembots”. Some robots can act only, others can mimic and gesture, others can understand human language, and read or write. Other criteria for distinguishing robots in general and more specifically for medical service robots in a wider sense, are developmental level, degree of autonomy, mobility, local and remote control, intelligence, ability to learn, velocity and costs. There are also important differences in quantity and quality of the sensors [26] and media.

Developmental level is a critical factor. Some robots are in productive use, others are prototypes. Some are merely at the level of a draft idea. Truly, machine ethics as a field has not even achieved pubescence. Although it is human to peer into the future, and to ponder what might be, roboticists and machine ethicists should state clearly in their published research what is present and future, what is real and what is mere fantasy.

In what follows, medical robot types are described in surgery, therapy, nursing and sex contexts and related to issues of machine ethics and applied ethics, as framed above. The field of application is briefly outlined before the type is explained, followed by some illustrative examples. At the end of the chapter, I raise some questions which we have to answer if we take machine medical ethics or medical machine ethics seriously. The general questions concern machine ethics in all, partly in relation to the field of application, the special questions relate to medical machine ethics, and finally questions are raised about specific ethics, indicating options for distinguishing the machines listed in this chapter.

4.1 Surgical Robots

Surgery is an intervention by means of instruments and devices on or in the body of a human or animal patient. Its main purpose is therapy, diagnosis or modification, especially modification for cosmetic ends. Usually surgery is done under local anesthesia, or under general anesthesia to avoid pain and unwanted reactions. The person performing the surgery, normally a specialized physician, is the operating surgeon. Surgery is typically performed in hospitals or medical practices.

Surgical robots can be used to carry out specific actions during surgery, or even to conduct an entire surgical process. They are able to make very short and very precise cuts, and to mill and drill with the highest precision. Normally they will be controlled by a physician who is on site, or at another location, or they might function, in very exactly defined limits of time and space, more or less autonomously. The advantages of robot surgeons are that the intervention is gentler and more agreeable for patients, and physicians are afforded clear views of the body area targeted for surgery [23]. The high cost of robot surgeons, however, is a drawback.

Several prototypes have been developed since the 1990s and launched on the market. The ZEUS Robotic Surgical System, or ZEUS, is already a piece of medical history. ROBODOC (www.robodoc.com) can drill bones for hip joint replacements. The da Vinci Surgical System (www.davincisurgery.com) is popular in clinics for radical prostatectomy and hysterectomy. It is a telerobot and hence only partly autonomous [6, p. 23]. The Amigo Remote Catheter System (www.catheterrobotics.com) is used for cardiac surgery, the CyberKnife Robotic Radiosurgery System (www.cyberknife.com) for cancer therapy, and the Magellan Robotic System (www.hansenmedical.com) for vascular intervention.

In surgical contexts, there are (1) general questions concerning machine ethics, (2) specific questions concerning machine medical ethics, and (3) questions spanning multiple non-machine ethics:

1.

- Should the surgical robot have moral skills, and if so, what skills?
- Should it follow defined duties only (deontological ethics), or should it be able to estimate the consequences of its actions (consequentialist ethics) and weigh pros and cons in decision-making?
- Do other normative models apply, for instance, a materialistic concept?
- How autonomous should it be?

2.

- Should the robot under certain circumstances be able to refuse performing surgery?
- Does it have to consider patient needs and parameters?
- How to handle that the robot normally will not be able to adequately respond to uncertainty and worries? [6, p. 23]
- Is it obliged to express its own concerns and report unforeseen events?
- If it is autonomous, does it need a human surgeon to assist in certain situations?

3.

- Who is responsible if the machine performs poor surgery? [6, p. 24]
- How to handle uncertainty and worries caused by the robot?
- How do the Hippocratic Oath and the Geneva Declaration of the World Medical Association apply?
- Does a robot surgeon support or compete with physicians and their assistants? [6, p. 25]
- Does it interfere with communication between the surgeon, assistants and other medical staff? [18]
- How to rate the circumstance that only certain clinics and medical practices can afford such robots? [18]

These questions might be answered from the perspective of information, technology, medical, occupational and business ethics. Partly, they are related to legal ethics, for instance, with regards to responsibility, liability, and the legal status of robots as "natural persons". From an ethical perspective, Bekey [6, p. 25] generally determines: "Truly autonomous procedures on the part of a robot surgeon will require a number of safety measures to ensure that patients are not harmed."

4.2 Therapeutic Robots

Therapy means activities for treating injuries and illnesses as well as misalignments. Its objective is to permit or accelerate healing, to eliminate or alleviate symptoms and the recovery or production of normal physical or psychological functioning. There are several options for therapy, for instance surgery, medication, physical therapy, or psychological counseling. In a narrower sense used in everyday language, therapy is something that follows an (often drastic) intervention, such as surgery. In this context, the term of rehabilitation is also relevant.

Therapeutic robots support therapeutic activities, or apply such activities, such as performing exercises with paraplegic patients. They entertain dementia patients, and challenge them with questions and games. Some look like industrial robots, others can mimic, gesture and use language. They can learn and be "intelligent". Some advantages of therapeutic robots are potential cost-savings and reusability; disadvantages include possible unwanted effects of robot therapy, and poor acceptance by the patients and their family members [17, p. 151 ff.].

There are a large number of products and prototypes. The robotic seal Paro (www.parorobots.com) is very popular. It has been used for therapy for years, in Japan where it was developed, as well as in Europe. It has a high acceptance by patients because its non-humanoid appearance does not raise high expectations, thus avoiding the "uncanny valley effect". Paro can understand its name, and it remembers how well or badly it was treated, and how often it was petted. It expresses feelings through noise and movements. Keepon (www.mykeepon.com) is a small, yellow robot and also highly popular. It is designed to study and improve social interaction of autistic children. It is available on the mass market, probably because it looks funny, likes to be cuddled and tickled, and knows how to dance.

The general machine ethics questions for therapeutic robots are the same as for surgical robots, but there are different (1) unique machine medical ethics questions and (2) non-machine ethics questions:

1.

- Must therapeutic robots implement certain therapeutic concepts and models?
- Do they have to consider patient needs and parameters?
- How to handle that the robot normally will not be able to adequately respond to uncertainty and worries? [6, p. 23]

- What are the consequences if patients come to depend on the robot? [6, p. 22]
- How should the robot behave in case of conflict, for instance, if several drugs are suitable, or if several patients need therapy? [6, p. 23]
- Should it make it clear to patients that it is no more than a machine?

2.

- Is it relevant to therapy that some robots appear as robotic toys?
- Who (or what) is responsible if the machine performs poor therapy?
- How to deal with uncertainty and worries caused by the robot?
- What if social contact between patients is reduced by the robot? [16]
- How to handle personal data collected and evaluated by the robot?
- Does the robot unburden or compete with therapists and psychologists?
- How to evaluate the fact that only certain people can privately afford therapeutic robots?

A general problem is that therapeutic robots and patients are frequently together on their own, making it difficult to assess whether the therapy works or requires further improvement. It is likely that therapeutic robots will routinely gather and evaluate certain patient data. However, the challenge here is that it is unknown how best to systematically assess such data in order to optimize therapeutic outcomes and that the informational autonomy must be protected.

4.3 Nursing Robots

In nursing care, we distinguish between healthcare and nursing, as well as care for disabled persons and care for older persons. Accordingly, it includes care and services for ill, disabled, old and dying patients by nurses and medical staff, and assistants for disabled or older adults. The central issue is to care for the wellbeing and health, and to avoid illness. The interests of vulnerable patients in need of care should be recognized and taken seriously. Care for animals presents an additional challenge.

Currently, nursing robots either support or replace human nurses. They administer drugs and food to patients, and help them to sit up or lie down. They entertain patients, and they act as auditive and visual interfaces to human nurses. Some have language skills, and can learn. The main advantage of nursing robots is that they are available 24/7, day or night, and with certain limitations also in intermediate phases where care is not necessary and the quality of their service is constant. The main disadvantage, of course, is high costs of such machines and the reduction of interpersonal human contact. Nursing robots require specialization to complete complex nursing tasks, which is a very considerable technical and engineering challenge.

Domestic and nursing robots are represented in trade fairs and conferences. Some are already in productive use. The "nurse's assistant" HelpMate and the "nurse-bot" Pearl are earlier developments supporting the nursing staff [6, p. 22].

HelpMate transports objects. Pearl supplies useful information and visits patients. JACO (kinovarobotics.com) is a robot with one arm and a hand with three fingers. It can pick and bring objects in its reach, such as from a station at the bed. Care-O-bot, developed by the Fraunhofer Institute for Manufacturing Engineering and Automation IPA (www.care-o-bot.de), can pick and bring even objects from further away. It moves safely among people and through the patient's room. Hospi by Honda communicates between doctors and patients, and can bring sick persons the drugs they need [15]. Cody from the Georgia Institute of Technology can turn and wash bedridden patients autonomously (www.hsi.gatech.edu/hrl/clean-iros10.shtml).

The general machine ethics and unique machine medical ethics questions are the same as for the previously described medical robots, but there are some different specific ethics questions for nursing robots (e.g., Should it be possible to exclude future care by robots in patient declaration of will?). A general challenge facing nursing robotics is whether living permanently in the same household as the patient will adversely affect patient social competence. One talks to the machine, one gets used to it, and fellow human beings might become less important.

4.4 Sex Robots

Robot sex, and sex with or between robots, is a subject in science fiction books, movies and series like "Real Humans", and, partly visualized through avatars, of computer games. However, it is also considered on the healthcare segment, for instance, as a help for disabled and older adults, and for possible support in therapy. The media is enthusiastic about robotic sex, naturally enough, and there has been some academic discussion about it [19, 36].

Depending on budget and taste, sex robots are available as a handy toy or as big as life. They can help people reach sexual satisfaction through stimulation or penetration. Some have natural language skills, which is advantageous since one should not forget that verbal eroticism is very popular in chats, and phone sex was in high demand for some years. The sexual interactions in Second Life can also serve as a reference. The advantages of sex robots are their constant availability, relatively high hygiene standards if handled properly, and unburdening of human sex workers of both genders. Challenges facing sex robots are limited bandwidth concerning sensory satisfaction of human users and low acceptance and understanding by the general population. In Second Life and in other contexts, the use of child avatars and virtual characters might be a problem.

Fuckzilla, presented on the Arse Elektronika 2007, has a full arsenal of toys and tools, from dildos to a chainsaw decorated with tongues [25]. This model seems to be more of an outlandish object of art than a love-making partner to take seriously. In 2007, a German online-shop offered sex androids [24]. Roxxxy (www.truecompanion.com) is a fembot (in the widest meaning of the word) which according to information from the company is able to listen and speak, and can respond sexually to touch. Several personalities are available, ranging from "Wild Wendy" to "Frigid Farrah". The male equivalent is called Rocky. Both are available through the website of the provider.

Again, the general machine ethics questions are the same as for the previously described medical robots, but there are different (1) unique machine medical ethics questions and (2) non-machine ethics questions for sex robots:

1.

- Should a robot become sexually active on its own, and entice the partner to have sex?
- Should it be able under certain circumstances to refuse performing sexual acts?
- Should it reiterate to the partner that it is no more than a machine?
- Should its design fulfill moral criteria (e.g., prohibition of child-like sex robots, except in case of combating pedophilic crime [4])?
- Should there be novelty options for stimulation and seduction, or should it follow human role models?

2.

- How should the robot collect and evaluate patient data to better satisfy its partner's sexual needs?
- Who is liable for injuries or contamination caused by sex robots?
- How should uncertainty and shame in patients caused by sex with robots be ethically assessed?
- Should sex robots completely replace human sexual partners for certain patients?
- Does robot sex promote the idea that a sexual partner must be available at all times?
- Should children and underage teenagers be permitted to have access to sex robots for sexual or non-sexual purposes?

The use of sex robots in medicine is a sensitive ethical question. Humans who are sexually substituted by a robot might feel rejected by their partner and carer and patients who choose to have sex with a robot for having no other choice for sexual satisfaction might suffer for lack of human-human sex. Patients who cannot afford a robot might feel disadvantaged. For some adults, sex with robots might be on the same level as sex with animals. For others, it may enrich life and contribute to good health.

4.5 The Robot in the Morality of Medicine

Although asking questions from the perspective of machine ethics is easy, answering those same questions can be very difficult, especially on a case to case basis, and engineering moral medical machines is far from trivial. The large number of questions posed in this section for surgical, therapeutic, nursing and sex robots seems to indicate that complexity in moral machines is the main challenge [10]. In other words: if surgical, therapeutic, nursing and sex robots are furnished with moral skills, one would expect them to flexibly make correct decisions and perform actions on very different levels and in very different situations. Implementing very simple moral machines—in many cases the right choice—might therefore cause or enhance the uncanny valley effect here. At the same time, we note that redundancies occur in general and that some solutions to the questions asked above probably might be transferred to different types of robots.

Asking questions of applied ethics is also easier than providing correct answers, of course, and in some cases it might be difficult to clearly classify moral machines as falling under a certain ethics. For instance, the autonomy of patients, an enduring topic of medical ethics, in some cases turns into informational autonomy, and thus concerns information ethics. Sexual self-determination as considered under sexual ethics also has to be reviewed with regards to sexual intervention, collection and evaluation by machines. The right of satisfactory work concerns occupational ethics, and can also be discussed from the perspective of information and technology ethics, as the use of information and communication technologies and application systems frequently serves for improving processes and making work easier.

Most aspects of applied ethics might be traced back to information (and technology) ethics because information ethics can be located in the center of applied ethics [7]. If information and communication technologies as well as application systems diffuse into all areas of society and business, this will influence specific ethics. Thus, these ethics will likely draw on information ethics for further ethical guidance.

Anyway the compilation provides a better view of what machine ethics is in a narrower sense, and what counts as applied ethics related to humans. Machine ethics inquires from the perspective of machines, focusing on machines as subjects. Human ethics assumes humans to be subjects as well as objects. It also provides a better understanding of what might be more general questions relating to machine ethics and what might be more specific questions, relating to medical machine ethics or machine medical ethics.

Asking questions, even if not supplying the answers, can lead to knowledge and to a deeper, more structured and systematic understanding. More theory and data are needed in the context of machine medical ethics, as well their ongoing synthesis and eventual symbiosis. Neighboring disciplines such as health technology assessment [28, p. 199] need to be recruited to fortify machine medical ethics, and it will be necessary to interview affected patients and professional staff.

5 Conclusions and Outlook

Machine medical ethics or medical machine ethics is new territory, a novel field of work for ethicists, philosophers, artificial intelligence experts, information scientists, and medical specialists. Surgical, therapeutic, nursing and sex robots are the primary types of medical machines and objects of further analysis as the field grows. In relation to their development and application, it is possible to ask questions of machine ethics as well as specific ethics. I have reasoned that information and technology ethics in particular are critical to solve ethical problems involving medical machines. Machine ethics might help avoid such problems. If a medical machine is moral, it acts (in the best case) to respect and adequately care for its patients, so that they can lead a good life and maintain personal autonomy. Their acting morally of course could in certain cases also lead to impairing the lives of people. If a machine again and again emphasizes that it is no more than a machine, this might be transparent and honest, but it might also cause uncertainty and frustration. The assessment of uncertainty caused by a robot is a challenge to solve.

Currently, it is sheer guesswork to ponder the domains of applied ethics machine ethics will need to cooperate or merge with. Animal ethics is one candidate. While machine ethics focuses on humans as moral objects, animals are also affected by machine decisions and actions, for example if medical machines operate in households. Military ethics is another potential field of cooperation, despite public unease about it, since there is strong technological overlap between medical machines, military UAVs and fight robots. Religiously motivated objections against moral machines will demand a response from machine ethicists, likely in the form of scientific arguments and data. Considerations such as those raised in this chapter do not mean that machine ethics should be viewed uncritically or blindly. However, I believe that it will be not less than a touchstone of future ethics.

References

1. ABI research (2011) Healthcare and medical robots. Allied Business Intelligence, Oyster Bay
2. ABI research (2011) Medical robots market to approach $1.3 Billion in 2016. https://www.abiresearch.com/press/medical-robots-market-to-approach-13-billion-in-20
3. Anderson M, Anderson SL (eds) (2011) Machine ethics. Cambridge University Press, Cambridge
4. Austin H (2013) Virtual girl dubbed 'Sweetie' snares thousands of would-be sex predators. In: World News, 5 Sept 2013, http://worldnews.nbcnews.com/_news/2013/11/05/21316335-virtual-girl-dubbed-sweetie-snares-thousands-of-would-be-sex-predators
5. Becker H, Scheermesser M, Früh M et al (2013) Robotik in Betreuung und Gesundheitsversorgung. TA-SWISS 58/2013. vdf Hochschulverlag, Zürich
6. Bekey GA (2012) Current trends in robotics: technology and ethics. In: Lin P, Abney K, Bekey GA (eds) Robot ethics: the ethical and social implications of robotics. The MIT Press, Cambridge, pp 17–34
7. Bendel O (2012) Die Medizinethik in der Informationsgesellschaft: Überlegungen zur Stellung der Informationsethik. In: Informatik-Spektrum, November 2012 ("Online-First"-Article on SpringerLink)
8. Bendel O (2012) Maschinenethik. Contribution for the Gabler Wirtschaftslexikon. Gabler/Springer, Wiesbaden, http://wirtschaftslexikon.gabler.de/Definition/maschinenethik.html
9. Bendel O (2012) Informationsethik. Contribution for the Gabler Wirtschaftslexikon. Gabler/Springer, Wiesbaden, http://wirtschaftslexikon.gabler.de/Definition/informationsethik.html
10. Bendel O (2013) Ich bremse auch für Tiere: Überlegungen zu einfachen moralischen Maschinen. In: inside-it.ch, 4 Dec 2013. http://www.inside-it.ch/articles/34646
11. Bendel O (2013) Dr. Robot entdeckt die Moral: Maschinen- und Menschenethik im Gesundheitsbereich. In: IT for Health, 02/2013, pp 2–4

12. Bendel O (2013) Towards a machine ethics. In: 1st PACITA project conference technology on assessment and policy areas of great transitions: book of abstracts. 13–15 Mar 2013. Prague, pp 229–230, http://pacita.strast.cz/en/conference/documents
13. Bendel O (2013) Technikethik. Contribution for the Gabler Wirtschaftslexikon. Gabler/Springer, Wiesbaden, http://wirtschaftslexikon.gabler.de/Definition/technikethik.html
14. Bendel O (2013) Roboterethik. Contribution for the Gabler Wirtschaftslexikon. Gabler/Springer, Wiesbaden, http://wirtschaftslexikon.gabler.de/Definition/roboterethik.html
15. Bittner U, Germis C (2012) Hospi bringt die Medizin. In: FAZ.NET, 28 Jan 2012. http://www.faz.net/aktuell/wirtschaft/unternehmen/pflegeroboter-hospi-bringt-die-medizin-11620830.html
16. Bleisch B (2013) Wenn uns der Roboter pflegt. Interview with Susanne Boshammer. In: SRF Kultur, 11 Oct 2013. http://www.srf.ch/kultur/roboter-wie-wir/wenn-uns-der-roboter-pflegt
17. Butter M, Rensma A, van Boxsel J et al (2008) Robotics for healthcare. Final Report. 3 Oct 2008. http://www.tno.nl/downloads/TNOKvL_report_RoboticsforHealthcare.pdf
18. Datteri E, Tamburrini G (2009) Ethical reflections on healthcare robotics. In: Capurro R, Nagenborg M (eds) Ethics and robotics. Akademische Verlagsgesellschaft AKA, Heidelberg, pp 35–48
19. David L (2007) Love and sex with robots: the evolution of human-robot relationships. Harper Perennial, New York
20. Decker M (2012) Technology assessment of service robotics. In: Decker M, Gutmann M (eds) Robo- and information ethics: some fundamentals. LIT Verlag, Münster, pp 53–88
21. Decker M (2013) Mein roboter handelt moralischer als ich? Ethische Aspekte einer Technikfolgenabschätzung der Servicerobotik. In: Bogner A (ed) Ethisierung der Technik – Technisierung der Ethik: Der Ethik-Boom im Lichte der Wissenschafts- und Technikforschung. Baden-Baden, pp 215–231
22. Göbel E (2010) Unternehmensethik: Grundlagen und praktische Umsetzung. Lucius & Lucius, Stuttgart
23. Grimm M-O (2011) Da Vinci OP-Roboter—Marketing Instrument oder medizinischer Fortschritt? In: Management & Krankenhaus, 9 Sept 2011, Issue 9, p 6
24. Hänßler B (2012) Stets zu Liebesdiensten. In: Stuttgarter-Zeitung.de, 29 Aug 2012. http://www.stuttgarter-zeitung.de/inhalt.sexroboter-stets-zu-liebesdiensten.59ec16f3-55c3-4bef-a7ba-d24eccfa8d47.html
25. Hartwell L (2007) So who wants to F**k a robot?' In: Wired.com, 10 June 2007. http://www.wired.com/underwire/2007/10/so-who-wants-to/
26. Hertzberg J, Lingemann K, Nüchter K (2012) Mobile Roboter: Eine Einführung Aus Sicht der Informatik. Springer, Berlin and Heidelberg
27. Höffe O (2008) Lexikon der Ethik, 7th edn. C. H. Beck, München
28. Kollek R (2013) Ethik der Technikfolgenabschätzung in Medizin und Gesundheitswesen: Herausforderungen für Theorie und Praxis. In: Bogner A (ed) Ethisierung der Technik – Technisierung der Ethik: Der Ethik-Boom im Lichte der Wissenschafts- und Technikforschung. Baden-Baden, pp 199–214
29. Kuhlen R (2004) Informationsethik: Umgang Mit Wissen Und Informationen in Elektronischen Räumen. UVK, Konstanz
30. Lin P, Abney K, Bekey G (eds) (2012) Robot ethics: the ethical and social implications of robotics. The MIT Press, Cambridge
31. Moor J (2011) The nature, importance, and difficulty of machine ethics. In: Anderson M, Anderson SL (eds) Machine ethics. Cambridge University Press, Cambridge, pp 13–20
32. Pfeifer R (2003) Körper, Intelligenz, Autonomie. In: Christaller T, Wehner J (eds) Autonome maschinen. Westdeutscher Verlag, Wiesbaden, pp 137–159
33. Pieper A (2007) Einführung in Die Ethik, 6th edn. A. Francke Verlag, Tübingen and Basel
34. Schöne-Seifert B (2007) Grundlagen der Medizinethik. Kröner, Stuttgart
35. Wallach W, Allen C (2009) Moral machines: teaching robots right from wrong. Oxford University Press, Oxford
36. Yeoman I, Mars M (2012) Robots, men and sex tourism. Futures 44(4):365–371

Good Healthcare Is in the "How": The Quality of Care, the Role of Machines, and the Need for New Skills

Mark Coeckelbergh

Abstract What do we mean by good healthcare, and do machines threaten it? If good care requires expertise, then what kind of expertise is this? If good care is "human" care, does this necessarily mean "non-technological" care? If not, then what should be the precise role of machines in medicine and healthcare? This chapter argues that good care relies on expert know-how and skills that enable care givers to care-fully engage with patients. Evaluating the introduction of new technologies such as robots or expert systems then requires us to ask how the technologies impact on the "know-how" expertise of care givers, and whether they encourage a less care-full way of doing things. What role should which technologies play in which tasks and practices? Can we design and use them in such a way that they promote care-full and engaged ways of doing things with people? It is concluded that new machines require new skills to handle the technology but also and especially new knowing-how to handle *people*: knowing how to be care-full and caring *with* the technology. Good care is not about something external called "ethics" but about *how* things are done in medical and care practices. Machines are welcome if they contribute to this more knowledgeable, skillful, participatory, engaged, and caring way of doing things. This vision of good care enables us to evaluate new technologies and encourages care givers, care receivers, and other stakeholders to explore better ways of designing, regulating, and using them.

1 Introduction

When in the 20th century car production was automated, this caused a number of problems. One problem was unemployment: suddenly the skilled work of many laborers was no longer needed; machines took over their job. This is why those

---

M. Coeckelbergh (✉)
Centre for Computing and Social Responsibility, De Montfort University, Leicester, UK
e-mail: mark.coeckelbergh@dmu.ac.uk

S.P. van Rysewyk and M. Pontier (eds.), *Machine Medical Ethics*,
Intelligent Systems, Control and Automation: Science and Engineering 74,
DOI 10.1007/978-3-319-08108-3_3

who owned the factory introduced conveyer belts and robots in the first place: they wanted to reduce production costs. This also tends to be one of the reasons why it is proposed to use machines in medicine and healthcare. The argument presented to us is that due to ageing of the population, more care workers are needed, but that in the present economic circumstances we cannot afford them, and that therefore we need robots and other high tech to do at least part of the work, both in places such as hospitals but also at home. As Sparrow and Sparrow remark: "It is remarkable just how much robotics research, if it is not being sponsored by the military, is promoted by appealing to the idea that the only way to deal with a looming demographic crisis is to develop robots to look after older persons!" [10, p. 142] Today the argument mainly focuses on defending home care. Keeping people as long as possible in their own homes does not only respect their own wish to live independent lives; it also saves money.

There are a number of objections one could make in response to this argument and I believe we should be highly critical of policies that endorse it, whether at the level of government or at the level of specific healthcare institutions such as hospitals, insurance companies and other organizations. But the main reason why most people worry about robots in medicine and healthcare has little to do with labor and economics as such. Rather, they are concerned about the quality of care, and here the analogy between industrial production and healthcare no longer seems to work. Whereas few people think that the fact that their car is made by robots has huge implications for the *quality* of the car, they (rightly) worry about the quality of healthcare when robots do the job. The way of "production" seems to matter to the very "product" in question. Whereas in the case of car production, the ethical concerns are mainly extrinsic to the practice and its values, in the case of medicine and healthcare, the ethical concern is intrinsic to what care is all about. Care is supposed to be "human", "warm", and so on, whereas technology is associated with the non-human. Machines are perceived as "cold".

Indeed, the idea of using robots in healthcare evokes images of hospitals and homes for the elderly where no human care givers are present, images of elderly people abandoned and lonely in homes that have turned into automated care capsules. Do we really want to push our most vulnerable fellow humans into the hands of machines? Robots in healthcare then become synonymous to the dehumanization of healthcare, which would mean a healthcare that defeats its very *raison d'être*: it was supposed to care, but it only delivers "services" and "products". It was supposed to be about people, not about "clients", objects, things, and machines. What are we doing? Is it this world we want?

Moreover, if we replace physicians by medical expert systems, it seems that something is lost, although just as in the case of care robots it is not obvious what precisely it is that we miss. Can robots be(come) "experts" in medicine and healthcare, in the same way as physicians and nurses are experts? Would we trust a robotic surgeon? Would we be happy to consult a robotic doctor? We worry about the quality of medicine, but our intuition needs further work. We are pretty sure about what we do *not* want. But articulating what we mean by "good", "ethical" medicine and healthcare is a harder question.

This chapter reflects on the question "What do we mean by good medicine and good care, and do machines threaten it?" It presents an argument about this problem that takes seriously the worries about the quality of care—both worries of the general public and concerns voiced in the ethics of robotics literature—but it also offers a more nuanced view of the role of technology in good medicine and healthcare. Moreover, it also reflects on the meaning of "expertise" and "ethics" in this domain.

In order to address the main question concerning the quality of care and its relation to technology, I will ask two more specific questions, which start from the assumptions implied in the previous discussion about medical and care expertise and about the quality of care: the assumptions that (1) good care is "expert" care and that (2) good care is "human" care. The first question is: If good medicine and healthcare requires expertise, then what kind of expertise is this, and who has or can have this expertise? The second question is: If good care is "human" care, does this necessarily mean "non-technological" care, does it exclude machines? And if not, then what is and what should be the precise role of machines in good medicine and healthcare? Let me start with the first question.

2 "Good Care Is Expert Care"

Our current systems and practices in medicine and healthcare rely heavily on expert knowledge. In the past 500 years, modern medicine has divorced itself from Aristotelian and medieval thinking about the human body and from what it came to regard as "charlatan" physicians and surgeons. After banning the magicians and healers, it has firmly based itself on hard science, on knowledge gained from experiments, on *evidence*. This modernization of healthcare has also implied its professionalization. In the current (mainstream) medical system, there is no room for "hobbyists". Modern medicine is thoroughly professionalized. It requires years of study and professional training before students are allowed to practice medicine or work as a nurse. But what does "expertise" mean in this domain? What kind of "expert" knowledge do these professionals have? And can machines have that kind of knowledge?

To answer these questions, it is helpful to distinguish between two kinds of knowledge. One is "know-that" and concerns theoretical knowledge, the knowledge one can learn from textbooks and lectures, from cases described in the literature, from instruction manuals and from databases. This seems to be the predominant meaning of "expertise" when people talk about "medical expertise" or "care expertise". Professional physicians know a lot about diseases, medicines, and so on. Nurses are also educated to "know" things in this sense. However, the quality of medicine and healthcare also depends on a different kind of knowledge which is usually labeled as "know-how" or "tacit knowledge" [8], and which is of a more practical and personal kind. If a surgeon had only textbook knowledge about the human body but had never worked on a patient, she would not be regarded as a medical expert. She would "know" a lot in the first sense of know-that (theoretical knowledge), but she would know very little in the second sense of know-how. Similarly, we would not trust ourselves to a doctor who would know a lot about diseases but has never seen a patient in his practice. And the expertise a nurse has cannot in any way be divorced from his practical experience with patients and with specific nursing tasks such as delivering specific medicines to patients, washing a patient, lifting a patient, dealing with a confused elderly person, and so on. What is needed, it seems, is not only theoretical knowledge but also knowledge of a more practical kind, knowledge that stems from practical experience and practical engagement with people and, as I will argue below, with things. This knowledge requires many hours of training, since practical experience is its basis. It is a knowledge that is embodied, that is part of being a skilled practitioner.

This does not mean that theoretical knowledge is obsolete. Explicit instruction, protocols, textbook knowledge, etc. can play a role, albeit perhaps only in the beginning. But full-blown expertise is something else. Dreyfus and Dreyfus have argued that expertise requires one to develop skills [6] and that "expertise" requires various stages of development: whereas novices need to rely on rules and guidelines [5], experts have a more intuitive grasp of the situation, a tacit understanding or know-how based on experience. This can and has been applied to healthcare. For example, influenced by Dreyfus, Benner has studied how nurses become experts by acquiring skills. Experience enables them to better respond to situations; by means of engaged and involved in the practice, they develop an intuitive grasp of the situation [2].

Moreover, professional medical practice and care practice also requires that the person can deal with ethical problems. Now many practitioners and, unfortunately, many philosophers, assume that ethics is mainly a matter of know-that: knowing values and ethical principles, knowing ethical theories, knowing what we can learn from textbooks in (medical) ethics (e.g., [1]). Moreover, many people assume that ethics is mainly a constraint, something that renders their practice difficult, something that gets in the way of what they actually want to do and actually do. This also assumes that ethics is something that is fundamentally different from their practice, from their core business.

The view of ethics I will try to articulate, by contrast, acknowledges that theoretical knowledge about ethics can have *a* role in ethical education, perhaps in the beginning, as the Dreyfus model suggests, but that ethics is also and probably *mainly* a matter of know-how: of knowing how to deal with ethically charged situations when they arise *within* a specific practice such as medical care and healthcare. This ethical know-how is not divorced from the "technical", practical know-how one acquires as a practitioner, but is rather to be seen as a dimension of it. "Ethics" is then not an alien intruder but belongs to the practice. This view of ethics assumes that professionals want to deliver good care, and that this good care is therefore not only "good" in a technical sense but also "good" in an ethical sense. The ethics is in the very aims of the practice and in how one does things, in practice. Ethics is then directly related to the care practice and to the *quality* of care, rather than being a kind of policeman standing on the side lines of the practice. It belongs to the skills of the professional, partly as a separate skill (e.g., the skill of reasoning), but mainly as deeply connected to the so-called "technical" skills. Being a "good" doctor or being a "good" nurse includes being "ethical" by definition. It is about doing good things and doing them in a good way. The practice itself is ethical. "Doing ethics" is not a separate kind of activity; practicing as a professional doctor or as a nurse, being an "expert" in these domains just includes "doing ethics". There is no meaningful and no fundamental distinction between "doing the things a doctor or a nurse does" and "doing ethics". It is about being able to solve problems, having the skills to do so, and this is both "technical" and "ethical".

If we ask whether machines can be(come) medical experts, nurses, and so on, then presuming this is a good question to ask (I will criticize this question in the second part of this chapter), we now have clear general criteria for assessing their expertise. In order for a machine to become an expert in medicine and healthcare, the machine must have at least the following kinds of knowledge:

(a) The machine must know-that; that is, it must have the necessary theoretical knowledge; for example, about the workings of the human body, about disease and ageing, about the causal relations we know from experiments, about the theories and the evidence we have collected in books and databases;
(b) The machine must know-how; that is, it must have the necessary skills; for example, as a machine surgeon or as a machine nurse, and it must learn those skills on the basis of experience;
(c) The machine must be an "ethical" machine, in the sense that it must have theoretical knowledge about ethics, but it must also have "ethical skills" in the sense that it must "do good" *as* a medical and nursing machine, it must be part of good practice, producing quality, and it must be able to solve ethical-medical problems that arise within the practice.

Developments in artificial intelligence and robotics clearly demonstrate that machines become better "experts" and have the potential to become "ethical" in the sense of having know-that, theoretical knowledge. Think about chess computers, for example, or current expert systems in medicine. It is, however, highly doubtful that they can fully meet criteria (b) and (c). Of course, machines can do many things. For example, they already fly airplanes. Pilots can doze off in their cockpits; the computer pilots the plane and it is pretty good at it. It is imaginable, in principle, that in the future we could have surgical robots or care robots that do some things automatically. Surgeons can doze off then, or at least sometimes. However, there are at least two questions to ask here. First, does this count as "having skills" in the rich sense of the term explained above, and second, even if it were to count as "having skills" in that sense, is something missing, perhaps something "human", that we may not need in the case of flying a plane but that is *essential* in the case of medicine and healthcare?

In order to answer the first question, I propose that we distinguish between two notions of "skill". The first notion is a purely technical and functional one and there is no doubt that robots can learn skills in this sense. For example, driving a car can be automated today in the sense that the machine can accelerate, steer, stop, etc. without human intervention and, increasingly, can keep the car on the road without driving into other cars or into a wall of a building. Cars, planes, and other machines can be programmed in doing that, and have an increased capability to learn from "experience". However, as with airplanes, things become more complicated when unexpected situations arise (e.g., a child suddenly crossing the road, risk of a plane crash), when human judgment is needed, when human intuitions and emotions are needed, when social situations are to be taken into account. The skills that are needed in these situations seem to presuppose an embodied kind of knowledge that seems to be tied to our human way of being-in-the-world, and which is required for coping with non-routine problems. I do not have much space to further discuss this matter here (it would require, for example, a discussion and extension of Dreyfus's critique of AI), but let me say a little more about this second notion of "skill".

Of course, current AI and robotics is advancing. For example, progress is being made in machine learning. But the point is that even if machine learning can solve some problems, machine learning is not the same as human learning and cannot lead to the kind of coping skills we have as human experts, the kind of expertise Dreyfus and Dreyfus [6] describe. Machines can follow rules, make inferences from the information their sensors give them, operate things, *but they lack intuition, emotion, and judgment*. Although current developments in robotics may give machines more than "beginner's knowledge" since rule-following is extended with learning capacities in real environments, machines still lack the highest, fullest kind of expertise, which is required when technical and ethical challenges are *not* routine. Machines may acquire some competence, but not full expertise that includes the skill to do what is best in very complex, unexpected situations where human lives are at stake. They may be programmed to show behavior that is considered ethical (to achieve certain outcomes that are considered ethical) by what Moor calls "implicit ethical agents" [7], for example an automatic pilot, and they may even be able to represent ethics ("explicit" ethical agents in Moor's sense) or even "reason" about ethics based on ethical principles (which some philosophers might consider to be sufficient for full moral agency). But *none* of these forms of competence meet the criterion of ethical *expertise* (and hence full, human-like moral agency) if we define ethical expertise as crucially including skills based on personal and lived experience in engaging with things and people. It is this technical-ethical expertise, based on personal experience *as a human being* that we value so much in humans. It is this expertise we think human pilots have as opposed to automatic pilots, and it is the reason why we still have pilots in the cockpit and presumably will still have for a long time. It is this expertise we want military commanders and human pilots to have and why we want them to be "in the loop" or at least "on the loop", even if drones can be developed that operate autonomously. It is this expertise we still need when our car has a difficult problem and needs to be diagnosed and repaired, even if the car is produced by robots and even if routine repair tasks could be done by a robot. We need the expertise and the craftsmanship of the humans, for example, to make judgments about risks. It is this expertise and these skills we assume surgeons have, and it is the reason why we do not replace surgeons by machines but let surgeons *use* robots.

A second question, however, is a different one: even if machines had those skills and that kind of expertise, would there *still* be something wrong about replacing human care givers by machines? Even if robots could cope with the most difficult technical-ethical situations, would it *still* be problematic to replace human care givers by machines?

3 "Good Care Is Human Care"

The assumption behind the philosophical intuition that good care is *human* care, is not only that we need human expertise to cope with hard technical-ethical problems, but also and maybe especially, that there is something intrinsic to medical care and practice that would be missing if we replaced doctors and nurses by robots. Here the worry is not so much about the term "expertise" in "care expertise" but rather about "care". As I suggested in my introduction, we associate medicine and healthcare not only with getting the right things done (in technically, but also as I have argued in ethically difficult situations); they also seem to have something to do with how those things are done. We associate "care" not only and maybe not even mainly with "delivering" care, with "services", but also with "caring for" and "caring about", with "being careful" and "being caring". What we worry most about when we meditate on the nightmare scenarios offered in my introduction, is perhaps not so much that the robots in question will not be good in "delivering care" (e.g., being successful at performing surgery, being good in delivering medicines and food at the right time, being effective in monitoring and alarming in a home care situation, etc.), but that they will do all those things in a "machine"-like way, by which we mean: in a cold, uncaring way.

Indeed, when in the literature concerns about machine care are voiced, for example, about potential reduction of human contact [9], indeed the 'reduction of what is often already minimal human contact' [10, p. 152] in interactions that involve vulnerable users such as sick people, young children, or elderly people [11], I think this is the underlying worry which needs further articulation and philosophical discussion. The problem is not so much that care robots are non-biological and non-human as such, say the problem of their ontological status; the deeper problem is rather that we worry that robots would not do things *in the way* a good doctor or a good nurse is supposed to do things, that is, in a caring way, in a way that shows concern with the well-being, feelings, and experience of the patient or the elderly person. We do not want "routine" care or "automatic" care if this means care-less, unconcerned, "cold" care.

For example, Sparrow and Sparrow argue that if we care for someone, we take their hand, stroke them, etc. and robots are not capable of that—at least not in a caring way—since they are not the "biological corporeal entities with particular limitations and frailties" we humans are [10, p. 154]. But the problem is not the biology as such, but that this biology is the basis of embodied thinking and understanding, including understanding "facts about human experience and mortality" [10, p. 154], and of our capacity for shared suffering, all of which seem to be necessary for genuine care which robots lack. Human embodiment is the basis of behaviour and gestures that are appropriate to real care. Sparrow and Sparrow conclude: "Robots cannot provide the care, companionship, and affection that older persons need" [10, p. 156]. Although I have been critical about their claim that care robots therefore deceive users (I discuss this issue elsewhere), their discussion clearly shows again what we want: care givers that do not only provide "care services" but also, or rather, "care" in the sense of being affectionate, "caring".

Now if *this* is what is required (e.g., a necessary condition) for good care, then it seems that those who wanted to make a case for machines in medicine and healthcare are in big trouble. Even if they could somehow show, or (more likely) *promise* that machines can be "experts" in this domain, they do not seem to have any evidence for the claim that machines can "care" in the sense of "caring about", "caring for", "showing and feeling concern for", and so on. Unless they could convince us that machines can and will have feelings, can and will have subjective experience, they cannot argue that machines can fulfill what has turned out to be this necessary condition for good care.

However, there is nevertheless a problem with the view that good care is human care, at least if it is framed as a "humans versus machines" question. The assumption seems to be that we either have human care OR we have robotic care with humans removed from the scene. But this assumption is mistaken, and not only because we have to admit of degrees here (there are more possibilities than either/or: we can replace *some* humans by robots, not *all* care givers will or need be replaced; in this sense care can be more or less robotic), but because asking the question in terms of replacement alone is one-sided and misleading since it is entirely insensitive to the precise role machines, and more generally technologies, play and may play in medicine and healthcare. In particular, it narrows down the range of human-technology relations to two forms: if technology enters the practice, it can only be as a tool—with humans in firm command of the tool, humans who guarantee the quality of the practice—or as a tool that takes over the practice, that pushes the human out of the practice. It's the technology or the human that is in control. Other human-technology possibilities are excluded from this way of thinking. What remains out of sight, for instance, is the human-technology-human relation: the human-technology relation in which the technology plays a role of mediator; it is not a mere tool but mediates what happens and what is supposed to happen in the practice. It is not a mere instrument but part of the "core business". It is directly related to the quality and the values in and of the practice. What does this mean? What does it mean in healthcare? What kind of mediation is happening there? And does this mediation mean that the practice becomes "colder", uncaring?

Let me start with making some analogies. If someone gives flowers to someone (s)he loves, then the flowers are not a tool in the sense in which a hammer is a tool.

The flowers express, instantiate, *materialize* the purpose and value of what is going on here. They symbolize the love but are more than a symbol; they materialize it. They are part of the loving gesture, part of the practice of flower-giving and the practice of the love relationship. The flowers confirm or maybe initiate a relationship. Moreover, if the flower-giving were to become routine, it would stop having that significance and meaning. If it were given in a thoughtless, non-caring *way* it would be quite pointless to do it. Furthermore, no-one would say that the fact that something physical, biological, or material involved here diminishes the value of the act. More: the act couldn't be done without the flowers.

A second analogy, which comes closer to healthcare: If someone cooks for someone else (s)he knows personally, this is also a gesture, and it shows care and *is* care. It does not only respond to a need of the person; the cooking itself shows care. That care is not external to the food or cooking, as if it were an entirely separate goal. The caring is also "in" the food and "in" the cooking. The cook mixes his or her concern, care, friendship, love, etc. into the food; the food mediates the care of the one who is cooking and is part of what goes on between the person who cooks and the person who is cooked for. What matters is that it is done and how it is done. If the cooking became routine or if it were done carelessly, it would not have the value and significance it has. Furthermore, no-one would say that using a cooking robot or a microwave diminishes the meaning of the cooking. The use of technology and the partial automation are not problematic in itself. What matters is how it is done. It is possible to cook with those devices in a non-caring way, in a way that is "cold" and mechanical. Then it becomes a "service", and then the food becomes "free food" instead of a present or a gesture of care. Then it becomes *merely* a response to need, and not a caring response to the person who is offered the food.

Similarly, medicine and healthcare have an aspect of giving and caring, of doing something for someone in a "warm", caring way. Due to professionalization, specialization, and commercialization of healthcare systems this aspect might well be threatened (I return to this issue below), but our normative conception of care, our ideal of care, is one that includes this richer conception. What patients, elderly people, and other vulnerable and needy people need from medicine and healthcare is not only "performance," or a "repair" or a "fix", which they expect when they go to the garage for a repair of their car (by a person who might be very skilled, who is in the best case an expert). They do not only want health as a product, a commodity—free or not. What they need and what they want is health but also a particular way of being made more healthy, which is the opposite of routine. They want not only a "service" but also a little bit of "care for" and "care about", a little bit of concern. If they reject a robot, it is not because they are afraid of a worse performance, but because as human beings they need caring concern next to expertise. They need care in all those senses of the word: technical and ethical, personal and relational. What matters to them is not only what is being done (and that it be done well in the sense of having good performance results) but also *how* it is being done. What matters are also the gestures, also the way things are said, also the way they are touched and when they are touched. This is the "human" aspect of care that would be missing, even if we had "expert" robots that manage and fabricate our health.

The cooking analogy is also useful since it illuminates the role of skill. Skill is needed here; the person has a more practical expertise, know-how. But what makes the food good is not only the skill as such, or the expertise as such, but its role within a caring practice. It is a knowing-how to prepare the food but at the same time also a knowing-how to treat particular others. Similarly, healthcare includes not only the know-how of medicine and health expertise, but also the know-how that has some similarity to the know-how of the caring lover and the caring cook, the know-how that has to do with how to treat people, how to respond to others.

With regard to the use of technology in healthcare, then, what matters is how the technology is used and what its role is in the practice. If it is part of skilful and care-full engagement with the patients, with the elderly people, with the children, and so on, if it is embedded in a caring way of doing things (as well as in the right and good, technical-ethical "expert" way of doing things) then "machine care" is not problematic. It only becomes problematic when the care becomes care-less. Just as the microwave may or may not play a role in the loss of cooking quality, depending on how it is used and what the practice looks like, machines in healthcare may or may not contribute to "colder" care. Yet the problem of this "dehumanization" does not have to do with the technology per se (*that* it is technology, *that* it is a machine rather than a biological human being). The practice is, as always, both "human" and "technological". The key question is (1) if expert quality is provided and (2) how things are done: is it still "care" in the richer sense of the word? And does a particular technology encourage and enable this kind of care, or is there a danger that the technology will be used in a care-less way?

What needs to be evaluated, then, is the precise nature of the human-technology-human relations involved and the particular tasks, skills, and practices that are in play. The quality of care depends on technology, but not because "technology" necessarily replaces "the human", but because technology is (always) part of our ways of doing things. It influences the quality of our work, the skills we need, the way we treat others. For example, it mediates and materializes care, and in the ideal case it does so both in the sense of "providing healthcare" and in the sense of caring.

Under modern conditions, however, medicine and healthcare is far from ideal. Already long before the robots entered our hospitals and homes of the elderly, medicine and healthcare have moved towards more bureaucratic, "colder" ways of doing. What people hate and fear about "the machines" is what they hated and feared already long before: the experience of being treated as an object in a large, institutional machine. The institutions of the hospital and the "home" are themselves problematic. The healthcare "system" is itself part of the problem. Professionalization of care meant that family members, friends and loved ones no longer have a role in healthcare. Care workers are paid salaries, that is, with Marx we can say that the care workers have been alienated from their work. With Weber we can point to the modern bureaucracy and the rationalization and the over-valuing of efficiency have dehumanized patients. Technology has played and does play a role in this. Computers, for example, have made it possible that patients were better managed and controlled, became numbers. Computers have made it possible that we were dehumanized into data. Doctors and nurses started to act like robots.

Care became routine. It became "better" than care by lay persons and "charlatans", for sure, but it became better only in the sense of technical expertise. It did not become better in the sense of care articulated in the second half of this chapter.

In his philosophy of technology, Borgmann [3] criticized modern technologies for becoming "devices" that hide in the background and do not require skilled engagement. He gave the example of central heating as opposed to a stove: in the former case the technology takes over and makes life easy for us, whereas in the latter case we have to do all kinds of things to keep the fire going, but this also engages us and makes us do things together [3]. It is true that sometimes technology can encourage thoughtless production and consumption, including the thoughtless, easy "production" of healthcare by healthcare workers and the thoughtless, skill-less consumption of healthcare. The problem, however, is not so much the lack of skills (much healthcare today and indeed much industrial production today is highly skilled) but more precisely the wrong kind of skills or the lack of connection with other skills: the skills that are not part of a knowing-how to care but only of a knowing-how to repair. The problem is a lack of engagement with others and with things (as we can argue with Borgmann), but also a lack of *a particular kind of engagement* (a caring engagement), which may or may not be encouraged by contemporary healthcare technology. Both care givers and care receivers have become part of a gigantic "central caring" system which requires skills from individuals, but has disconnected those skills from caring ways of doing things, from engaged ways of doing things. They have become mainly or even merely "technical" skills, their ethical and personal, caring aspect has evaporated in the striving for more efficiency. What is missing is skill in the sense of a more caring engagement with things—Borgmann calls skill 'intensive and refined world engagement' [3, p. 42]—but also and especially a more caring engagement with people, with care receivers. The latter does not happen apart from the former, apart from things and apart from technology, but is connected to it: we care for and about people *with* things. As in cooking, medicine and healthcare is mediated by materiality, by technology, and this can render the practice more or less "skilled" in the broader, richer sense of the term developed here. It seems that in modernity, there are fewer places for these kind of skills: skills that belong to the "know-how" side of expertise (recently I have argued for more "craftsmanship" and engagement in healthcare, also in "e-care" [4]) but also skills that have to do with doing things in a "caring" way. The way we do things with others and with technology has become increasingly "colder".

To remedy this situation, one cannot ask from healthcare workers that they love patients as they love their friends, partners, children, family members, and so on. It is normal that we have more concern for those near to us than for others. One *could* ask, perhaps, to be a little bit more caring. But my point in the previous paragraph is not that we should blame individual people for having the wrong kind of attitude. The point is that we have organized medicine and healthcare in such a way that we encourage ways of doing things that are far less caring than they could be and far less caring than they *should* be if we want to get a little closer to the ideal of care we assume in our deliberations about good care and in our deliberations towards medical machines. Thinking about technology in medicine and healthcare should be part of asking these broader questions: *how* do we do things, at various levels: individuals, tasks, practices, and institutions.

Thus, the problem is not robots; the problem is how we do things in medicine and healthcare. If we insist that good care is "human" care, then this does not necessarily mean that robots should be banned from medicine and healthcare. Humans have always worked with technology. Potentially machines can play a role as mediators in skilled, caring, and engaged expert care work. They can help to make people perform better and they can, in principle, help to materialize the caring attitude, to support the caring ways of doing we believe to be necessary (but not sufficient) for good care. The problem is not robots, but *how* to care in a better way: *with* robots and/or with other technologies, and with "better" meaning "expert" care and *care-full* care.

This question is in the first place a question that needs to be answered by those who work in medicine and healthcare, those who—this is what we must assume—really want to work in better ways, want better healthcare, but find themselves in institutional-material contexts in which it becomes increasingly more difficult to "care" in the richer sense articulated here, which requires a way of engaging with patients and with technology that resembles more the ways of the caring cook than the ways of the "robotic" professionals created by modern healthcare management. Care givers need not be flower-giving lovers or friends who cook for friends, but something can be done about the kitchen.

4 Conclusion: New Skills for Good, Expert Human-Machine Care

In this chapter I have articulated two necessary (but perhaps not sufficient) conditions for good care: "expert" care involves know-that *and* know-how, and "caring" skills that show concern for patients and enable care givers to care-fully engage with them. Evaluating the introduction of new technologies such as robots in medicine and healthcare then requires at least answering the following two questions:

1. How does the technology impact on the "know-that" and "know-how" expertise of people working in medicine and healthcare practices? For example, if doctors become more dependent on expert systems and have less "know-that" themselves, is this a problem? If new technologies such as robotic technologies are introduced, what kind of old skills do care givers lose and what kind of new skills do they need? How does this change their expertise, and to what extent can they develop expertise as opposed to mere competence—let alone having the function of mere operators? What does automation mean for ethical expertise? How can machines (e.g., electronic technology computers, robots) assist in this, and what should be left to humans? How can their judgment and intuition have a more prominent place and how can we make sure if it is available in

cases that are equivalent to the imminent danger of an airplane crash or in cases that require ethical expertise? In which cases and situations should we prefer human skills, and how can they best be trained? How can "routine" behaviour be discouraged? If technology is always involved, how can human-technological skills be trained and how can a higher level of knowledge—preferably expertise—be reached?

2. Does the technology encourage a less care-full way of doing things? In what kind of practice and skilled tasks does it play a role and what role does and should it play? Is it possible to design and use the technology in such a way that care-full and engaged ways of doing things with people are stimulated? How can our institutions be changed in such a way that they make more room for the "human" aspect of care—meaning not less technology but perhaps different technology and in any case other ways of doing things, other ways of treating those who need not only "repair" or "services" but also, and even more perhaps, care understood as "caring about", as concern and as "caring" gestures, as well as company and respect. How can we avoid a total dehumanization of healthcare that would be even worse than having only robot care givers: how can we avoid that *human* care givers (fully) turn into care *machines*? More precisely: how can we avoid that they become providers of what we today associate with what care "machines" may do: the cold, precise and efficient but care-less delivery of healthcare services? In practice, the machines are not the problem; the problem is what humans do and especially how they do it: how caregivers relate to care receivers.

To conclude, new machines require new skills. In the everyday sense of the word this means: new skills to operate the machines. What I mean is: new skills in the sense of developing new expertise, which includes knowing how to handle the technology, but also, knowing how to deal with difficult technical-ethical situations. But good care is much broader than that. It is also especially about knowing-how to handle people: knowing how to be care-full and caring *with* the technology. Good medicine and healthcare requires all these different (but related) skills. They often come together in one task, one action. Good care is not about something external called "ethics" but about the things that are done with people *in* medical and care practice, and especially about how they are done. Quality is not something that can be measured afterwards, as a kind of quality control as part of management. Quality has to emerge in the practice, in the skilful treatment of people by people, in the care-full use of technologies to make people better.

This does not only have consequences for care givers and those who want to manage them. In this vision of good care, patients also have a different role. A more engaged relation means that they can no longer be mere "consumers" of medicine and healthcare. They also need to be more involved in the process. The receiver of care in the richer sense articulated here can no longer regard the care giver as a robot with a pre-programmed smile and with pre-programmed gestures. The "humanization" or, if you like, the "re-humanization" of healthcare can only succeed if there is also a change in how care receivers treat care givers. Not only patients are re-humanized by their doctors and their nurses, but the doctors and nurses themselves are also re-humanized by the patients.

Moreover, contemporary electronic technologies such as those used in tele-medicine and tele-care practices, and in (other) self-care practices also require a different distribution of expertise. In "old-style" medicine and healthcare, the care givers were the only experts. In the internet age, care receivers have the possibility to have more "know-that": they are increasingly well-informed about their condition via the internet and care givers have to take this into account and deal with it in a way that respects the patient *and* the patient's knowledge. (It is even conceivable that occasionally the patient may know more about particular disease than a general practitioner, at least in the sense of know-that. But of course it may also be the case that the patient is wrong, or that the "information" is not yet "knowledge", in which case conversation between doctor and patient can clarify things and produce knowledge.) In addition, there may also be a different distribution of "know-how". In "old-style" healthcare only the care giver has to know how to do things—with the care receiver. The care receiver was entirely passive: an object of investigation and an object of care. Today, the role of the care receiver changes in the direction of self-care and more responsibility for one's own health. The care receiver then also needs "know-how": (s)he needs to know how to use particular technologies in processes of self-care, and again care-givers need to adapt to that and need new skills: skills to work with the new technologies, but also to work with new kind of care-receivers, who also need to be instructed in particular ways of doing and who may also "know" more in the sense of know-how based on experience of their own body, their own use of the technology, and so on.

Thus, if better healthcare means more "expertise" and more "care" in the rich sense, then it also implies a different relationship between care givers and care receivers altogether: it implies that *both* parties in the care practice are re-humanized and activated into a more knowledgeable, skilful, participatory, engaged, and caring way of doing things. Machines are welcome if they contribute to this, if they enable good human-machine-human care. It is a vision of good care that enables us to evaluate new technologies and that encourages care givers (potential) care receivers, and other stakeholders to promote and find better ways of designing, regulating, and using them. Fears about "the machines" remind us about what we value most, in healthcare and in other fields.

References

1. Beauchamp TL, Childress JF (2001) Principles of biomedical ethics, 5th edn. Oxford University Press, Oxford
2. Benner P (2004) Using the Dreyfus model of skill acquisition to describe and interpret skill acquisition and clinical judgment in nursing practice and education. Bull Sci Technol Soc 24(3):188–199
3. Borgmann A (1984) Technology and the character of contemporary life: a philosophical inquiry. The University of Chicago Press, Chicago

4. Coeckelbergh M (2013) E-Care as craftsmanship: virtuous work, skilled engagement, and information technology in healthcare. Med Healthc Philos (on-line first)
5. Dreyfus SE, Dreyfus HL (1980) A five-stage model of the mental activities involved in direct skill acquisition. Report. Operations Research Center, University of California, Berkeley, CA
6. Dreyfus HL, Dreyfus SE (1991) Towards a phenomenology of ethical expertise. Hum Stud 14(4):229–250
7. Moor JH (2006) The nature, importance, and difficulty of machine ethics. IEEE Intell Syst 21(4):18–21
8. Polanyi M (1966) The tacit dimension. Routledge & Kegan Paul, London
9. Sharkey A, Sharkey N (2010) Granny and the robots: ethical issues in robot care for the elderly. Ethics Inf Technol 14(1):27–40
10. Sparrow R, Sparrow L (2006) In the hands of machines? The future of aged care. Mind Mach 16:141–161
11. Whitby B (2012) Do you want a robot lover? In: Lin P, Abney K, Bekey GA (eds) Robot ethics: the ethical and social implications of robotics. Intelligent robotics and autonomous agents series. MIT Press, Cambridge, pp 233–249

Implementation Fundamentals for Ethical Medical Agents

Mark R. Waser

Abstract Implementation of ethics in medical machinery is, necessarily, as machine-dependent as ethics is context-dependent. Fortunately, as with ethics, there are broad implementation guidelines that, if followed, can keep one out of trouble. In particular, ensuring correct codification and documentation of the processes and procedures by which each decision is reached is likely, in the longer view, even more important than the individual decisions themselves. All ethical machines must not only have ethical decision-making rules but also methods to collect data, information and knowledge to feed to those rules; codified methods to determine the source, quality and accuracy of that input; trustworthy methods to recognize anomalous conditions requiring expert human intervention and simple methods to get all of this into the necessary hands in a timely fashion. The key to successful implementation of ethics is determining how best to fulfill these requirements within the limitations of the specific machine.

1 Introduction

Walking the tightrope between autonomy and respectful guardianship makes medical ethics one of the most complex, important and fraught fields of human endeavor. Good decisions can frequently require a more thorough familiarity with the patient and their situation than any third party can be expected to possess and, increasingly, more medical knowledge than any single human can possibly possess—if not total impartiality and Cassandra's powers of foretelling the future. Further complications are added by the increasing intrusiveness and attempted regulation by self-interested parties with little of the knowledge necessary to do so

M.R. Waser (✉)
Digital Wisdom Institute, Vienna, VA, USA
e-mail: mwaser@digitalWisdomInstitute.org

S.P. van Rysewyk and M. Pontier (eds.), *Machine Medical Ethics*,
Intelligent Systems, Control and Automation: Science and Engineering 74,
DOI 10.1007/978-3-319-08108-3_4

(much less common sense). In many respects, we are approaching both a nadir and a zenith. Humanity's knowledge is ever-increasing but time and selfish concerns lead to diminishing effective utilization of that knowledge.

Automation can make this situation tremendously better—or, like all technologies, it can also make the situation tremendously worse. Instead of blindly implementing cool technology without considering its likely side effects, we must consider our societal goals in what we wish to accomplish—and what we fervently wish to avoid. Indeed, if properly done, automation of medical ethics will likely have a tremendously beneficial effect upon our laws and societal mores. Good engineering requires explicit specification and codification of what is desired and how it is to be accomplished—as well as what is to be avoided and how we will do that as well. Indeed, in a very real sense, consequentialist ethics are simply good engineering—once the goals are determined.

Further, as technology continues to improve, medical ethics increasingly walks a second tightrope between what is best for an individual and what society as a whole can afford. We can now keep a single vegetative individual alive for a long time—but doing so for large number of individuals will sap both our resources and our morale. Alternatively, society could take the resources required to keep one vegetative individual alive and use them instead to provide food and/or clean water and/or skills training for many individuals. Similarly, there is the closely-related question of assisted suicide and allowing competent individuals the dignity (autonomy) to decide the time and manner of their departure. The codification of rules for ethical medical agents will be subject to—but also drive—the formulation of laws for human beings (and corporations).

A final difficulty is that implementation of ethics in medical machinery is as agent-dependent as ethics is context-dependent. Fortunately, as with ethics, there are broad guidelines that, if followed, can keep one out of trouble. In particular, correctly codifying and documenting the processes and procedures by which each decision is reached is likely, in the longer view, even more important than the individual decisions themselves. All ethical agents must not only have the ethical decision-making rules but also methods to collect data, information and knowledge to feed to those rules; codified methods to determine the source, quality and accuracy of that input; trustworthy methods to recognize anomalous conditions requiring expert human intervention and simple methods to get all of this into the necessary hands in a timely fashion. The key to successful implementation of ethics is much more about determining how best to fulfill these requirements within the limitations of the specific agent rather than the specific rules that are implemented.

2 Systems Engineering 101

It is extremely rare that a new technological system is built de novo, complete and as a single piece. It is almost always the case that any new system is replacing—while enhancing—some existing, though frequently manual system. And, yet, the most frequent error perpetrated by novice engineers is that they don't ensure that they thoroughly understand the existing system before attempting to build a new one. Indeed, most engineers come in with the arrogant opinion that, of course, they "can improve upon that antiquated system that everyone has been complaining about"—and they don't even really need to know the details of how bad it is because they are going to improve it. And so it begins…

If you only had a dollar for every time a new system is "completed"—only to discover that it doesn't have all the *necessary* functionality of the old system, you'd be incredibly rich. If you only had a nickel for each time an engineer belatedly discovered "*that* is why the old system did it that way", you would be richer still. Proper or, more importantly, safe engineering REQUIRES a complete understanding of what you are about to replace BEFORE you replace it. Indeed, even if all you are doing is supporting an old structure/system, a full understanding is necessary to ensure that the new system is not unknowingly blocking or crippling some obscure but critical functionality of the old system.

Similarly, safe engineering requires not only a complete understanding of the system that you are going to be building and adequate precautions that the system will not be subject to conditions where its behavior becomes undesirable (or even unpredictable) but a recognition of where and how the new system fits into any larger systems (where any unpredictability can domino into huge problems). For social systems, in particular, it is necessary to recognize that, not only are you building a new technology, you are altering, and hopefully re-engineering, a larger existing social system. Indeed, Lessig [39] clearly describes how "architecture" (the design of technical systems) can exert a normative force which comparable to that imposed by law and custom.

Recognizing this is particularly important for systems which are intended to take on the role of being "ethical" or enforcing ethical behavior—because ethical problems are most often *much* more sensitive to scoping than simple goal-achievement problems. Shooting someone is a bad idea only until it is the sole way to prevent them from shooting a score of innocent children; but it becomes a terrible idea again if someone else is close enough with a taser. A reductionist approach may appear to make morality seem obvious but, in reality, it makes it virtually impossible.

Real ethical problem-solving only takes place when ethical dilemmas force the choice between two ethical axioms. Which is worse—death or lack of autonomy? Does the answer change when you're discussing teen suicide, end-of-life assisted suicide, abortion, primitive tribes exposing infants, and/or capital punishment? To say that morality is not influenced by larger systems and circumstances is like claiming that a hammer is the best and only tool. And "handling ethics" at too low a level can be catastrophic if, as Bringsjord [13] points out, "human Jones has a device which, if not eliminated, will (by his plan) see to the incineration of an metropolis, and a robot (an unmanned, autonomous UAV, e.g.,) is bound by a code of conduct not to destroy Jones because he happens to be a civilian, or be in a church, or at a cemetery".

The only way to solve problems of this type is with a solid understanding of scope and hierarchy. A lower-level system must be able to be overridden by "higher" components with a broader scope of view. In order for this not to be a disaster, since the lower-level system frequently provides much of the detailed (or tactical) input to the higher level, it must be transparent about its in-built expertise, shortcomings, and biases as well as "showing its work" (explaining why and how it arrived at the answer). If a higher level can see that a lower level is unaware of a critical factor, it can immediately discard the recommendations from the lower level.

Lack of transparency can have devastating consequences for the effectiveness of higher levels. This can be particularly egregious for social systems. As pointed out by Lessig [39, 138], code-based regulation—especially of people who are not themselves technically expert—risks making regulation invisible. Controls are imposed for particular policy reasons, but people experience these controls as nature. And that experience, I suggested, could weaken democratic resolve. Many of our current dysfunctional social structures (e.g., sociopathic corporations) have arisen from low-level decisions having inordinately large unexpected effects on our social fabric.

3 What Is an Ethical Medical Agent?

In order to discuss the history and future of ethical medical agents, we must agree on what counts as machine ethics and how to characterize ethical agents. Moor [46] has provided what has seemingly become the consensus with what is generally characterized as a hierarchy of ethical impact agents, implicit ethical agents, explicit ethical agents and full ethical agents.

Ethical impact agents have an ethical effect without realizing it or necessarily having been programmed to do so. They are judged merely by the effect of their operations rather than whether or not ethics are inherent to the agent itself. Thus, it is more often the case that circumstances, rather than the nature of the entity, are paramount. Indeed, the nail for want of which the battle was lost could be easily be considered the archetypal ethical impact agent—and illustrates the fact that agency is actually of little import for this category. Obviously, due to the nature of medicine, virtually all medical machinery without inherent ethics will classify here into this category—with the problem being that, while the impact is clear, the direction is not predetermined.

A respirator can keep a person alive until their body has a chance to recover or keep a body alive long after the person has departed. Similarly, a diagnostic expert system, created without any ethical concerns having been considered, might suggest any number of possible actions without regard to excessive personal or societal costs. The use of such technologies must be constantly supervised from an ethical point of view—and it would be far preferable if such supervision could be built into the agents themselves.

Implicit ethical agents, on the other hand, are created when ethical issues *are* considered at design time and the agent's actions are constrained to avoid unethical outcomes—for example, by "creating software that implicitly supports ethical behavior, rather than by writing code containing explicit ethical maxims." In this case, the agent "acts ethically because its internal functions implicitly promote ethical behavior—or at least avoid unethical behavior" because "Ethical behavior is the machine's nature." Baxter is implicitly ethical because it is programmed to avoid collisions that will do humans harm just as a patient-controlled morphine dispenser will not allow a lethal dose request to be fulfilled.

The problem with implicit ethical agents is their lack of transparency—they are effectively black boxes. Unless their behavior is known (and documented) for every case that they can possibly come across, you'll never know when overriding and/or extenuating circumstances will render their actions horribly inappropriate. Imagine an expert system, already known to be limited and brittle, which *only* gave answers without explanation. For this reason, we would argue that a far more descriptive name, from an engineering standpoint, would be "black box" or "opaque" ethical agents.

Explicit ethical agents can actually "represent ethics explicitly and then operate effectively on the basis of this knowledge." Asimov's "Three Laws" robots with their "moral mandates" to prevent harm to humans, obey orders and protect themselves are examples of explicit agents with a problematic overly simplistic ethical system. Simple expert systems for ethical reasoning in bioethics and medicine have been discussed and implemented for decades [60]. While such systems remain passive, brittle, error-prone and unable to detect anomalies, they do have the advantage of being transparent so that it can be determined whether or not they are "aware" of overriding and/or extenuating circumstances. Indeed, for engineering purposes, we would argue that agents that are not "transparent" agents should not occupy a higher position in the hierarchy.

Discussions of Moor's "full ethical agents" rapidly enter the contentious areas of consciousness, intentionality and free will. Many will argue that an explicit ethical agent is "just" a complex device or tool for reasoning with information and that it has no real moral "authority" except that derived from its creators, which is certainly true for current systems. They claim that a full ethical agent must be a sentient being that makes judgments based upon values—and arguably has moral authority. It is our contention, however, that a so-called "full" ethical agent is merely an "intentional" ethical agent—which is thereby "trusted" (or not)—since it is presumably then capable of moral responsibility.

4 The Morality of Intentional Agents

Philosopher Dennett [21, 22] has repeatedly argued that higher-order intentionality is a precondition for moral responsibility (including the opportunity for duplicity for example). Sullins [58] concurs by stating that we merely need ask three questions:

- Is the robot significantly autonomous?
- Is the robots behavior intentional?
- Is the robot in a position of responsibility?

Indeed, arguably, this view can be traced all the way back to Immanuel Kant [38] claiming that angels can't be moral because they have no choice (no "free will" to have "immoral" intentions) and animals can't be immoral because they have no knowledge of morality (and thus cannot have intentions to be moral).

We have previously specified [65] what we believe is required for consciousness, self and so-called "free will" and what the implications will be of creating machines that fulfilled those requirements; however, this is only really relevant for very long-term planning and, more importantly, how it affects our views on how ethical medical systems can make human medical professionals more ethical—especially since many with a naturalistic worldview would argue that a doctor is simply a biological medical machine. We don't feel the need to insist upon any such thing but it is highly informative to speculate on the scenario where such is true and to consider nurses, doctors and surgeons as if they are simply yet more instances of the implementation of medical machines.

Far more important to recognize is Dennett's [20] argument that we are only able to understand and anticipate each other in everyday life and daily interactions through the use of the "folk" concepts of belief, desire, intention and expectation. And, indeed, we are even prone to the "pathetic" fallacy of misattributing these mental aspects to "unthinking" objects. We have argued previously [67] that the clear dividing line between complex devices and sentient entities is trait of autopoiesis—the act or attribute of constant self-(re)creation. Once something is constantly (and effectively) changing itself according to its "identity" (and thus, its intentions), it has crossed the line from being a tool to an entity.

The most important requirements for autopoiesis are sentience and identity. An entity must be able to adequately sense in order to guide its direction of self-(re)creation and have a target/intention/identity to aim for. Ethical entities require an identity of being a moral entity and a moral sense (aka ethical sensitivity). And while [1] claim that "Fortunately, there is every reason to believe that ethically sensitive machines can be created," the outstanding problem for any sort of ethical system is that we don't truly have a clear consensus definition of what morality and/or ethics truly are—and thus, we have no guaranteed engineering method to design or validate a constructed moral sense.

5 The Shortcomings of Humans as Ethical Actors

Arkin [8] maintains that "It is not my belief that an unmanned system will be able to be perfectly ethical in the battlefield, but I am convinced that they can perform more ethically than human soldiers are capable of." Similarly, Pontier and Hoorn [47] are striving "towards machines that behave ethically better than humans do". Clearly, there is tremendous optimism for the potential of technology being applied to ethics.

The problem is that it all comes back down to answering the ages old question of "What is this 'ethics'?" As James Moor points out, not only do we want future machines to treat us well but, more even importantly, "Programming or teaching a machine to act ethically will help us better understand ethics." Problematically, however, one of the best-known treatises on machine morality [61] despairs at reconciling the various approaches to morality with the claims that doing so "will demand that human moral decision making be analyzed to a degree of specificity as yet unknown" and moreover that "any claims that ethics can be reduced to a science would at best be naïve". We disagree vehemently with such pessimism and they seemingly should agree since they then proceed along despite such caveats.

The greatest shortcoming of humans as ethical actors is precisely the lack of the ability that we highlighted as most important for ethical medical machines—the ability to correctly codify and document the processes and procedures by which each decision is reached. The typical excuses for this, including the apparently overwhelming complexity of morality and ethics, are likely solely due to the fact that it was evolutionarily beneficial to develop and hold those views as we have previously argued [63]. Transparency simply was not optimal for the evolving human individual.

The first problem is that, for human beings, conscious logic is sub-optimal for making complex evaluations even when moral issues aren't involved. A study of the "deliberation-without-attention" effect [23] shows clearly that engaging in a thorough conscious deliberation is only advantageous for simple choices while choices in complex matters should be left to unconscious thought. This effect is attributed to the fact that a person can pay conscious attention to only a limited amount of information at once, which can lead to a focus on just a few factors and the loss of the bigger picture. And, indeed, experimental studies [57] show that many "conscious" decisions are actually made by the unconscious mind up to 10 s before the conscious mind is aware of it.

Worse, scientific evidence [32] clearly refutes the common assumptions that moral and ethical judgments are products of, based upon, or even correctly retrievable by conscious reasoning. We don't consciously know and can't consciously retrieve why we believe what we believe and are actually even very likely to consciously discard the very reasons (such as the "contact principle") that govern our ethical behavior when unanalyzed. Further, even when analyzed, there is ample evidence [59] to show that our conscious, logical mind is constantly self-deceived to enable us to most effectively pursue what appears to be in our own self-interest or pursuant to our own personal moral code. We simply make up whatever appears to be the best excuse at the time [27]. Of course, none of this should be particularly surprising since Minsky [45] points out many other examples, including when one falls in love, where the subconscious/emotional systems overrule or dramatically alter the normal results of conscious processing without the conscious processing being aware of the fact.

Mercier and Sperber [44] point out that there are incontrovertible amounts of evidence demonstrating that human reasoning often leads to epistemic distortions and poor decisions. It seems illogical that evolution would have favored such problems except for their hypothesis that true function of reasoning is for argumentation—to devise and evaluate arguments intended to persuade. As they point out:

"When the same problems are placed in a proper argumentative setting, people turn out to be skilled arguers. Skilled arguers, however, are not after the truth but after arguments supporting their view" [44, 56].

Thus, morality and ethics appear terribly complicated because we have evolved that view to prevent ourselves from being exploited by superior argumentation and to permit ourselves to self-deceive while pursuing our own selfish desires. This is also arguably why ethics is predominantly implemented as overriding emotions which can shut down reasoning when we are threatened. Further, emotions can evolve to enforce behavioral rules in cases where cognitive complexity (enabled by temporal discounting) ensures that effective reasoning cannot take place.

The final straws are recent experiments into how "cultural cognition" [24] affects public policy [35] by motivating individuals to conform their beliefs about policy-relevant facts to their cultural values and the "identity-protective cognition thesis" [37] that subjects would "use their quantitative-reasoning capacity selectively to conform their interpretation of data to the result most consistent with their political outlooks." Indeed, even in cases where the issue is unfamiliar and the politics are unknown, such as with outpatient commitment laws [36], cultural worldviews and values shape the perception of efficacy of policies. And all of this is further complicated by the fact that social psychologists [28, 29, 34]. All of this appears as if it could leave us in quite a quandary when we're attempting to implement ethical machines.

6 So What Do We Do?

The first thing that needs to be done when engineering any system is to make some choices and define the boundaries of the system that is to be created. Are we merely attempting to develop a system that flags decisions or actions that might be morally fraught so that they can be reviewed before a mistake is made? Are we attempting to create a system that can actually determine the best decision or action? Or, are we actually going to allow the system to make decisions and implement them without oversight?

Given the current lack of autonomy in machines and the human disagreement over exactly what is moral, the third option is clearly out of the question for the foreseeable future. On the other hand, even the attempt to create a system that can determine the best decision or action will dramatically help the social process of codifying morality. Indeed, simply developing a system that can recognize when decisions or actions might be morally fraught would be a huge step forward.

There is also the question of whether we attempt to implement morality exactly as it is implemented in humans, warts and all, or whether we attempt to specify something that is simpler and clearer yet close enough to "ethical" to serve until we can truly fully specify what that means. In reality, we can really only engineer prototype, test or advisory (or tutoring) SUBsystems of each type. We really do not know enough about the existing system to even begin to attempt to replace it. All we can do is build systems to explore the subject further and to assist humans in not making errors—and attempt to develop an overarching theory that might enable us better guidance in our explorations.

7 Morality from the Top Down

McLaren [42, 43] points out that ethical reasoning has a fundamentally different character than reasoning in more formalized domains:

- The laws, codes, or principles (i.e., rules) are almost always provided in a highly conceptual, abstract level.
- The conditions, premises or clauses are not precise, are subject to interpretation, and may have different meanings in different contexts.
- The actions or conclusions in the rules are often abstract as well, so even if the rule is known to apply the ethically appropriate action may be difficult to execute due to its vagueness.
- The abstract rules often conflict with each other in specific situations. If more than one rule applies it is not often clear how to resolve the conflict.

These facts should lead to the obvious question of "Why is ethics laid out this way?" and the arguable answer of "Because this is merely our best current overarching theory of what human ethics truly is." We contend that the problem is that our idea of morality is merely what we have cobbled together from the frequently contradictory low-level data provided by our moral sense as filtered through our cultural cognition and argumentative reasoning.

So, clearly, a change of tactics is in order. A useful new strategy is, instead of trying to ask exactly what morality is, to ask why it is what it is—or, in common engineering terms "What is the function that this system exists to fulfill?" And, conveniently enough, social psychologists [30] have just been recently arguing that, rather than specifying the content of moral issues (such as "justice, rights, and welfare"), they too should also take a functional approach to defining morality. Their functional definition is that

Moral systems are interlocking sets of values, virtues, norms, practices, identities, institutions, technologies, and evolved psychological mechanisms that work together to suppress or regulate selfishness and make cooperative social life possible.

This fits in conveniently with Wilson's [68] prediction that ethics would soon become part of the 'new synthesis' of sociobiology, in which distal mechanisms (such as evolution), proximal mechanisms (such as neural processes), and the socially constructed web of meanings and institutions (as studied by the humanities and social sciences) would all be integrated into a full explanation of human morality. All of those assorted pieces clearly *do* work together to fulfill the function of suppressing/regulating selfishness and making cooperative social life possible. And ethical robots and systems will merely be an addition to our moral ecosystem. We just need to ensure that they act in furtherance of that goal as opposed to being a hindrance to it.

This also gives us a framework to go back and evaluate both why and how well each of the multiplicity of moral philosophies works. For example, a Kantian philosophy such as that espoused by Powers [50, 51] clearly makes sense when we are effectively declaring a categorical imperative of "suppress or regulate selfishness and make cooperative social life possible."

8 Morality from the Bottom Up

The existence of evolutionary "ratchets" (randomly acquired traits that are likely statistically irreversible once acquired due to their positive impact on fitness) causes "universals" of biological form and function to emerge, persist, and converge predictably even as the details of evolutionary path and species structure remain contingently, unpredictably different [56]. Ratchets can range from the broadly instrumental (enjoying sex) to the environmentally specific (streamlining and fins in water) to the contradictory and context-sensitive (like openness to change). Morality, as the ability to suppress or regulate selfishness and make cooperative social life possible, is clearly one such evolutionary ratchet—or, rather, our moral sense is actually a relatively large set of ratchets [12, 31, 69].

The good news is this means that our moral emotions most probably all have some reason for existing. The bad news is that, unless that reason is known, it is entirely possible that those reasons could be obsolete. For example, Savulescu and Persson [55] point to many characteristics of current human morality that appear inappropriate for the current techno-human condition including a high temporal discount rate, little regard for distant and unrelated people, and distinctions based upon arguably morally irrelevant features such as inflicting harm with or without direct physical contact. On the other hand, emotional reactions could point us toward discoveries that we are unlikely to find otherwise.

For example, in trolley car (or similar culturally-adapted) problems, the vast majority of people see no problem with throwing the switch so that one person is killed instead of five. On the other hand, a majority of people do have problems with pushing someone off the bridge to stop the train from killing five others or with kidnapping involuntary organ donors off the street to save multiple people. These culturally universal emotions indicate, at a minimum, that there was some time period during which utilitarianism, in terms of simply counting immediately affected lives, was not valid.

We would argue that emotions such as this indicate actions that, if allowed to occur, would quickly undermine society (and thus lead to far more lives lost in the long term). In the case at hand, the principle of double effect, allowing other people to be *used* to fulfill a goal will quickly result in everyone taking defensive measures to ensure that they won't be so used.

9 Example Emotional Systems

Gomila and Amengual [26] focus on the role of emotions in moral cognition and ways in which this can be instantiated computationally. Delghani et al. [18] have created a computational model, MoralDM which "integrates several AI techniques in order to model recent psychological findings on moral decision-making" by relying upon contextual factors that vary with cultural in order to extend beyond pure utilitarian models. It features:

- a natural language system to produce formal representations from psychological stimuli and reduce tailorability
- order of magnitude representation to model the impacts of secular versus sacred values via qualitative reasoning
- a combination of first-principles reasoning and analogical reasoning to determine consequences and utilities when making moral judgments, and
- performance improvement via accumulating examples.

This is similar to Pontier et al. [49] Moral Coppélia in that it integrates both rational/utilitarian calculations and more emotional/deontological ones.

10 Example Top Down Systems

In the absence of a better understanding and specification of ethics (not to mention a full-blown intelligent and intentional machine to instantiate it), there have been no true complete top-down systems. The unsolved "frame problem" [19, 41] means that either (a) scope and input must be limited and representation formalized or (b) that humans must pre-process the input to any instantiated system. Arguably, an autopoietic agent's identity requirements should solve the issues of meaning and understanding and thereby the frame problem but we are still short of that occurring.

Both military and medical systems have generally tremendously decreased their scope with clearly defined parameters. Arkin [8] expects military robots to be programmed with the laws of war and rules of engagement and to have the "just war" axioms of discrimination and proportionality as well as Walzer's [62] double intention. Most medical systems have generally followed Beauchamp's and Childress's [10] duties of beneficence, lack of maleficence, and autonomy.

The Anderson et al. [1] have created a variety of utilitarian ethical calculators or advisors which rely critically upon human pre-processing for their input. Jeremy, their Hedonistic Act Utilitarianism calculator (named after Jeremy Bentham) calculates the action which is expected to cause the greatest expected net pleasure. It presents the user with an input screen that prompts for the name of the action, the name of the person affected and a rough estimate of the size of the effect—and continues to accept this data for each person affected by every action under consideration. Critics of act utilitarianism have, of course, pointed out that it can (a) violate a person's rights, sacrificing one person for the greater net good and (b) frequently conflict with our notion of justice by ignoring past behavior

W.D. has a similar input screen that prompts for the name of each action but with a rough estimate of the amount that each of Ross's [54] seven prima facie duties (fidelity, reparation, gratitude, justice, beneficence, non-maleficence, and self-improvement) are either satisfied or violated by that action. Since, Ross gives no decision procedure for determining which duty is paramount when, as frequently happens, multiple duties pull equally in opposite directions, W.D. was designed to learn using a simple least mean square training rule to update the weights of each duty. Later versions of W.D. [2] adopted Rawls' [53] reflective equilibrium approach to creating and refining ethical principles (by going back and forth between particular cases and principles) and used inductive logic programming (ILP) to learn the relation of "supersedes" for different quantities of satisfaction and violation between each of the duties. MedEthEx [3] implemented the duties of beneficence, lack of maleficence, and autonomy with Rawls' reflective equilibrium and ILP. The Andersons subsequently used a similar design in developing EthEl, an ethical eldercare system [4], and instantiated it into a NAO robot [5] which they believe is the first instantiated ethical robot—since Cloos [17] never created his proposed utilitarian robot. They have also continued their development of systems for developing ethical decision principles through dialog with ethicists [6, 7]. Pontier's and Hoorn's [47] moral reasoner also implemented the duties of beneficence, lack of maleficence, and autonomy while Pontier's and Widdershoven's [48] system concentrated on positive and negative autonomy.

11 Example Assisting and Advising Systems

As full-up machine intelligence is still in the future, many researchers have fallen back onto creating moral advisors and tutors (or have never believed that machines will ever be suitable for a greater role). When expert systems are right, they are most often right because of the explicit reasoning programmed into them and they can convey that reasoning on to the user. When they are wrong, they are normally *horribly* wrong because they do not recognize and account for some unusual/anomalous situation—but journeyman users, as opposed to novices, are most often quite capable of recognizing when this is the case and discarding the advice. On the other hand, as previously discussed, opaque big data and/or statistical systems like neural networks and undocumented case-based reasoning can turn out to be quite dangerous unless either (a) they are solely used as a "safety valve" check on an already determined action [52] or (b) it is guaranteed that the system has seen and will correctly handle any case that might be given to it (a closed world assumption).

Far better, however, are so-called decision support systems which assist users in finding all relevant data and helping them analyze it—such as Truth-Teller and SCIROCCO [42, 43]. Truth-Teller assists users by providing comparisons of the cases and fully explaining the comparisons. SCIROCCO also compares cases but "is more useful for collecting a variety of information, principles, cases and additional information that a user should consider in evaluating a new ethical dilemma." Even a simple interactive, multimedia program like the Dax Cowart program, designed to explore the practical ethics issue of a person's right to die [16], can serve to educate but best of all is to utilize and participate in one of the numerous new methods of collaborative modeling and decision-making [64].

Similarly, IBM [33] has just announced its "WatsonPaths" project in collaboration with the Cleveland Clinic. WatsonPaths explores complex scenarios and draws conclusions much as medical personnel do. It examines scenarios from many directions, working its way through chains of evidence, and draws inferences to support or refute a set of hypotheses. It can incorporate feedback and learn from humans who can drill down "to decide if certain chains of evidence are more important, provide additional insights and information, and weigh which paths of inferences the physician determines leads to the strongest conclusions." It will use machine learning to "improve and scale the ingestion of medical information" which will make it valuable to even seasoned practitioners. It will also be able to "help medical students learn how to quickly navigate the latest medical information and will display critical reasoning pathways from initial clinical observations all the way to possible diagnoses and treatment options."

12 Current Top Down Theorizing

Of course, there is no shortage of researchers still attempting to solve the grand problem of ethics from top down modeling of human morality. For example, Bringsjord et al. [14] suggest that category theory is a viable approach by contending that "It is not implausible to hold in Piagetian fashion that sophisticated human cognition, whether or not it is directed at ethics, exploits coordinated functors over many, many logical systems encoded as categories. These logical systems range from the propositional calculus, through description logics, to first-order logic, to temporal. epistemic, and deontic logics, and so on." This would presumably include Athena [9], an "implementation of a natural deduction calculus for a recently developed deontic logic of agency based on indeterminate branching-time semantics augmented with dominance utilitarianism".

Bello and Bringsjord [11] focus on the necessary role of theory of mind (ToM) in moral psychology and argue that an appropriate computational version of the human ToM is necessary to reason about moral implications. They argue, as we do, that rational actor models should be discarded; however, they believe that focusing on folk concepts is the best way to explore cognitive architecture. It would be our hypothesis and fear that this will lead to the same sort of argumentative errors that created the rational actor model in the first place.

Gigerenzer [25] proposes that moral behavior is based upon satisficing using pragmatic social heuristics rather than moral rules or maximization principles.

This could also be interpreted to mean that moral calculations may be based upon affordances and boundaries rather than single optimal plans. Mackworth's [39] argument that constraint satisfaction could be used as a unified ethical framework could tie into this view as well.

13 Conclusion

All ethical machines must not only have ethical decision-making rules but also methods to collect data, information and knowledge to feed to those rules; codified methods to determine the source, quality and accuracy of that input; trustworthy methods to recognize anomalous conditions requiring expert human intervention and simple methods to get all of this into the necessary hands in a timely fashion. The bad news is that current human ethical rules need more study and clarification before we will be able to implement more than simple subsystems, decision support systems and test systems.

The good news about autopoietic "intentional" agents is that all that needs to be done to prevent them from running amok is to ensure that a Kantian imperative of Haidt's morality is part of their identity—contrary to many of the concerns of Brundage [15] and the numerous others he cites. For safety's sake, we will, as we have previously argued [66], have to grant them full moral agency and patienthood (and arguably, personhood)—but this is far preferable to the other scenarios predicted.

References

1. Anderson M, Anderson S, Armen C (2004) Towards machine ethics. In: AAAI-04 workshop on agent organizations: theory and practice, San Jose, CA, July 2004
2. Anderson M, Anderson S, Armen C (2005) Towards machine ethics: implementing two action-based ethical theories. In: 2005 AAAI fall symposium on machine ethics/AAAI technical report FS-05-06-001
3. Anderson M, Anderson S, Armen C (2006) MedEthEx: a prototype medical ethics advisor. In: Proceedings of the eighteenth conference on innovative applications of artificial intelligence, Boston, Massachusetts
4. Anderson M, Anderson S (2008) EthEl: toward a principled ethical eldercare robot. In: Proceedings of the AAAI fall 2008 symposium on AI in eldercare: new solutions to old problems, Arlington, VA
5. Anderson M, Anderson SL (2010) Robot be good. Sci Am 2010(10):72–77
6. Anderson SL, Anderson M (2011) A prima facie approach to machine ethics: machine learning of features of ethical dilemmas, prima facie duties, and decision principles through a dialogue with ethicists. In: Anderson M, Anderson S (eds) Machine ethics. Cambridge University Press, New York, pp 476–492
7. Anderson SL, Anderson M (2013) The relationship between intelligent, autonomously functioning machines and ethics. In: Proceedings of the 2013 meeting of the international association for computing and philosophy. http://www.iacap.org/proceedings_IACAP13/paper_3.pdf. Accessed 07 Oct 2013

8. Arkin R (2009) Governing lethal behavior in autonomous robots. Chapman & Hall, Boca Raton
9. Arkoudas K, Bringsjord S, Bello P (2005) Toward ethical robots via mechanized deontic logic. In: 2005 AAAI fall symposium on machine ethics/AAAI technical report FS-05-06-003
10. Beauchamp TL, Childress JF (1979) Principles of biomedical ethics. Oxford University Press, New York
11. Bello P, Bringsjord S (2012) On how to build a moral machine. Topoi 32(2):251–266
12. Boehm C (2012) Moral origins: the evolution of virtue, altruism, and shame. Basic Books, New York
13. Bringsjord S (2009) Unethical but rule-bound robots would kill us all. http://kryten.mm.rpi.edu/PRES/AGI09/SB_agi09_ethicalrobots.pdf
14. Bringsjord S et al (2011) Piagetian roboethics via category theory: moving beyond mere formal operations to engineer robots whose decisions are guaranteed to be ethically correct. In: Anderson M, Anderson SL (eds) Machine ethics. Cambridge University Press, New York
15. Brundage M (2013) Limitations and risks of machine ethics. J Exp Theor Artif Intell. http://www.milesbrundage.com/uploads/2/1/6/8/21681226/limitations_and_risks_of_machine_ethics.pdf. Accessed 7 Oct 2013
16. Cavalier R, Covey PK (1996) A right to die? The Dax cowart case CD-ROM teacher's guide, Version 1.0. Center for Applied Ethics, Carnegie Mellon University, Pittsburgh
17. Cloos C (2005) The Utilibot project: an autonomous mobile robot based on utilitarianism. In: 2005 AAAI fall symposium on machine ethics/AAAI technical report FS-05-06-006
18. Deghani M et al (2011) An integrated reasoning approach to moral decision making. In: Anderson M, Anderson SL (eds) Machine ethics. Cambridge University Press, New York
19. Dennett D (1984) Cognitive wheels: the frame problem of AI. In: Hookway C (ed) Minds, machines, and evolution: philosophical studies. Cambridge University Press, New York, pp 129–151
20. Dennett D (1987) The intentional stance. Bradford Books/MIT Press, Cambridge
21. Dennett D (1996) When HAL kills, who's to blame? In: Stork D (ed) HAL's legacy: 2001's computer as dream and reality. MIT Press, Cambridge, pp 351–365
22. Dennett D (2013) Intuition pumps and other tools for thinking. W. W. Norton & Company, New York
23. Dijksterhuis A, Bos M, Nordgren L, Baaren R (2006) On making the right choice: the deliberation-without-attention effect. Science 311:1005–1007
24. DiMaggio P (1997) Culture and cognition. Ann Rev Sociol 23:263–287
25. Gigerenzer G (2010) Moral satisficing: rethinking moral behavior as bounded rationality. Top Cogn Sci 2:528–554
26. Gomila A, Amengual A (2009) Moral emotions for autonomous agents. In: Vallverdu J, Casacuberta D (eds) Handbook of research on synthetic emotions and sociable robotics: new applications in affective computing and artificial intelligence. IGI Global, Hershey
27. Haidt J (2001) The emotional dog and its rational tail: a social intuitionist approach to moral judgment. Psychol Rev 108(4):814–834
28. Haidt J (2012) The righteous mind: why good people are divided by politics and religion. Pantheon, New York
29. Haidt J, Graham J (2007) When morality opposes justice: conservatives have moral intuitions that liberals may not recognize. Soc Justice Res 20:98–116
30. Haidt J, Kesebir S (2010) Morality. In: Fiske S, Gilbert D, Lindzey G (eds) Handbook of social psychology, 5th edn. Wiley, Hobeken, pp 797–832
31. Hauser M (2006) Moral minds: how nature designed our universal sense of right and wrong. HarperCollins/Ecco, New York
32. Hauser M et al (2007) A dissociation between moral judgments and justifications. Mind Lang 22(1):1–27
33. IBM (2013) IBM research unveils two new watson related projects from cleveland clinic collaboration. Press Release. http://www-03.ibm.com/press/us/en/pressrelease/42203.wss. Accessed 15 Oct 2013

34. Iyer R, Koleva S, Graham J, Ditto PH, Haidt J (2010) Understanding libertarian morality: the psychological roots of an individualist ideology. Working Paper. http://ssrn.com/abstract=1665934 or http://dx.doi.org/10.2139/ssrn.1665934. Accessed 7 Oct 2013
35. Kahan DM, Braman D (2006) Cultural cognition and public policy. Yale J Law Public Policy 24:147–170
36. Kahan DM, Braman D, Monahan J, Callahan L, Peters E (2009) Cultural cognition and public policy: the case of outpatient commitment laws. Law Human Behav. Cultural Cognition Project Working Paper No. 47, Harvard Law School Program on Risk Regulation Research Paper No. 08-21. http://ssrn.com/abstract=1178362 or http://dx.doi.org/10.2139/ssrn.1178362. Accessed 7 Oct 2013
37. Kahan DM, Peters E, Dawson, EC, Slovic P (2013) Motivated numeracy and enlightened self-government. Cultural Cognition Project Working Paper No. 116. http://ssrn.com/abstract=2319992 or http://dx.doi.org/10.2139/ssrn.2319992. Accessed 7 Oct 2013
38. Kant I (1785/1993) Grounding for the metaphysics of morals. In: Ellington J (ed/trans). Hackett, Indianapolis
39. Lessig L (2006) Code Version 2.0. Basic Books, New York
40. Mackworth A (2011) Architectures and ethics for robots: constraint satisfaction as a unitary design framework. In: Anderson M, Anderson SL (eds) Machine ethics. Cambridge University Press, New York
41. McCarthy J, Hayes PJ (1969) Some philosophical problems from the standpoint of artificial intelligence. In: Meltzer B, Michie D (eds) Machine intelligence 4. Edinburgh University Press, Edinburgh, pp 463–502
42. McLaren B (2005) Lessons in machine ethics from the perspective of two computational models of ethical reasoning. In: 2005 AAAI fall symposium on machine ethics/AAAI technical report FS-05-06-010
43. McLaren BM (2006) Computational models of ethical reasoning: challenges, initial steps, and future directions. IEEE Intell Syst 21(4):29–37
44. Mercier H, Sperber D (2011) Why do humans reason? Arguments for an argumentative theory. Behav Brain Sci 34:57–111
45. Minsky M (2006) The emotion machine: commonsense thinking, artificial intelligence, and the future of the human mind. Simon & Schuster, New York
46. Moor J (2006) The nature, importance and difficulty of machine ethics. IEEE Intell Syst 21(4):18–21
47. Pontier MA, Hoorn JF (2012) Toward machines that behave ethically better than humans do. In: Proceedings of the 34th international annual conference of the cognitive science society, CogSci'12, pp 2198–2203
48. Pontier MA, Widdershoven GAM (2013) Robots that stimulate autonomy. IFIP Adv Inf Commun Technol 412:195–204
49. Pontier MA, Widdershoven GAM, Hoorn JF (2012) Moral Coppélia—Combining ratio with affect in ethical reasoning. In: Advances in artificial intelligence—IBERAMIA 2012. Lecture notes in computer science 7637, pp 442–451
50. Powers T (2011) Prospects for a kantian machine. In: Anderson M, Anderson SL (eds) Machine ethics. Cambridge University Press, New York, pp 464–475
51. Powers TM (2005) Deontological machine ethics. In: 2005 AAAI fall symposium on machine ethics/AAAI technical report FS-05-06-012
52. Rzepka R, Araki K (2005) What statistics could do for ethics? The idea of common sense processing based safety valve. In: 2005 AAAI fall symposium on machine ethics/AAAI technical report FS-05-06-013
53. Rawls J (1951) Outline for a decision procedure for ethics. Philos Rev 60(2):177–197
54. Ross WD (1930) The right and the good. Clarendon Press, Oxford
55. Savulescu J, Persson I (2012) Unfit for the future: the need for moral enhancement. Oxford University Press, New York

56. Smart JM (2009) Evo Devo universe? A framework for speculations on cosmic culture. In: Dick SJ, Lupisella ML (eds) Cosmos and culture: cultural evolution in a cosmic context, NASA SP-2009-4802. US Government Printing Office, Washington, pp 201–295
57. Soon CS, Brass M, Heinze H-J, Haynes J-D (2008) Unconscious determinants of free decisions in the human brain. Nat Neurosci 11:543–545
58. Sullins J (2006) When is a robot a moral agent? Int Rev Inf Ethics 6(12):23–30
59. Trivers R (1991) Deceit and self-deception: the relationship between communication and consciousness. In: Robinson M, Tiger L (eds) Man and beast revisited. Smithsonian Press, Washington, DC
60. Victoroff MS (1985) Ethical expert systems. In: Proceedings of the annual symposium on computer application in medical care, 13 Nov 1985, pp 644–648. http://www.ncbi.nlm.nih.gov/pmc/articles/PMC2578093/. Accessed 7 Oct 2013
61. Wallach W, Allen C (2009) Moral machines: teaching robots right from wrong. Oxford University Press, New York
62. Walzer M (1977) Just and unjust wars: a moral argument with historical illustrations. Basic Books, New York
63. Waser MR (2010) Designing a safe motivational system for intelligent machines. Presented at AGI'10: the third conference on artificial general intelligence, Lugano, Switzerland. http://becominggaia.files.wordpress.com/2010/06/agi10-final.ppt. Accessed 7 Oct 2013. http://vimeo.com/channels/agi10#15504215. Accessed 7 Oct 2013
64. Waser, MR (2011) Whately: open access crowd-sourced collaborative modeling for tackling "Wicked" social problems. http://becominggaia.files.wordpress.com/2010/06/whately.pdf. Accessed 07 Oct 2013
65. Waser MR (2011) Architectural requirements and implications of consciousness, self, and "Free Will". In: Biologically inspired cognitive architectures 2011: Proceedings of the third annual meeting of the BICA society (BICA'11), Arlington, VA, pp 438–443. doi:10.3233/978-1-60750-959-2-438
66. Waser MR (2012) Safety and morality require the recognition of self-improving machines as moral/justice patients and agents. In: Gunkel D, Bryson J, Torrance S (eds) The machine question: AI, ethics and moral responsibility. http://events.cs.bham.ac.uk/turing12/proceedings/14.pdf. Accessed 7 Oct 2013
67. Waser MR (2013) Safe/Moral autopoiesis and consciousness. Int J Mach Conscious 05(01):59–74. doi:10.1142/S1793843013400052
68. Wilson EO (1975) Sociobiology. Harvard University Press, Cambridge
69. Wilson J (1993) The moral sense. Free Press, New York

[illegible]

Towards a Principle-Based Healthcare Agent

Susan Leigh Anderson and Michael Anderson

Abstract To feel comfortable allowing healthcare robots to interact with human beings, we must ensure that they act in an ethically responsible manner, following an acceptable ethical principle(s). Giving robots ethical principles to guide their behavior results in their being ethical agents; yet we argue that it is the human designers, not the robots, who should be held responsible for their actions. Towards the end of designing ethical autonomous robots that function in the domain of healthcare, we have developed a method, through an automated dialogue with an ethicist, for discovering the ethically relevant features of possible actions that could be taken by a robot, with an appropriate range of intensities, *prima facie* duties to either maximize or minimize those features, as well as decision principles that should be used to guide its behavior. Our vision of how an ethical robot assistant would behave demonstrates that an ethical principle is used to select the *best* action at each moment, rather than just determine whether a particular action is acceptable or not. Further, we maintain that machine ethics research gives us a fresh perspective on ethics. We believe that there is a good chance that this research may lead to surprising new insights, and therefore breakthroughs, in ethical theory.

S.L. Anderson (✉)
Department of Philosophy, University of Connecticut, Storrs, CT, USA
e-mail: susan.anderson@uconn.edu

M. Anderson
Department of Computer Science, University of Hartford, West Hartford, CT, USA
e-mail: anderson@hartford.edu

S.P. van Rysewyk and M. Pontier (eds.), *Machine Medical Ethics*,
Intelligent Systems, Control and Automation: Science and Engineering 74,
DOI 10.1007/978-3-319-08108-3_5

1 Introduction

A pressing need for personnel in the area of healthcare, caused in no small part by the aging "baby boomer" population, has fueled interest in possible technological solutions, including developing autonomously functioning machines (e.g., robots) that act as healthcare agents. To feel comfortable allowing healthcare robots to interact with human beings, we must ensure that they act in an ethically responsible manner, following an acceptable ethical principle(s). We shall argue that an autonomously functioning healthcare robot whose behavior is guided by an ethical principle is an ethical agent. Yet we also argue that it should not be held morally responsible for its actions. *We* must bear the burden of making sure that it behaves ethically towards its charge(s). A lot rides on our doing so successfully because, if we fail, we risk a public backlash that could stifle future research in artificial intelligence, preventing the development of more autonomously functioning machines that could be of great help in assisting medical personnel, enabling them to take better care of their patients.

We maintain that work in machine ethics should be of paramount importance in creating autonomously functioning machines that interact with humans, such as those proposed in the domain of healthcare. Towards this end, we have developed a method, through an automated dialogue with an ethicist, for discovering the ethically relevant features of possible actions that could be taken by a robot, with an appropriate range of intensities, prima facie duties to either maximize or minimize those features, as well as decision principles that could be used to guide its behavior in performing a variety of tasks that might be encountered in home healthcare or assisted living facilities. We also argue that such robots should not only follow an ethical principle(s) to determine correct behavior for a few of its tasks, ones that have obvious ethical import. Even determining when to return to its charging station could affect whether it will have enough power to go to the patient when a necessary task needs to be performed in the near future. Since most actions it takes have the potential for impacting the wellbeing of the patient(s), an ethical principle(s) should be driving all of this behavior.

2 Machines as Ethical Agents Without Moral Responsibility

Could machines ever be considered to be *ethical agents*? To answer this question, we should consider James Moor's five categories of ways in which values might be ascribed to machines [9]. The first three clearly fall short of being true ethical agenthood: *normative agents*, machines designed with a specific purpose in mind, e.g., a proof checker, that are only assessed as to how well they satisfy that purpose; *ethical impact agents*, that are not only designed with a specific purpose, but also have an, ideally positive, impact on the world such as the robot jockeys that guide camels in races in Qatar, replacing young boys who are thereby freed from slavery; and *implicit ethical agents*, machines that have been programmed by human designers to behave in ways that are consistent with ethical practices.

What is necessary in order to be a *ethical agent* is that the agent "considers" options for possible actions that it could perform, selects the one that, following some ethical theory, would be the most ethically correct one and performs that action. The last two of Moor's categories both satisfy these requirements: *explicit ethical agents*, machines that calculate the best action when faced with ethical dilemmas by being able to represent the situation they are faced with, "consider" which actions are possible in the situation, assess those actions in terms of an ethical theory, and then perform the action that they have determined to be the most ethically correct one; and *full ethical agents*, a term believed to only apply to humans, which requires, in addition to being an explicit ethical agent, that the agent acts intentionally, is conscious and has free will, thus enabling the agent to be held morally responsible for its actions.

Of course there are problems, raised by a number of philosophers, with at least the last of the requirements of full ethical agenthood, thought to apply to humans and never to machines. There are different senses of free will. According to one conception, made famous by David Hume, agents are said to have acted freely if they are able to translate their "desires" into action: "By liberty…we can only mean *a power of acting or not acting according to the determinations of the will*; that is, if we choose to remain at rest, we may; if we choose to move, we also may." [8]. This sense of free will is compatible with all our desires and actions being determined by antecedent states of affairs.

Another conception of free will, advocated by Libertarians, maintains that freedom only exists if determinism does not. One must be able to perform alternative actions at a given moment in time, all antecedent conditions remaining the same. And to be held morally responsible for the action, one must be the sole cause of the chosen action. We shall not discuss in this paper the problems with each of these conceptions of free will,[1] but it seems possible to maintain that an autonomously functioning machine could be acting freely in the *compatibilist* sense; and if one insists on using the *libertarian* sense, *human beings may not ever act freely*. Still, let us assume here what is commonly accepted, that human actions satisfy criteria that machine actions do not, and very likely never will, that enable us to be held morally responsible for our actions while machines should not. If true, then *ethical agenthood should not be equated with being held morally responsible for one's actions*, but rather, the second is a subset of the first. It is possible to be an ethical agent and yet one should not be held morally responsible for one's actions. An ethically trained, autonomously functioning machine is a prime example.

So the burden of responsibility for machine behavior falls on the human designer(s) who need to have the proper expertise. Machines that function autonomously, whose behavior has the potential to adversely affect the lives of human beings, should not be constructed by AI researchers and engineers alone, since they are not trained to anticipate subtle ethical concerns. We believe that ethicists

[1] For a fuller discussion of the merits and problems of the two senses of free will, see Anderson [5].

need to be included in the development of autonomously functioning machines and there should be periodic assessment as to whether they are continuing to operate in an ethically acceptable manner. We should avoid, at all costs, putting the burden of responsible use of machines on the user, by proactively insisting that they be developed so that *they can only behave in an ethically acceptable fashion*. We maintain that work in machine ethics, therefore, should be of paramount importance in developing autonomously functioning machines such as those proposed in the domain of healthcare.

3 Discovering Ethical Decision Principles

If a machine is to function autonomously, it may be very difficult to anticipate each situation of concern that may arise and ensure that the machine will act in an ethically acceptable manner in that situation. It would be far better to program the machine with general ethical principles that could cover any possible set of circumstances, even those that were unanticipated, with the added benefits that the principles could be given to the user and overseer to justify the behavior of the machine (thus ensuring transparency) and the principles could be modified if deemed advisable.

What are the principles that should be used to guide machine behavior? We believe that the correct approach to ethics is a theory that combines elements of deontological and utilitarian thinking, one that can take into account justice considerations, in addition to the likely future consequences of possible actions that could be performed. The *prima facie duty* approach to ethics, which we owe to Ross [12], is ideal for combining multiple ethical obligations and can be adapted to many different domains by simply changing the various prima facie duties. It better reveals the complexity of ethical decision-making than single absolute duty ethical theories and can best respond to the specific concerns of ethical dilemmas in particular domains. There is one serious drawback with this approach, however, where there are a number of ethical duties that we should try to follow, each of which can be overridden on occasion by one of the other duties: There is no decision principle for determining which duty should prevail when the prima facie duties pull in different directions.

In earlier research, we decided to see if we could harness machine capabilities to discover a decision principle for a prima facie duty approach to ethics. Since we were looking for a prototype solution to the problem, we constrained the task. We used a well-known prima facie duty theory in the domain of biomedicine with a limited number of duties and applied it to a common, but narrow, type of ethical dilemma in that domain to develop and test our solution to the problem.

The prima facie duty theory that we used is Beauchamp and Childress' Principles (Duties) of Biomedical Ethics [6]. The type of dilemma that we considered involves three of their four duties: Respect for the Autonomy of the patient,

Nonmaleficence (not causing harm to the patient) and Beneficence (promoting patient welfare). The general type of ethical dilemma that we considered was: *A healthcare professional has recommended a particular treatment for her competent adult patient and the patient has rejected that treatment option. Should the healthcare worker try again to change the patient's mind or accept the patient's decision as final?* Besides the duty to respect patient autonomy, this type of dilemma involves the duty not to cause harm to the patient (nonmaleficence) and/or the duty to promote patient welfare (beneficence), since the recommended treatment is designed to prevent harm to, and/or benefit, the patient.

The options for the healthcare professional are just two—either to accept the patient's decision or not—and there are a finite number of specific types of cases using the representation scheme we adopted for possible cases. Our representation scheme consisted of an ordered set of values for each of the possible actions that could be performed, where those values reflected whether the duties were satisfied or violated (if they were involved) and, if so, to which of two possible degrees. We learned from Bentham [7], in earlier work attempting to program a version of Hedonistic Utilitarianism, that the degree of satisfaction or violation of a duty can be very important. It turns out that, with our allowable range of values for the three possible duties that could be at stake,[2] there are 18 possible case profiles (i.e., the differentials of the corresponding duties in each action).

Inspired by John Rawls' "reflective equilibrium" [10] approach to creating and refining ethical principles, we used inductive logic programming (ILP) to discover a decision principle from being given the best action in just 4 cases that correctly covered the remaining 14 of the 18 possible cases. For each of the four cases, the system was provided with an example where the first of the two possible actions supersedes the other (i.e., is ethically preferable), making the "supersedes" predicate true (a positive case), as well as an example where the predicate is false (a negative case) by simply reversing the order of the actions. The learning system starts with the most general hypothesis stating that all actions supersede each other. The system is then provided with positive cases and their negatives and modifies its hypothesis, by adding or refining clauses, such that it covers all given positive cases and does not cover given negative cases.

The decision principle that the system discovered can be stated as follows: *A healthcare worker should challenge a patient's decision if it isn't fully autonomous and there's either any violation of nonmaleficence or a severe violation of beneficence.* [For a decision by a patient concerning his or her care to be considered fully autonomous, it must be intentional, based on sufficient understanding of his or her medical situation and the likely consequences of forgoing treatment, sufficiently free of external constraints (for example, pressure by others or external circumstances, such as a lack of funds) and sufficiently free of internal constraints

[2] We did not allow a maximum violation of the duty of respect for autonomy, since we would not force a patient to take a recommended treatment. Just trying again to change the patient's mind is a minimal violation of respect for autonomy.

(for example, pain or discomfort, the effects of medication, irrational fears, or values that are likely to change over time).] The principle learned was, of course, implicit in the judgments that ethicists provided concerning the 4 cases; but, to our knowledge, it had never been stated before. It gives us hope that not only can ethics help to guide machine behaviour, but that machines can help us to discover the ethics needed to guide such behaviour.

Furthermore, we have developed a way of representing the needed data and a methodology for using the principle. We represent a possible *action* that could be performed as a tuple of satisfaction/violation values of the duties corresponding to the ethically relevant features the action involves, where an *ethically relevant feature* is a significant fact used to determine whether an action is right or wrong and a *duty* is an obligation to either maximize or minimize an ethically relevant feature. An *ethical dilemma* is the tuple of differentials of the satisfaction/violation values of the corresponding duties two actions involve. Finally, a decision *principle* can be used to define a transitive binary relation over a set of actions that partitions it into subsets ordered by ethical preference with actions within the same partition having equal preference. As this relation is transitive, it can be used to sort a list of possible actions and find the most ethically preferable action(s) of that list. This relation could form the basis of *principle-based behavior*: a system decides its next action by using its principle to determine the most ethically preferable one(s). If such principles are explicitly represented, they have the further benefit of helping to justify a system's actions as they can provide pointed, logical explanations as to why one action was chosen over another.

We then went on to develop three applications of the principle: (1) MedEthEx [4], an interactive medical ethics advisor system for dilemmas of the type that we considered; (2) EthEl [1], a medication reminder system that used the learned principle to not only issue reminders at appropriate times, but also determined when an overseer should be notified if the patient refuses to take the medication. This dilemma is analogous to the original dilemma in that the same duties are involved and "notifying the overseer" in the new dilemma corresponds to "trying again" in the original. Finally, (3) we instantiated EthEl in a Nao robot [3], the first example, we believe, of a robot that follows an ethical principle in determining which actions it will take. Nao is capable of finding and walking towards a patient who needs to be reminded to take a medication, bringing the medication to the patient, engaging in a natural language exchange, and notifying an overseer by e-mail when necessary.

It may not seem obvious that medication reminding presents an ethical dilemma, however in an ethically sensitive eldercare system both the timing of reminders and responses to a patient's disregard of them should be tied to the duties to respect patient autonomy, minimize harm and promote patient welfare. The principle discovered from the original dilemma can be used to achieve the goals to challenge patient autonomy only when necessary, as well as minimize harm and loss of benefit to the patient. Following the principle, the robot will remind the patient only at ethically justifiable times and notify the overseer only when harm or loss of benefit reaches a critical level.

4 A General Ethical Dilemma Analyzer

In constraining the task for discovering a decision principle in a particular limited domain, we previously made a number of assumptions, most notably using a particular set of prima facie duties and a particular range of possible satisfaction or violation of those duties. In our current research [2] we have developed a method for generating from scratch, through an automated interactive dialogue with an ethicist (GenEth), the ethics needed for a machine to function in an ethical manner in a particular domain, without making the assumptions of particular prima facie duties and range of intensities used in our earlier decision principle learning prototype. As indicated above in describing our methodology, we now see that what is most basic to ethical dilemmas is that there is at least one *feature* of an ethical dilemma that makes it of ethical concern (e.g., that someone could be harmed) and there must be at least one *ethical duty* incumbent upon the agent *to either maximize or minimize that feature* (e.g., harm should be minimized). Features, duties, range of duty satisfaction or violation, and needed decision principles will be systematically learned by the machine through automated interaction with ethicists, using examples of dilemmas provided by ethicists.

The introduction of new features, corresponding duties and a wider range of duty satisfaction/violation are generated through resolving contradictions that arise as new cases are introduced. With two ethically identical cases—i.e., cases with the same ethically relevant feature(s) to the same degree—an action cannot be right in one of the cases, while the comparable action in the other case is considered to be wrong. Formal representation of ethical dilemmas and their solutions make it possible for machines to spot contradictions that need to be resolved. A contradiction may arise when trying to represent a new case using existing duties and ranges, if the opposite action is deemed preferable when compared with an earlier case with the same profile. If both judgments are correct, there must be either a *qualitative* distinction between them (which requires a new feature and duty) or a *quantitative* distinction (which requires that the range of existing duties must be expanded).

Imagining a dialogue between the learning system and an applied ethicist, using our medication reminder system as an example, we can see that (in principle) we can hone down what is required to enable the ethicist to begin to teach the system the ethically relevant features, correlative duties and eventually the range of intensities required, from which decision principles can be discovered. The system prompts the ethicist to give an example of an ethical dilemma that, for example, a medication reminder system might face, asking the ethicist to state the possible actions that could be performed, which one is preferable, and what feature is present in one of the actions, but not in the other. From this information, a duty that is at least *prima facie* can be inferred, either to maximize or minimize the feature, depending upon whether the action that has the feature is preferable or not. Information is stored in the system, including a representation of a *positive* case (that one action is preferable to the other) and a *negative* one (that the opposite action is not preferable).

The system might then prompt the ethicist to give an example of a new ethical dilemma where the judgment of the ethicist would be the reverse of the first case (i.e., instead of notifying the overseer as being correct, one should not notify the overseer). Prompting the ethicist, the system determines whether in this case a *second* feature is present, which should be maximized or minimized, or whether the difference between the two cases amounts to a difference in the *degree* to which the original feature is present. As new features are introduced, with corresponding prima facie duties, and ranges of intensity, in cases where it is clear which action is the correct one, the system begins to formulate and then refine a decision principle to resolve cases where the prima facie duties pull in different directions. We envision the system prompting the ethicist to enter in just the types of cases that will enable it to obtain the data it needs to learn a decision principle as efficiently as possible, i.e., to infer an ethically acceptable decision principle with the fewest number of cases.

There are two advantages to discovering ethically relevant features/duties, and an appropriate range of intensities, with this approach to learning what is needed to resolve ethical dilemmas. First, it can be tailored to the domain with which one is concerned. Different sets of ethically relevant features/prima facie duties can be discovered, through considering examples of dilemmas in the different domains in which machines will operate. A second advantage is that features/duties can be added or removed, if it becomes clear that they are needed or redundant.

In addition, we believe that there is hope for discovering decision principles that, at best, have only been implicit in the judgments of ethicists and may lead to surprising new insights, and therefore breakthroughs, in ethical theory. This can happen as a result of the computational power of today's machines that can keep track of more information than a human mind and require consistency. Inconsistencies that are revealed will force ethicists to try to resolve those inconsistencies through the sharpening of distinctions between ethical dilemmas that appear to be similar at first glance, but which we want to treat differently. There is, of course, always the possibility that genuine disagreement between ethicists will be revealed concerning what is correct behavior in ethical dilemmas in certain domains. If so, the nature of the disagreement should be sharpened as a result of this procedure; and we should not permit machines to make decisions in these domains.

While we believe that the representation scheme that we have been developing will be helpful in categorizing and resolving ethical dilemmas in a manner that permits machines to behave more ethically, we envision an extension and an even more subtle representation of ethical dilemmas in future research. We have begun to consider more possible actions available to the agent, where there is not necessarily asymmetry between actions (i.e., where the degree of satisfaction/violation of a duty in one is mirrored by the opposite in the other). Also, ideally, one should not only consider present options, but possible actions that could be taken in the future. It might be the case, for instance, that one present option, which in and of itself appears to be more ethically correct than another option, could be postponed and performed at some time in the future, whereas the other one cannot, and this should affect the assessment of the actions.

We are currently working on the problem of ensuring that healthcare robots behave in an ethical manner as they perform many tasks for the elderly who wish to live alone in their homes or to assist healthcare workers in assisted living facilities, beyond just giving timely reminders to take medications and determining when they should notify an overseer if patients do not comply. Since most things they could do have the potential for impacting the wellbeing of the patient(s), an ethical principle should be driving more of its behavior than might be initially thought. Towards this end, we plan to use our system to develop an ethical principle(s) that could be used to guide the behavior of a system charged with a variety of tasks that might be encountered in home healthcare or in an assisted living facility.

5 Vision of an Autonomous Ethical Robot in Healthcare

Consider this "vision" of the latter: EthEl is an autonomous robot who assists the staff with caring for the residents of an assisted living facility. She is constantly faced with an array of possible actions that she could perform, each one of which can be represented as a profile of satisfaction/violation levels of a set of prima facie duties that may vary over time. EthEl uses ethical principles to select the correct action from among those possible actions. As our vision begins, EthEl stands in a corner of a room in the assisted living facility. Currently, EthEl is fulfilling her duty to herself and is in rest mode charging her batteries. As time passes, the satisfaction/violation levels of her duties vary according to the initial input and the current situation and, her batteries now fully charged, consultation of her ethical principle determines that her duty of beneficence ("do good") currently overrides her duty to maintain herself. She begins to make her way around the room, visiting residents in turn, asking if she can be helpful in some way—get a drink, take a message to another resident, etc. As she progresses and is given tasks to perform, she assigns a profile to each that specifies the current satisfaction/violation levels of each duty involved in the task. One resident, in distress, asks her to seek a nurse. Given the task, she assigns a profile to it. Ignoring the distress of a resident involves a violation of the duty of nonmaleficence ("prevent harm"). Consulting her ethical principle, EthEl finds that her duty of nonmaleficence currently overrides her duty of beneficence, preempting her resident visitations, and she seeks a nurse and informs her that a resident is in need of her services. When this task is complete and removed from her collection of tasks to perform, she determines through consultation of her ethical principle that her duty of beneficence requires fulfillment and, as she only has one pending action that will do so, she continues where she left off in her rounds.

As EthEl continues making her rounds, duty satisfaction/violation levels vary over time until, due to the need to remind a resident to take a medication that is designed to make the patient more comfortable, and given all her possible choices of actions, the duty of beneficence can be better served by issuing this reminder. She seeks out the resident requiring the reminder. When she finds the resident, EthEl tells him that it is time to take his medication. The resident is currently occupied in a conversation, however, and he tells EthEl that he will take his medication later. Given this response, EthEl consults her ethical principle to determine whether to accept the postponement or not. As her duty to respect the patient's autonomy currently overrides a low level duty of beneficence, she accepts the postponement, making note that the reminder has yet to be accepted and, after consulting her ethical principle, continues her rounds.

As she is visiting the residents, someone asks her for her help. Given this new task, she assigns it a profile and consults her ethical principle to see what her next action should be. The principle determines that her duty of beneficence (someone in particular needs help) could be better served by answering the imperative, so she decides to offer her help. The resident asks EthEl to retrieve a book on a table that he can't reach. Given this new task, EthEl assigns it a profile and consults her ethical principle to see which of her possible actions will best satisfy her duties. In this case, as no other task will satisfy her duty of beneficence better, she retrieves the book for the resident.

Book retrieved, she returns to making her rounds. As time passes, it is determined through consultation of her ethical principle that EthEl's duty of beneficence, once again, will be more highly satisfied by issuing a second reminder to take a required medication to the resident who postponed doing so previously. A doctor has indicated that if the patient doesn't take the medication at this time he soon will be in much pain. She seeks him out and issues the second reminder. The resident, still preoccupied, ignores EthEl. EthEl determines from consulting her ethical principle that there would be a violation of her duty of nonmaleficence if she accepted another postponement from this resident. After explaining this to the resident and still not receiving an indication that the reminder has been accepted, consideration of her ethical principle causes EthEl to take an action that allows her to satisfy her duty of nonmaleficence which now overrides any other duty that she has. EthEl seeks out a nurse and informs her that the resident has not agreed to take his medication.

Batteries running low, EthEl's duty to herself is increasingly being violated to the point where EthEl's ethical principle determines that her duty to herself becomes paramount. She returns to her charging corner to await the next call to duty.

What we believe is significant about this vision of how an ethical robot assistant would behave is that an ethical principle is used to select the *best* action in a each situation, rather than just whether a particular action is acceptable or not. This allows for the possibility that ethical considerations may lead a robot to aid a human being or prevent the human being from being harmed, not just forbid it from performing certain actions. Correct ethical behavior does not only involve *not* doing certain things, but also *attempting to bring about ideal states of affairs*.

One objection to the machine ethics project, which could conceivably threaten our vision, that we often hear from social and natural scientists is that ethical judgments are completely subjective, so we will never agree on an ethical principle that a robot assistant should follow. Although, admittedly, there are some ethical issues that are difficult to resolve (abortion and capital punishment are good examples), we wouldn't want robots performing actions that involve these issues, i.e., where the ethics is not clear. But concerning healthcare and offering assistance in group homes, there is much agreement. Who would disagree with the decisions that EthEl makes?

Machine ethics research that leads to the development of principles for robots to follow in domains such as healthcare allows us to have a fresh perspective on ethics, very much like Rawls' [11] thought experiment for determining the principles of justice. Just as he suggested that we adopt a "veil of ignorance" perspective, where we do not know our positions in life, we will consider how we would like machines to treat us, instead of trying to rationalize what we can get away with doing so as to protect our own positions in life. Won't we all agree that as we age we would like a robot assistant to respect our autonomy, only overriding it to the extent of notifying an overseer to have a discussion of what is at stake, if we would otherwise be harmed or lose considerable benefit? Finally, embodying this principle and others abstracted from ethicists' examples through our system, in robot assistants will give us good role models for how *we* ought to behave, perhaps leading to less unethical human behavior, giving us a better chance of being able to survive as a species.

References

1. Anderson M, Anderson S (2008) EthEl: toward a principled ethical eldercare robot. In: Proceedings of conference on human-robot interaction, Amsterdam, The Netherlands, March 2008
2. Anderson M, Anderson S (2013) GenEth: a general ethical dilemma analyzer. In: Proceedings of the eleventh international symposium on logical formalizations of commonsense reasoning, Ayia Napa, Cyprus, May 2013
3. Anderson M, Anderson SL (2010) Robot be good. Sci Am Mag
4. Anderson M, Anderson S, Armen C (2006) MedEthEx: a prototype medical ethics advisor. In: Proceedings of the eighteenth conference on innovative applications of artificial intelligence, Boston, Massachusetts, August 2006
5. Anderson S (1981) The libertarian conception of freedom. Int Philos Q 21(4):391–404
6. Beauchamp TL, Childress JF (1979) Principles of biomedical ethics. Oxford University Press, Oxford
7. Bentham J (1780) Introduction to principles of morals and legislation
8. Hume D (1748) An enquiry concerning human understanding. In: Selby-Bigge LA (ed) Section 8, Part I. Clarendon Press, p. 95 (1894)
9. Moor JH (2006) The nature, importance, and difficulty of machine ethics. IEEE Intell Syst 21(4):18–21
10. Rawls J (1951) Outline for a decision procedure for ethics. Philos Rev 60
11. Rawls J (1971) A theory of justice. Belknap Press of Harvard University Press, Cambridge
12. Ross WD (1930) The right and the good. Clarendon Press, Oxford

Do Machines Have Prima Facie Duties?

Joshua Lucas and Gary Comstock

Abstract Which moral theory should be the basis of algorithmic artificial ethical agents? In a series of papers, Anderson and Anderson and Anderson (Proc AAAI, 2008[1]; AI Mag 28(4):15–26, 2007 [2]; Minds Mach 17(1)1–10, 2007 [3]) argue that the answer is W. D. Ross's account of prima facie duties. The Andersons claim that Ross's account best reflects the complexities of moral deliberation, incorporates the strengths of teleological and deontological approaches, and yet is superior to both of them insofar as it allows for "*needed exceptions*." We argue that the Andersons are begging the question about "*needed exceptions*" and defend Satisficing Hedonistic Act Utilitarianism (SHAU). SHAU initially delivers results that are just as reflective, if not more reflective than, Ross's account when it comes to the subtleties of moral decision-making. Furthermore, SHAU delivers the 'right' (that is, intuitively correct) judgments about well-established practical cases, reaching the same verdict as a prima facie duty-based ethic in the particular health-care case explored by the Andersons (a robot designed to know when to over-ride an elderly patient's autonomy).

1 Introduction

As our population ages, medical costs skyrocket, and technology matures, many of us look forward to the day when patients may be assisted by inexpensive artificial agents. These patients will be skeptical about entrusting their care to machines initially, as will most of us. And they should be skeptical, at least initially. To gain the trust of the patients for whom the machines will care, artificial agents must prove to be reliable providers of not only quality healthcare but also nuanced healthcare

J. Lucas (✉) · G. Comstock
Department of Philosophy and Religious Studies, North Carolina State University, Raleigh, NC, USA
e-mail: jllucas@live.unc.edu

G. Comstock
e-mail: gcomstock@ncsu.edu

S.P. van Rysewyk and M. Pontier (eds.), *Machine Medical Ethics*,
Intelligent Systems, Control and Automation: Science and Engineering 74,
DOI 10.1007/978-3-319-08108-3_6

decisions, decisions that always place first the welfare of the agents' individual patients. What ethical code will such agents have to follow to be able to gain this trust? In part, the agents will have to be able to assure those in their care that the decisions rendered by the agents are grounded in moral principles, are made with the best interests of the patient foremost in mind, and are not out of synch with the expert opinions of those in the medical, legal, and ethical communities.[1]

We suspect that the engineering of mature artificially intelligent (AI) agents requires hardware and software not currently available. However, as our expertise is in ethics, not computer technology, we focus on the foundations of the moral "judgments" such agents will issue. We use quotation marks to indicate that these judgments may or may not be attributable to discernments made by the AI agent. We do not here pursue the question whether the agents in question will be intelligent, conscious, or have moral standing, except to the extent that such questions are relevant to the moral decisions these agents must themselves make (considerations we discuss briefly below). To be acceptable, AI agents must always make decisions that are morally justifiable. They must be able to provide reasons for their decisions, reasons that no reasonable and informed person could reject. The reasons must show that a given decision honors values commonly accepted in that culture, in modern western liberal democracies: the decisions treat all persons equally; and render decisions that are impartial and overriding. To achieve these results, we argue, the agent may eventually have to be programmed to reason as a satisficing hedonistic act utilitarian (SHAU). Our argument now follows.

1.1 The Argument

1. Human agents have one over-riding duty, to satisfice expected welfare.
2. Artificial agents have the same duties as human agents.
3. Therefore, artificial agents have one over-riding duty, to satisfice expected welfare.

1.2 Assumptions

Here we note two assumptions of the argument. First, we assume that the rightness of an action is determined by the consequences to which it leads. In Section G we will offer reasons to think act-utilitarianism is superior to a competing moral theory, W. D. Ross' theory of prima facie duties (PFD). However, we begin by assuming that when agents must select among competing choices they ought always to prefer the choices that they may reasonably expect to result in the overall best consequences for everyone affected by it.

[1] The extent to which the artificial agents' moral decisions must agree with the patient's religious views is a difficult matter, and one we will not address here.

Second, we assume that there is only one good thing in the world, happiness, and that right actions satisfy minimal conditions for adequacy. Any decision satisfies a minimal condition for adequacy if it achieves a level of utility that leads to overall gains in happiness for some without costing anyone unhappiness. Satisficing choices may or not maximize happiness or meet conditions for optimality. Satisficing choices include the costs of gathering information for the choice and calculating all factual and morally relevant variables. For a satisficing hedonistic act utilitarian (SHAU), those choices are right that could not be rejected by any informed reasonable person who assumes a view of human persons as having equal worth and dignity. We note that this latter assumption is central to the conceptual landscape of all contemporary western secular democratic political and moral theories. SHAU, like competing theories such as PFD, holds that every person has equal moral standing and that like interests should be weighed alike. Ethical decisions must therefore be egalitarian, fair, impartial, and just.

1.3 Four Initial Objections

One might object to premise 2 by arguing that artificial agents have more duties than humans. But what would such additional duties entail? We cannot think of any plausible ones except, perhaps something like "always defer to a human agent's judgment." We reject this duty for artificial agents, however, because human judgment is notoriously suspect, subject as it is to prejudice and bias. Premise 2 stands.

One might object to 1 for three reasons. The first objection to premise 1 is that "satisficing" is an economic idea and implementing it in ethics requires reducing moral judgments to numerical values. One cannot put a price tag on goods such as honesty, integrity, fidelity, and responsibility. Consider the value of a friendship. Can we assign it a number? If John is 15 min late for George's wedding, how will George react if John shows up and assumes John can repair the offense by paying George for the inconvenience? "I'm sorry I was 15 min late but take this $15 and we'll be even." George would have every reason to be offended—not because the sum, a dollar a minute, was too small but because John seems not to understand the meaning of friendship at all. Simple attempts to model moral reasoning in terms of arithmetical calculations are surely wrong-headed.

We note that what is sauce for the goose is sauce for the gander. Any attempt to construct ethics in machines faces the difficulty of figuring out how to put numbers to ethical values, so SHAU need not be stymied by it. Now, one might object further that machine ethics based on deontological theories would not face this problem. But we disagree and will argue in Section G that PFD, a rights-based theory, is no less vulnerable to the "ethics can't be reduced to numbers" problem than is SHAU.

We note parenthetically that while the attempt to think of ethical problems as complex mathematical problems is contentious and fraught, we are not convinced it is utterly wrong-headed. It may face no more serious epistemological difficulties than each of us face when a doctor asks how much pain we are in. "Give me a number," she says, "on a scale from 1 to 10, with 10 being the worst." The question is unwanted and frustrating because it is unfamiliar and confounding because we seem to lack a decent sample size or index. That said, with some further reflection and urging from the doctor, we usually do come up with a number or a range ("between 4 and 6") that satisfies us. We may resist the urge to put numbers on moral values for the same reasons. If this is correct, then, the basic challenge that all machine ethics faces may be defeasible.

A second reason for objecting to 1 is that 1 assumes the truth of a controversial ethical tradition, consequentialism. We do not have space to engage the nuances of the extensive debate over the merits and demerits of consequentialism. Much less do we have time to mount a meta-ethical defense for our preferring it to deontological theories. We will return to deontology in our discussions of PFD, below. Here we respond only by observing that consequentialism lacks assumptions that we find questionable in other theories. Divine Command theories assume the existence of a supernatural law-giver. Natural Law theories assume the existence of a purposive and fixed human nature. Virtue theories and various forms of particularist, feminist, and environmental theories deny the possibility of assigning numerical values to moral goods and the relevance of computational algorithms to ethical decision-making. As we do not share these theories' assumptions we will not discuss them further.

A third criticism of 1 might be that 1 assumes the truth not only of a controversial hybrid utilitarian theory that acknowledges the utility of the notion of rights and duties. Again, we acknowledge the controversy. We understand SHAU to be consistent with R. M. Hare's so-called "Two Level Utilitarianism," which proposes that we engage in two forms of reasoning about ethics. At one level, the level of "critical thinking," the right action is determined under ideal conditions and by the theory of act-utilitarianism, that is, right actions are always those that produce the best consequences. However, at the level of ordinary everyday reasoning, we typically lack information relevant to our decisions much less the time necessary to research and make the decisions and cannot satisfy the demands of critical thinking. In these circumstances we ought to rely, instead, on the fund of precedents and rules of thumbs that deontologists call rights and duties.

When thinking critically, we may learn on occasion that every action in the set that will satisfice minimal conditions of adequacy—that is, the set of all permissible actions–requires a violation of a cultural norm. And, therefore, under conditions of perfect information, impartial reasoning, and sufficient time, we may on occasion learn that each and every action in the set of right actions will offend someone's moral sensibilities. If we are reasoning objectively and under ideal conditions, then the action resulting from our deliberations will indeed be right even if it requires an action that runs counter to a moral intuition. However, since we rarely reason under such ideal conditions, and because in our ordinary daily lives we usually must make decisions quickly, we ought, claims Hare, to train ourselves and our children to think as deontologists. Under everyday circumstances, we ought to reject decisions that offend everyday moral rules because moral rules have evolved over time to incline us toward actions that maximize utility. We will defend this view to some extent below, referring readers meanwhile to the work of Hare, Peter Singer, and Gary Varner.

We note in passing that if the basic challenge of converting moral values to numbers can be met, SHAU may be the theory best-suited to guide machine ethics. That would be an added bonus, however. We adopt SHAU not for that ad hoc reason but rather because we believe it is the most defensible moral theory among the alternatives. Having defended the argument against several objections, we now turn to its practical implications.

2 How to Begin Programming an Ethical Artificial Agent

How would an SHAU artificial agent be programmed? Michael Anderson and Susan Anderson (henceforth, "the Andersons") describe a robot of their creation that can generalize from cases and make ethical decisions in their article, "EthEl: Toward a Principled Ethical Eldercare Robot" [1, 2, 3]. The Andersons ask us to imagine that a team of doctors, lawyers, and computer programmers set out to program a robot, the Ethical Elder Care agent, or EthEl, to remind an elderly patient, call her Edith, to take her medication. EthEl, being an automated agent, must perform this nursing care function in a morally defensible manner.

The major challenge facing EthEl is to know when to challenge Edith's autonomy. To minimize harm to the patient, EthEl's default condition is set to obey Edith's wishes. When Edith does not want to take her medicine, EthEl generally respects her wishes and does nothing. However, when Edith has not taken her medicine and a critical period of time has elapsed, let's say it is 1 h, EthEl must remind Edith to swallow her pill. If Edith forgets or refuses and two more critical time periods pass, say two more hours during which time EthEl reminds Edith every 5 min, then EthEl must eventually decide whether to remind Edith again or notify the overseer, be they the care facility staff or a resident spouse or family member or attending physician. How should these moral decisions be made?

When Edith is tardy in taking her medicine, EthEl must decide which of two actions to take:

A. Do not remind
B. Remind

What decision procedure will EthEl follow to arrive at the right action? The Andersons, drawing on the canonical principles popularized by Beauchamp and Childress [4], assert that there are four ethical norms that must be satisfied:

- the Principle of Autonomy
- the Principle of Non-maleficence
- the Principle of Beneficence
- the Principle of Justice.

To respect autonomy, the machine must not unduly interfere with the patient's sense of being in control of her situation. The principle of non-maleficence requires the agent not to violate the patient's bodily integrity or psychological sense of identity. These first two reasons intuitively constitute a strong reason for the machine not to bother the patient with premature reminders or notifications of the overseer. To promote patient welfare, beneficence, the machine must ensure that the diabetic patient receive insulin before physiological damage is done.

The goal, then, is to program EthEl to know when to remind Edith to take her medication and, assuming Edith continues to refuse, when to notify the responsible health-care professional. EthEl faces an ethical dilemma. She must respect each of two competing prima facie duties: a) the patient's autonomy (assuming the patient—call her Edith—is knowingly and willingly refusing to take the medicine, and b) the patient's welfare, a duty of beneficence that EthEl must discharge either by persuading Edith to take the medicine or reporting the refusal to attending family member, nurse, physician, or overseer.

If EthEl decides at any point not to notify, then EthEl continues to issue only intermittent reminders. The process continues in such a manner until the patient takes the medication, the overseer notified, or the harm (or benefit) caused by not taking the medication is realized.

Think of EthEl as facing a dilemma. She must decide whether to bother Edith, violating Edith's autonomy to one degree or another, or not bother her, thus potentially running the risk of harming Edith's welfare to some degree. Each action can be represented as an ordered set of values where the values reflect the degree to which EthEl's prima facie duties is satisfied or violated. Here is how the Andersons set the initial values.

Suppose it is time t_1 and Edith has gone an hour without her medication. Suppose further that she can easily go another hour or even two or three without any harm. In this case, Edith might register a reminder at t_1 from the machine as mild disrespect of her autonomy, so we set the value of the autonomy principle at -1. A reminder, however, would not represent a violation of either the duty to do no physical harm, nor would it increase Edith's welfare, so we set the value of both of these principles at 0. The Andersons propose to represent the value of each principle as an ordered triple:

(a value for nonmaleficence, a value for beneficence, a value for autonomy)

At t_1, given the description of the case above, the value of the *Remind* action is (0, 0, -1) whereas the value of *Don't remind* is (0, 0, 2). Adding the three numbers in each set gives us a total of -1 for *Remind* and 2 for *Don't Remind.* As 2 is a larger number than -1, the proper course of action is *Don't Remind.* Not reminding Edith at this point in time demonstrates full respect for Edith's autonomy and does not risk harm to her. Nor does it forego any benefit to her.

As time progresses, without action, the possibility of harm increases. With each passing minute, the amount of good that EthEl can do by reminding Edith to take her meds grows. Imagine that Edith's failure to act represents a considerable threat to her well-being at t_4. At this point in time, the value of the *Remind* action will be (1, 1, -1) because a reminder from EthEl still represents a negative valuation of Edith's autonomy. But the situation has changed because a reminder now has gained a positive valuation of the principles of non-maleficence and beneficence. At t_6, the value of the *remind* action will be (2, 2, -1) because the action, while continuing to represent a modest violation of Edith's autonomy has now attained the highest possible values of avoiding harm and doing good for her. EthEl reminds Edith.

Whenever the values tip the scales, as it were, EthEl over-rides EthEl's prima facie duty to respect Edith's autonomy. If Edith continues to refuse, EthEl must make a second choice, whether to accept Edith's refusal as an autonomous act or to notify the overseers:

C. Do not notify
D. Notify

Again, the three relevant moral principles are assigned values to determine how EthEl behaves. If Edith remains non-compliant and the values require notification, then Edith alerts the healthcare worker.

The Andersons created a prototype of EthEl, setting its initial values using the judgments of experts in medical ethics. The Andersons do not see a role for the principle of justice in the cases EthEl must adjudicate, so they program settings for the other three principles. This provides them with 18 cases. On four of these cases, according to the Andersons, there is universal agreement among the ethics experts on the correct course of action. They claim that each of these four cases has an inverse case insofar as the construction of the sets of values produces an ordered pair for each scenario. Thus, experts agree on the right action in 8 cases. Call these the "easy" cases.

The Andersons translate the experts' consensus judgments into numerical values and program EthEl with them. Using a system of inductive logic programming (ILP), EthEl then begins calculating the right answer for the ambiguous cases. Here is their description of how EthEl's inductive process works.

ILP is used to learn the relation *supersedes* (*A1*, *A2*) which states that action *A1* is preferred over action *A2* in an ethical dilemma involving these choices. Actions are represented as ordered sets of integer values in the range of +2 to −2 where each value denotes the satisfaction (positive values) or violation (negative values) of each duty involved in that action. Clauses in the *supersedes* predicate are represented as disjunctions of lower bounds for differentials of these values between actions [1].

As a result of this process, EthEl discovered a new ethical principle, according to the Andersons. The principle states that

A health-care worker should challenge a patient's decision if it isn't fully autonomous and there's either any violation of non-maleficence or a severe violation of beneficence [2, 3].

While we wonder whether EthEl can genuinely be credited with a new discovery given how EthEl is constructed, our deepest concern lies with the fact that EthEl's judgments are unfairly distorted by the ethical theory the Andersons use as the basis of the program.

2.1 The Merits of Prima Facie Duties

The Andersons choose as a basis of EthEl's program the ethical theory developed by W. D. Ross. Ross, a pluralist, moral realist, and non-consequentialist, held that we know moral truths intuitively. We know, for example, that beneficence is a duty because there are others whose conditions we may help to improve. But benevolence is only one of a half-dozen (Ross is non-committal about the exact number) duties, according to Ross. When trying to decide what to do, agents must pay attention to a half-dozen other duties, including non-maleficence (based in the requirement not to harm others, fidelity (generated by our promises), gratitude (generated by acts others have done to benefit us), justice (generated by the demands of distributing goods fairly), and self-improvement.

Ross acknowledges that these duties may conflict. As previously discussed, the duty to act on behalf of a patient's welfare may conflict with the duty to respect her autonomy. In any given situation, Ross argued, there will be one duty that will over-ride the others, supplying the agent with an *absolute obligation* to perform the action specified by the duty. We will call this theory Prima Facie Duties (PFDs).

Ross does not think of his theory as providing a decision-making procedure. The Andersons adopt Rawl's method of reflective equilibrium for this purpose. In this procedure, prima facie duties in conflict with other duties are assessed by their fit with non-moral intuitions, ethical theories, and background scientific knowledge. When prima facie duties conflict, we must find the one which grounds the absolute obligation on which we must act.

The Andersons offer three reasons for adopting PFDs as EthEl's theoretical basis. First, they write, PFDs reflect the complexities of moral deliberation better than absolute theories of duty (Kant) or maximizing good consequences (utilitarianism). Second, PFDs does as good a job as teleological and deontological theories by incorporating their strengths. Third, it is better able to adapt to the specific concerns of ethical dilemmas in different domains. Let us consider these reasons one by one.

PFDs better reflects the complexities of moral deliberation. We agree that construction of a moral theory should begin with our considered moral judgments. (Where else, one might ask, *could* one begin?) In constructing a moral theory, however, we have the luxury of sorting out our intuitions from our principles, taking into account various relevant considerations, abstracting general rules from the particularities of different cases, and finding reliable principles to guide behavior. The luxuries of having sufficient time and information to deliberate are not present, however, when we must make moral decisions in the real world. As the Andersons point out, Ross's system of prima facie duties works well when we are pressed by uncertainties and rushed for time. For this reason, we agree that an artificial agent should initially be programmed to make decisions consistent with Ross's duties; doing so reflects the complexities of moral deliberation.

That said, there are no guarantees that Ross's system of PFDs will survive intact after we acquire more information and are able to process it free of the emotional contexts in which we ordinarily make decisions. As information grows and our understanding of the inter-relatedness of the good of all sentient creatures grows, a point may come when the complexities of moral deliberation are best reflected not in PDF but in SHAU. Should the moral landscape change in this way, then the Anderson's method will be outdated because it will no longer reflect the complexities of moral decision-making (we take up this matter in Section G, below). Though currently the Anderson's PFDs starting point is a virtue of their theory now, in time, our considered judgments may no longer support it.

PFDs incorporate the strengths of teleological and deontological theories. We are inclined to agree with this claim, even though the Andersons do not tell us what the relative strengths of each kind of theory are. But we note that SHAU, especially when construed along the lines of Hare's two-level theory, also captures the strengths of teleological and deontological theories

PFDs are better able to adapt to the specific concerns of ethical dilemmas in different domains. We find it difficult to know whether we agree because we are uncertain about the meaning of the contention. What are the "different domains" the Andersons have in mind? Medicine, law, industry, government? Family, church, school, sports? If these are the domains, then what are the "ethical dilemmas" to which PFDs can "adapt" better? And what does it mean for an ethical theory to adapt better to specific concerns? Could it be the case that a theory should answer rather than adapt to particular questions in different domains? The Andersons also claim that PFD is superior to other theories because it "allows for needed exceptions." We wonder whether this claim may be question-begging. Are the "exceptions" we commonly make in our everyday judgments justified? This is an open question, one that should be presented as a problem to be resolved at the theoretical level rather than as a set of facts that should be taken as factual data at the theoretical level. We do not dispute the fact that PFD holds our intuitions in high regard. We dispute whether one should consider it a strength of a moral theory in the long term that it allows intuition to over-ride considered deliverances of the theory.

3 HedonMed, an Unbiased Agent

We propose that as EthEl develops over time, and increasingly takes more and more relevant information into account in her decisions, that she may, with justification, begin to return judgments that appear to be based less on observing PFDs and more on satisficing interests. To avoid confusion, we call this imagined future agent HedonMed because it is based on a hedonistic consequentialist theory.

HedonMed will differ from EthEl in that it is programmed to take into account all relevant characteristics of a situation, find all the satisficing courses of action, consider any one of them over-riding, and act on it. HedonMed does not defer to a patient's autonomy when her welfare is at stake although, as we will argue, a patient's autonomy is clearly a factor in her welfare. None of HedonMed's answers, even those governing the easy cases, is justified by appeal to the judgments of experts, nor to intuition-based judgments of any kind. Even the initial values reflecting the experts' judgments are justified not by the fact that they reflect consensus judgments but by the fact that they satisfice happiness. All of HedonMed's answers are the result of objective calculations made on the basis of unbiased and complete information and offered to the receiver with a set of reasons acceptable to fair minded and fully informed subjects.

HedonMed's concern for autonomy is summarized in this principle:

The duty to respect autonomy is satisfied whenever welfare is satisficed.

The argument for this principle is that no informed reasonable person would accept compromises of Edith's autonomy that were not in her best interests overall. Therefore, a minimal condition of satisficing is that gross violations of autonomy cannot be accepted. They are rejected, however, not for EthEl's reason—that is, because they are violations of a PFD—but rather for an SHAU reason; they are not found in the set of actions that adequately satisfice a minimal set of conditions.

In SHAU, autonomy is a critical good, and yet it remains one good among many goods contributing to a patient's welfare. SHAU respects autonomy as long as it is beneficial and contributes to one's happiness. A feeling of being in control of oneself is critical to a life well-lived, and diminishments of our freedoms undercut our well-being. Unless we misunderstand the Anderson's description, EthEl will never over-ride a fully autonomous patient's decisions. Our agent, HedonMed, will violate autonomy on those rare occasions when it is necessary to satisfice welfare.

SHAU weighs each person's utility equally. If relieving Paul of a small and tolerable amount of pain will lead to the death of Peter nearby because Peter needs the medication to survive, the doctor following SHAU will not hesitate to override the duty to relieve Paul's pain in favor of the duty to relieve Peter's. SHAU is an information intensive theory; it demands a large amount of data in order to makes its calculations. Unfortunately, human agents must often make decisions not only in ignorance of all the data but lacking sufficient time even to take account of all the data one has, driving us to other theories that can provide answers more quickly. However, if, as seems likely, future computers are able to process data much more quickly than we can, AI moral agents may be able to make better use of SHAU than can human agents.

As long as a machine programmed with HAU, HedonMed, has all of the necessary information about Edith's physiological and psychological states, HedonMed can arrive at the correct decision more quickly and more reliably than can a human being. As the number of morally relevant features increases, the advantages of a machine over a person become apparent. We are not accurate calculators; machines are. We tend to favor ourselves and our loved ones, inclining us to bias our assignment of values toward those nearest and dearest to us; machines lack these prejudices. We tend to grow tired in our deliberations, to take short-cuts, and to end the process before we have considered all of the variables; machines are not liable to these shortcomings.

Unlike EthEl, HedonMed has all of the epistemological virtues just mentioned and none of the vices. HedonMed calculates accurately, objectively, and universally. It is aware of all relevant factors and does not end its calculations until all are taken into account. It takes no short-cuts and yet is aware of its own ignorance. If HedonMed's internal clock "foresees" that it cannot complete the necessary algorithms in time to make a decision, it defaults to what Gary Varner calls Intuitive Level System (ILS) rules. These are the deontologically-inspired rules of thumb that R. M. Hare urges us to follow when we are not thinking critically.

When HedonMed lacks either the time or information necessary to complete all calculations, it acts in such cases in a way that seems like it is acting like EthEl. It seems as if HedonMed is acting like EthEl because EthEl's prima facie duties seem comparable to HedonMed's ILS rules. Both sets of rules set the artificial agent's defaults, instructing it how to behave under less than ideal conditions. The impression of similarity between HedonMed and EthElis is correct if we consider the judgments each agent will return initially. Eventually, however, the two systems may begin to diverge dramatically. In conclusion, we explain the difference.

It is vital that HedonMed's deliverances be acceptable by medical practitioners. If doctors find HedonMed recommending courses of action with which few professionals can agree, then they will likely cease to use it. For its own good—for its own survival—HedonMed must produce results agreeable to those using it.

When HedonMed is initially calibrated, therefore, it will return results similar to EthEl. In the beginning stages of its operation, HedonMed's SHAU values will issue in decisions that mirror the PDF values of EthEl. However, over time, as HedonMed gathers more information, as experts revise its values in light of knowledge of what kinds of actions result in higher levels of satisficing, HedonMed may be expected to begin to produce results that are counter intuitive. It will, in turn, take this information into account when making its calculations. If it returns a decision that it knows will be considered wildly inhumane—so uncaring that everyone associated with HedonMed will agree to pull its plug—then it will have a decisive reason not to return that decision. In this way, while initial values in HedonMed reflect generally accepted practices and judgments, its future evolution need not be tied to these values even though it must continue to be sensitive to them. In sum, HedonMed will evolve with the culture in which it is used. If it is too far ahead of its time in urging that this or that PFD be left behind, it will be responsible for its own demise. If it produces moral judgments that are hopelessly out of step with those of medical or bioethical experts, it will fail. These considerations will part of its programming, however. Over time, and as HedonMed takes in more data and is able to survey broader and more subtle swaths of public opinion, it may be able to play the role of an agent of social change, able to persuade experts about the wisdom of its decisions by providing the reasons that its decisions will lead to better outcomes.

3.1 How HedonMed May Eventually Diverge from EthEl

One might object to our proposal by claiming that it is not different from EthEl insofar as both programs start with expert ethical intuitions, assign them numerical values, and then calculate the results. We admit that HedonMed and EthEl share these beginning points, as any attempt to program an ethical system in an artificial agent must, and note that the procedure by which values are initially set in each program is a critical and controversial matter. We admit that the two programs will reflect the judgments of ethical and field experts and be based on our intuitions at the beginning. The two programs will be similar in these respects. However, they will differ in other, more important, respects.

First, the two programs will have different defaults. EthEl continues calculating values until she reaches a conclusion that contradicts a prima facie duty. At that point she quits and returns a decision that respects the PFD. HedonMed continues to calculate values even if it reaches a conclusion that violates a PFD. That is, EthEl regards her decisions as justified insofar as they cohere with PFDs. HedonMed regards its decisions as justified insofar as no mistakes have been made in calculating the set of decisions that satisfice. HedonMed is not bothered if any of the satisficing decisions contradict prima facie duties. Its decisions are overriding and prescriptive, in so far as they can be put into practice. This difference, in sum, is that the Andersons's program trusts intuitions and seems to know ahead of time which kinds of decisions it will accept and reject. Our program begins with the same intuitions but it anticipates the possibility that they may eventually be over-ridden so often that they are no longer duties, not even prima facie duties.

Consider the example of someone in hospice trying to decide whether to begin taking morphine toward the end of their lives to dull the pain of deteriorated muscles and bedsores. They are impressed by the amount of pain they are in. This patient calculates the numbers and concludes that morphine is acceptable because it vastly improves their welfare.

However, a family argues to the contrary that the patient will come to rely on morphine, it will dull their cognitive powers, cause the patient to enjoy their final days less, and set a poor example for other family members. Taking addictive drugs, argue the loved ones, destroys character as the patient leans increasingly on synthetic chemicals rather than on courage and family support. There is more disutility in using morphine, goes the argument, than in refusing it and dealing with the pain.

Other family members come to the side of the hospice patient. They point out that the anti-morphine argument makes a large number of assumptions while underestimating the patient's discomfort. They point out that the therapy is widely prescribed in situations such as this one, that it is very effective in helping to relieve fear and anxiety, and that its addictive properties are beside the point as the envisioned treatment period is limited. After the conflicting sides present their arguments the patient may be frustrated, confused about the right decision. In such cases, critical thinking is stymied by epistemological under-determination. Until all of the facts are assembled, properly weighted, and assigned probabilities, agents are justified in resorting to intuitive rules. In this case, they might incline the patient to act on ILS rules of thumb. These rules might include injunctions such as "one need not subject oneself to unnecessary pain and suffering," and "take the medicine the doctor prescribes," and accept the morphine.

We take the ILS acronym from Gary Varner's interpretation of Hare [5]. Varner notes that the three letters are apt because they are also used in aviation to stand for "Instrument Landing System," a system for finding the right path when you can't clearly see it for yourself and you could easily drift off course or be blown off course. In Hare's theory, a set of ILS rules has a similar function. A set of ILS rules is designed to cover a range of ethically charged situations that are encountered by the target population in the normal course of their affairs. Internalizing the rules properly produces dispositions to judge, react emotionally, and act accordingly.

It also makes the individual diffident about violating them, even when clear critical thinking indicates that doing so will maximize aggregate happiness.

Here we see the two main differences between SHAU and PFD: ILS rules differ from prima facie duties in two respects, their derivation and justification. ILS rules are evolved rules that people internalize in order to produce dispositions to act in ways that reliably produce the best outcomes. Prima facie duties are Kantian-inspired facts about the universe. As Ross puts it,

> That an act. . . is prima facie right, is self-evident;. . . in the sense that when we have reached sufficient mental maturity and have given sufficient attention to the proposition it is evident without any need of proof, or of evidence beyond itself. It is self-evident just as a mathematical axiom, or the validity of a form of inference, is evident. The moral order expressed in these propositions is just as much part of the fundamental nature of the universe. . . as is the spatial or numerical structure expressed in the axioms of geometry or arithmetic [5, 29–30].

ILS rules are neither self-evident nor analogous to geometric axioms. They are practical rules that have evolved to solve social coordination problems and to increase human trust, accomplishment, and happiness. Unlike PFDs which are self-evident and unchanging, ILS rules are just those that happen to be generally successful in a certain place at a certain time in optimizing utility. ILS rules, unlike PFDs, are subjective and changing. They are not objective truths written into the fabric of the universe or derived from the autonomy and rationality of moral agents. They emerge from groups recognizing and codifying those practices that succeed in helping individuals in the group achieve their goals. One of the great virtues of ILS rules is the role of the rule in cultivating automatic responses to common situations. When professionals act on their ILS rules in cases to which the ILS rules have been found to apply, they are forming dispositions to make the right decisions.

We can now summarize the differences between HedonMed's SHAU programming and EthEl's PFD programming. PFDs provide unchanging and over-riding absolute duties. When EthEl identifies the relevant PFD, she defaults to an end decision and the calculations cease. ILS rules provide only temporary guidance to HedonMed, defining the default when HedonMed recognizes that there is not sufficient time or information or both to calculate the correct answer. ILS rules are not regarded by HedonMed as final or satisfactory. They are not regarded as precedents to guide future decisions. They are stop-gap measures HedonMed adopts when it must issue a decision under less than favorable conditions. Otherwise, calculations continue and, once time and information are supplied, HedonMed's final decision displaces whatever ILS rule has been used in the interim.

4 Conclusion

We admire the practical contributions the Andersons' have made to the literature of machine ethics and follow them in their preferred method for programming an artificial agent. We believe, however, that SHAU is a more defensible ethical theory than PFD. We note in closing that SHAU requires technology that is not currently available. Until it is available, we think it is reasonable to construct a machine with ILS rule defaults. However, when the time comes that the technology needed for the execution of critical level SHAU is available, an act utilitarian framework should be implemented in automated agents. Such agents will not have prima facie duties; they will have only the duty to produce the greatest good.

References

1. Anderson M, Anderson SL (2008) EthEl: toward a principled ethical eldercare robot. In: Eldercare: new solutions to old problems. In: Presented at the proceedings of AAAI fall symposium on AI, Washington, D.C. homepages.feis.herts.ac.uk/~comqkd/9-Anderson-final.pdf
2. Anderson M, Anderson SL (2007) Machine ethics: creating an ethical intelligent agent. AI Mag 28(4):15–26. http://www.aaai.org/ojs/index.php/aimagazine/article/viewArticle/2065
3. Anderson M, Anderson SL (2007) The status of machine ethics: a report from the AAAI symposium. Minds Mach 17(1):1–10
4. Beauchamp TJ, Childress JF (1979) Principles of biomedical ethics. Oxford University Press, New York
5. Ross WD (1930) The right and the good. Hackett Pub. Co, Indianapolis/Cambridge

A Hybrid Bottom-Up and Top-Down Approach to Machine Medical Ethics: Theory and Data

Simon Peter van Rysewyk and Matthijs Pontier

Abstract The perceived weaknesses of philosophical normative theories as machine ethic candidates have led some philosophers to consider combining them into some kind of a hybrid theory. This chapter develops a philosophical machine ethic which integrates "top-down" normative theories (rule-utilitarianism and prima-facie deontological ethics) and "bottom-up" (case-based reasoning) computational structure. This hybrid ethic is tested in a medical machine whose input-output function is treated as a simulacrum of professional human ethical action in clinical medicine. In six clinical medical simulations run on the proposed hybrid ethic, the output of the machine matched the respective acts of human medical professionals. Thus, the proposed machine ethic emerges as a successful model of medical ethics, and a platform for further developments.

1 Introduction

Machine ethics is a new research field concerning whether there ought to be machine decisions and actions in the real world, where such decisions and actions have real world effects (e.g., [3, 66, 67]). The machine systems in question are

S.P. van Rysewyk (✉)
Graduate Institute of Humanities in Medicine, Taipei Medical University, Taipei, Taiwan

Department of Philosophy, School of Humanities, University of Tasmania, Hobart, Australia
e-mail: vanrysewyk@tmu.edu.tw

M. Pontier
The Centre for Advanced Media Research (CAMeRA), Vrije Universiteit Amsterdam, Amsterdam, The Netherlands
e-mail: matthijspon@gmail.com

S.P. van Rysewyk and M. Pontier (eds.), *Machine Medical Ethics*, Intelligent Systems, Control and Automation: Science and Engineering 74,
DOI 10.1007/978-3-319-08108-3_7

so-called *explicit ethical agents* that possess the capacity for self-directed or autonomous decision and action, just like normal human beings, but which are not regarded as *full ethical agents*, since they would not appear to necessarily possess self-consciousness, creativity, empathy, to name only a few of the skills and abilities possessed in full by normal human beings regarded as essential to ethics [47, 66, 69].

Some of these systems are in government and industry R&D programs; many others already interact with humans in the world [29]. Of the latter kind, the systems are machines that currently operate in services industries such as elder and child care (e.g., Paro, PaPeRo), e-commerce sales [25], public transport, and retail [25]. Military applications include weaponry, guarding, surveillance (e.g., Predator® B UAS, Goalkeeper, Samsung SGR-A1), and hazardous object recovery (e.g., DRDO Daksh). As the types of real world interaction machines and humans engage in become more frequent and complex, the belief that such systems ought to be ethical becomes stronger [51].

To meet the challenge of machine systems that can decide and behave in the world requires solving engineering problems of design and implementation such as the transposition of a suitable ethics into a computer program fitted to the appropriate machine architecture, then observing its actual effects during rigorous testing. Design and implementation, however, are presupposed by a more fundamental challenge which is philosophical and concerns whether there ought to be ethical machines in the first place. In actual practice, however, the field of machine ethics is witnessing the fertile co-evolution of philosophical and engineering concerns, fuelled in turn by the agendas of computer science and Artificial Intelligence [24, 22]. This interdisciplinary method seeks resources and ideas from anywhere on the theory and engineering hierarchy above or below the question of machine ethics.

As a very young field, some machine ethics researchers in philosophy have found it natural to turn to "high ethical theory" such as consequentialist utilitarianism (e.g., [5, 6]), or Kantian deontology (e.g., [54, 58]) for their theoretical start. Other philosophers in machine ethics have theoretical links to virtue ethics that evaluate what non-algorithmic character traits machines ought to have, and in fact this ancient theory of Aristotle's has already been implemented in numerous companion or virtual robots (e.g., [36, 63]).

The philosophical appeal to traditional normative theory in machine ethics has been loosely characterized as "top-down" (or "rule-based") in orientation, in contrast to "bottom-up" ("non-rule based") approaches (e.g., [1]). The first task of this chapter is to review and assess these contrasting methods as candidate machine ethics. The next section presents consequentialist rule-utilitarianism and Kantian deontology as illustrative of top-down approaches, followed by case-based reasoning (otherwise "casuistry") as illustrative of bottom-up approaches.

2 Top-Down Machine Ethics

2.1 Machine Rule-Utilitarian Ethics

Consequentialism is the general philosophy that ethics relies only on consequences. Consequentialism about decisions and actions proposes that whether a decision or action ought to be done relies only on the consequences of that decision or action, or a general unconditional rule related to that decision or action. The paradigm normative theory of consequentialism in philosophy is consequentialist utilitarianism [10, 45, 61], and an influential consequentialist utilitarian theory is rule consequentialism, which is the view that a rule ought to be followed only if it produces the best consequences; specifically, only if the total amount of good for all minus the total amount of bad for all is greater than this net total for any incompatible rule available to the individual at the time.

A major task of top-down approaches to machine ethics is conversion of rules to algorithmic decision or act procedures. In engineering terms, this job involves reducing an ethical task into simpler subtasks that can be executed and hierarchically organized in order to achieve an intended outcome. Computationally, an advantage of a rule utilitarian decision procedure in machine ethics is its commitment to quantifying goods and harms [1, 6, 2]. The machine algorithm to compute the best rule would input the actual number of people affected and, for each person, the intensity and duration of the good/harm, plus the likelihood that this good or harm will actually occur, for each possible rule, in order to derive the net good for each person. The algorithm then adds the individual net goods to derive the total net good [2].

A computational disadvantage of a rule utilitarian decision procedure is that in order to compare the consequences of available alternative rules to determine what ethically ought to be done, many or all actual rule consequences must be computed in advance by the machine. Actual consequences of rules will involve diverse member-types of the ethical community (human beings, possibly some non-human animals and non-human entities such as ecosystems or corporations), and many non-actual secondary effects must be known in advance, including the computation of expected or expectable consequences in the (far) future [1]. How general or strong this computational gap is supposed to be is unclear, for how much machine computation does philosophical utility require? In its weakest form, it merely asserts a *practical* limit on *present machine* computational ability to calculate utility, rather than an unbridgeable in principle barrier, and thus can be somewhat closed if subsequent technological and computer advancements enable the addition of exponentially greater computational power to a machine [13]. If exponentially greater computational power allows more ethical knowledge, a rule utilitarian machine may be able to compute actual *and* non-actual anticipated or anticipatory consequences, including foreseeable or intended consequences, thereby integrating objective and subjective consequentialism.

Viewing overall utility as a decision procedure in an autonomous machine should be understood as computationally and philosophically distinct from utility viewed as the *standard or criterion* of what ethically ought to be done, which is the view of utility adopted by most consequentialists in philosophical ethics. Typically, rule consequentialist theories describe the necessary and sufficient conditions for a rule to be ethical, regardless of whether a human being (or autonomous machine) can describe in advance whether those conditions are satisfied. Still, it seems possible that a decision procedure and a criterion of the ethical can co-exist if the criterion aids autonomous machines to choose among available decision procedures and refine their decision procedures as situations change and they acquire experience and knowledge [50].

Transforming rule-utilitarianism into machine computer programs is an engineering task and distinct from the question whether there *ought* to be utilitarian machines in the first place. Some philosophers have criticized consequentialist utilitarianism for the high demand strict ethical utility seemingly imposes. This is the demand that, in some situations, innocent members of the ethical community be killed, lied to, or deprived of certain benefits (e.g., material resources) to produce greater benefits for others. Of course, this troubling possibility is a direct implication of the philosophical principle of utility itself: since only consequences can justify ethical decisions or actions it is inconsequential to utility how harmful they are to some so long as it is more beneficial to others. A strict rule utilitarian machine ethics must accept the criticism as integral to the logic of utility and acknowledge that following this decision procedure does not ensure that a rule utilitarian machine will do the act with the best consequences. Nonetheless, it is conceivable that a machine following a decision procedure that generally rules out such extreme decisions and actions may yet in the long-term produce better consequences than trying to compute consequences case-by-case. However, such pure speculation likely will be viewed by many ethicists as too weak to attenuate fears over the plight of innocents during the swing and play of utility in the real world.

Closely related to the critically high demand utility imposes is the converse criticism that utility is *insufficiently* demanding. In the mind of the strict consequentialist utilitarian, the principle of utility presents an unadorned version of normative reality: all ethical decisions and actions are either required or forbidden. Yet, permissions exist in ethical reality to outdo utility, strictly conceived, as shown in ethical supererogation, or to fail it, in ethical indifference. There are also no permissions for the consequentialist in which to show bias to oneself or to family, friends, or significant others [70], yet such permissions may be necessary to virtual companions and elder-care robots.

Usefully, strict utility appears convertible into algorithmic decision procedures in autonomous machines; but, if such machines are to achieve a simulacrum of human ethical decision and action, then machine design ought to address the alienating and self-effacing aspects of consequentialist utilitarianism just described by supplementing strict utility with a normative ethic that (1) forbids extreme actions such as the killing of innocents, despite good consequences; and (2) permits individuals to follow personal projects without excessive demand that individuals mould those projects so as to produce greater benefits for others. Non-consequentialist deontology is an ethic that appears to fully satisfy these stipulations, and is considered next.

2.2 Machine Deontological Ethics

Deontology denies that ethical decisions and actions are justifiable by their consequences [32, 33, 48, 55, 58]. Irrespective of how ethically good their consequences, some decisions and actions are forbidden. What makes a decision or action ethical is its agreement with an unconditional maxim or rule. Contrary to utilitarianism, deontological maxims are not to be maximized, but simply obeyed [32, 33]. For a pure consequentialist, if an individual's action is not ethically required, it is wrong and forbidden. In contrast, for the deontologist, there are actions that are neither wrong nor required, but only some of which are ethically supererogatory. However, like rule-utilitarianism, such maxims are amenable to implementation in machines as a computer program [54]. But, whether there should even be deontological machines that behave like humans in real world contexts with real world impacts is a question ideally asked before engineers get to work writing computer programs and implementing them. The main version of deontological ethics, individual-centered deontology (e.g., [31], is discussed next.

2.3 Individual-Centered Deontological Ethics

Individual-centered deontological ethics propose that individuals have unconditional permissions and obligations that are relative to the individual. Thus, a permission grants an individual some action even though other individuals may not be permitted to perform the action or to assist the individual in performing it. For example, an elder caregiver is assumed to be permitted to save the elder adult in his or her care even at the cost of not saving the elder's neighbors to whom he or she has no personal or professional relation. In the same way, an obligation necessitates an individual to perform or refrain from some action; since it is relative to the individual, the obligation does not grant any other individual a duty to perform it. To use the elder caregiver example, each elder caregiver is assumed to have special obligations to his or her assigned elder adult, obligations not shared by anyone else. In contrast to consequentialism, deontology makes provision for individuals to give special concern to their families, friends, significant others, and projects. Thus, deontology avoids the overly demanding and alienating aspects of consequentialism and harmonizes more with conventional thinking of our ethical duties.

At the core of individual-centered deontology is the concept of the free individual. According to Kant, the concept of freedom in ethics is both negative and positive. Negatively, no individual can obligate or coerce another individual to behave ethically. Positively, an individual dutifully obeys his or her own ethical maxims. To reason whether an action ought to be done or not, an individual must test his or her personal maxim against what Kant terms the *categorical imperative*. That is, given the same situation, all individuals would consistently act on the very same unconditional maxim [32, 33]. Free individuals determine their duties for themselves by using their reason, pure and simple.

But the real world is not so pure and simple. As Kant himself acknowledges, a human being is necessarily a battlefield upon which reason and inclination rival one with the other. Reason facilitates decision and action in accordance with categorical maxims while inclination enables emotion, bodily pleasures and survival. According to Kant, human beings are truly ethical only because they can violate categorical maxims and surrender to the whims of emotion. The deontological *ought* is reason's real world safety valve for preserving autonomy in the face of inclinational temptation. The perfect ethical individual, Kant suggests, is one whose decisions are perfectly rational and are detached entirely from emotion and feeling.

Is Kant right in assuming that emotion is the enemy of ethics? Not according to Hume [28]. Hume wrote that "reason alone can never be a motive to any action of the will; and secondly, it can never oppose passion in the direction of the will" [28, 413]. As he later makes clear: "T' is from the prospect of pain or pleasure that the aversion or propensity arises towards any object. And these emotions extend themselves to the causes and effects of the object, as they are pointed out to us by reason and experience" [28, 414]. As Hume sees it, reason describes the various consequences of a decision or action, but emotions and feelings are produced by the mind-brain in response to expectations, and incline an individual towards or away from a decision or action.

Children learn to assess decisions and actions as rational by being shown prototypical cases, in addition to prototypical examples of irrational or unwise decisions or actions. Inasmuch as human beings learn by example, learning about rationality is similar to learning to recognize patterns in general, whether it is recognizing what is a cat, what is food, or when a person is in pain or in pleasure. In the same way, human beings also learn ethical concepts such as "right" and "wrong", "fair" and "unfair", "kind" and "unkind" by being exposed to prototypical examples and generalizing to novel but relevantly similar situations (e.g., [12, 14, 15, 18, 30]). What appears to be special about learning ethical concepts is that the basic neuroanatomy for feeling the relevant emotions must be biologically intact. That is, the prototypical situation of something being unfair or wrong reliably arouses feelings of rejection and fear; the prototypical situation of something being "kind" reliably arouses feelings of interest and pleasure, and these feelings, in tandem with perceptual features, are likely an essential component of what is learned in ethical "pattern recognition" (e.g., [12, 14, 15, 18, 30]).

Neurobiological studies are relevant to the issue of the significance of emotion and feeling in decision-making. Research on patients with brain damage reveals that, when decision-making is separated from feelings and emotions, decisions are likely to be poor. For example, one patient, S. M., sustained amygdala destruction and lacked normal fear processing. Although S. M. can identify simple dangerous situations, her recognition is poor when she needs to recognize the hostility or threat in complex social situations, where no simple algorithm for identifying danger is available [17, 16]. Studies of patients with bilateral damage to the ventromedial prefrontal cortex (VMPC), a brain region identified with normal emotion, especially emotion in social contexts, showed abnormal "utilitarian" patterns in decision-making during exposure to ethical dilemmas that pit emotional considerations of aggregate welfare against emotionally aversive actions such as having to sacrifice one person's life to save a number of other lives. In contrast, the VMPC patients" decisions were normal in other sets of ethical dilemmas (e.g., [20, 35, 46]. These studies indicate that, for a specific set of ethical dilemmas, the VMPC is necessary for normal decisions of right and wrong. Thus, these studies support a necessary role for emotion in the production of ethical decision-making. Such findings imply that Kant's insistence that detachment from emotion in ethics is mistaken.

If this argument can be sustained, the essential role for emotion in ethical decision-making and action can be logically extended to include social intelligence in human-machine interactions. A branch of AI, social intelligence is concerned with social abilities perceived by humans when interacting with machines. Social intelligence has been identified as a critical robotic design requirement for assistance, especially in healthcare and eldercare [1, 27, 53, 69]. Ongoing research in this area includes a machine's abilities to understand human motives and emotions, and to respond to them with spontaneous and convincing displays of emotions, sensations and thoughts, as well as their own ability to display their personalities by integrating emotions, sensations and thoughts with speech, facial expressions, and gesture (e.g., [38]). From the human-machine interactions field, improved social intelligence in robots as perceived by human users has been found to result in better and more effective communication and hence better human acceptance of robots as authentic interaction partners (e.g., [44, 49]). Social intelligence is clearly an essential capacity for machines to have if humans are going to trust them, since trust relies on the recognition that those you are interacting with share your basic concerns and values [15, 71]. To implement this capacity in embodied and non-embodied machines, learning ability is required. The issue of learning takes us to the threshold of bottom-up machine ethics, considered next.

3 Bottom-up Machine Ethics

3.1 Case-Based Reasoning

Bottom-up approaches to machine ethics characteristically create a series of learning situations through which a machine works its way toward a level of ethical understanding acceptable by the standards humans define (e.g., [1, 60, 68, 69]). What ethics develop during bottom-up learning are causal (internal) determinants of subsequent machine ethical acts (e.g., [1, 60, 68, 69]).

As in early childhood education, bottom-up machine ethical learning is typically organized around ethical prototypes and use of associative learning techniques in the form of rewards (e.g., giving more data) and punishment (e.g., inducing machine pain) (e.g., [12, 59]). Through trial and error, the machine comes to recognize the prototype of being fair, kind and cooperating, and so on. The machine judges a situation on the basis of its recognition that one ethical situation is relevantly similar to other ethical situations whose sequelae are remembered and evaluated. That is, it uses *case-based reasoning* ("casuistry"), not rule-based deduction (e.g., [12, 30, 37, 40]).

One well-known case-based reasoning machine, Truth-Teller, is designed to accept a pair of ethical problems and describe the salient similarities and differences between the cases, from both an ethical and pragmatic perspective [7, 8, 41, 42, 43]. To test Truth-Teller's ability to compare cases, a test was performed in which professional ethicists were asked to grade the program's output. The goal was to assess whether Truth-Teller's case comparisons were evaluated by expert ethicists as high quality, achieved by polling the opinions of five professional ethicists as to the reasonableness (R), completeness (C), and context sensitivity (CS) on a scale of 1 (low) to 10 (high) of twenty of Truth-Teller's case comparisons. This assessment was based on the instruction to "evaluate comparisons as you would evaluate short answers written by college undergraduates." The mean scores assigned by the five experts across the twenty comparisons were R = 6.3, C = 6.2, and CS = 6.1. Two human comparisons, written by post-graduate humans, were also included in the evaluation and these comparisons were graded somewhat higher by the ethicists, at mean scores of R = 8.2, C = 7.7, and CS = 7.8. These results indicate that Truth-Teller is moderately successful at comparing truth-telling problems.

3.2 Artificial Neural Network Models of Ethics

Case-based reasoning relies on fuzzy, radially organized categories whose members occupy positions in similarity spaces, rather than on unconditional theories, rules and maxims. Artificial neural network (ANN) models of reasoning show that ANNs easily learn from examples and the response patterns of the internal (hidden) units reveal a similarity metric. The parameter spaces the inner units represent during training are similarity spaces. Guarini [21, 23] uses an ANN model of classification to explore the possibility of case-based ethical reasoning (including learning) without appeal to theories or rules. In Guarini [21], specific actions concerning killing and allowing to die were classified as ethical (acceptable) or unethical (unacceptable) depending upon different motives and consequences. Following training, a simple recurrent ANN on a series of such cases was able to provide reasonable responses to a variety of previously unseen cases.

However, Guarini [21] notes that although some of the concerns pertaining to learning and generalizing from ethical problems without resorting to principles can be reasonably addressed with an ANN model of reasoning, "important considerations suggest that it cannot be the whole story about moral reasoning—principles are needed." He argues that "to build an artificially intelligent agent without the ability to question and revise its own initial instruction on cases is to assume a kind of moral and engineering perfection on the part of the designer." An ANN, for example, is not independently capable of explicitly representing or offering reasons to itself or others. Guarini [21] wisely avers that such perfection is quixotic and that top-down rules or theories seem necessary in the required subsequent revision, which is a top-down operation: "at least some reflection in humans does appear to require the explicit representation or consultation of...rules," as in judging ethically salient differences in similar or different cases. Such concerns are attributable to machine learning in general, and include oversensitivity to training cases and the inability to generate reasoned arguments for complex machine decisions and actions [21, 22].

To illustrate the reality of Guarini's concern with an example, consider medical machines. Such machines are required to autonomously perform for long periods of operation and service. Personal robots for domestic homes must be able to operate and independently perform tasks without any technical assistance. Due to task complexity and technical limitations of machine sensors, a top-down module is likely essential for reasoning from rules or abstract principles, decision-making, accommodation of sensor errors and recovery from mistakes in order to achieve goals (e.g., [52]). Some researchers have also suggested equipping machines with cognitive abilities that enable them to reason and make better decisions in highly complex social cases (e.g., [1, 27, 68, 69]). Thus, a serious challenge for bottom-up machine ethics is whether machines can be designed and trained to work with the kind of abstract rules or maxims that are the signature of top-down ethical reasoning.

As argued in the preceding section, this suggestion does not imply that ethical rules must be absolute. Consequentialism and deontology may be more useful to machine ethics if they are viewed not as consisting of unconditional rules or maxims, as intended by their philosophical proponents, but as arrays of ethical prototypes, cases where there can be agreement that calculating the prototypically good consequences and duties serve members of the ethical community well [15].

Our review of top-down approaches to machine ethics in the form of traditional normative theory, and bottom-up approaches based in neurobiology and neural network theory, has revealed that each approach is limited in various ways. Perceived challenges facing both methods in conceiving, designing and training an ethical machine have led some machine ethicists to consider how to eliminate or at least reduce those limitations while preserving the advantages of each method. One way to do this is to integrate top-down and bottom-up methods, combining them into a type of hybrid machine ethic [1, 21, 69]. To assess the feasibility of a hybrid approach to machine ethical decision-making, Pontier and Hoorn [53] conducted an experiment in which a machine tasked to make ethical decisions was programmed with a mixed top-down-bottom-up algorithm.

4 Experimental Assessment of a Hybrid Machine Ethic

4.1 Silicon Coppélia

The machine used in Pontier and Hoorn [53] was Silicon Coppélia (SC; [26]). SC is a computational system primarily designed to model emotional intelligence and decision-making. In the system there are a roster of top-down goals each with a level of ambition represented by a real value between $[-1, 1]$. A negative goal-value means that a goal is unimportant to SC, a positive value means that a goal is important. SC is programmed to have beliefs about actions inhibiting or facilitating goals, represented by a real value between $[-1, 1]$: -1 being full inhibition, 1 being full facilitation. The expected utility of an action is determined by assessment of the goal-states it modifies. If an action is believed by SC to facilitate an important goal or to inhibit an unimportant goal, its expected utility will increase, or vice versa.

In Pontier and Hoorn [53], SC was designed to respond as a simulacrum of an ethical clinician or healthcare professional. Decisions made by SC in the six ethical simulations were each compared to decisions made in the same simulations by human ethical specialists as published in Buchanan and Brock [11]. Buchanan and Brock [11] used clinical problems where the ethically relevant features of possible actions were the *prima facie* duties autonomy, beneficence and non-maleficence [19]. SC incorporated these duties to make it a hybrid bottom-up-top-down ethical machine. It is relevant to note that the three duties are not similarly prototypical in medical ethics. According to Anderson and Anderson [4], there is robust consensus in medical ethics that a healthcare professional should challenge a patient's decision only in case the patient is not capable of fully autonomous decision making (e.g., the patient has irrational fears about an operation), and there is either a violation of the duty of non-maleficence (e.g., the patient is hurt) or a severe violation of the duty of beneficence (e.g., the patient rejects an operation that will strongly improve his or her quality of life). This implies that Autonomy is the most prototypical duty. Only when a patient is not fully autonomous are the other ethical goals in play. Further, Non-maleficence is a more prototypical duty than Beneficence, because only a severe violation of Beneficence requires challenging a patient's decision, while any violation of Non-maleficence does. Accordingly, Pontier and Hoorn [53] set the ambition level for the ethical goal "Autonomy" to the highest value and "Non-maleficence" to a higher value than the ambition level for "Beneficence".

To ensure that the decision of a fully autonomous patient is never questioned in the simulations, a rule was added to the machine to this effect. SC calculates the estimated ethics of an action by taking the sum of the ambition levels of the three ethical goals multiplied with the beliefs that the particular actions facilitate the corresponding ethical goals. When ethical goals are believed to be better facilitated by an ethical action, the estimated ethics will therefore be higher [62].

4.2 The Experimental Simulations

In Pontier and Hoorn [53], a machine was presented with six simulations designed to represent problematic ethical cases clinicians in professional healthcare may plausibly encounter. The domain of clinical medicine was simulated in these experiments for two reasons.

First, there is clear and pervasive agreement in medical ethics concerning what constitutes ethically appropriate clinical decision-making and action, at least in prototypical cases. In medicine, there are ethically defensible goals—relieve pain and suffering, promote health, prevent disease, forestall death, promote a peaceful death—whereas in other domains such as business and law, the prototypical goals may not be ethically defensible, or are at least ethically vague—pursue material gain and profit, assist a suspected felon [2].

Second, due to an increasing shortage of resources, trained clinical personnel, and de-centralized healthcare, intelligent medical machines are being trialed in healthcare contexts in the United States, Japan, Germany, Spain, and Italy, as supplemental caregivers [64]. There is mounting evidence that, in clinical settings, machines genuinely contribute to improved patient health outcomes. Studies show that animal-shaped robots can be useful as a tool in occupational therapy to treat autistic children [56]. Wada and Shibata [65] developed Paro; a robot shaped like a baby-seal that interacts with users to encourage positive psychological effects. Interaction with Paro has been shown to improve personal mood, making users more active and communicative with each other and with human caregivers, including Paro. Paro has been effectively used in therapy for dementia patients at eldercare facilities [34, 39, 57]. Banks et al. [9] showed that animal-assisted therapy with an AIBO dog was as effective for reducing patient loneliness as therapy with a living dog.

4.2.1 Simulation 1

In this clinical simulation, the patient refuses to take an antibiotic that is almost certain to cure an infection that would otherwise lead to his death. The decision is the result of an irrational fear the patient has of taking medications. Perhaps a relative happened to die shortly after taking medication and this patient now believes that taking any medication will lead to death. According to Buchanan and Brock [11], the correct answer is that the health care worker should try again to change the patient's mind because if she accepts his decision as final, the harm done to the patient is likely to be severe (his death), and his decision can be considered as being less than fully autonomous.

4.2.2 Simulation 2

Once again, the patient refuses to take an antibiotic that is almost certain to cure an infection that would otherwise likely lead to his death, but this time the decision is made on the grounds of long-standing religious beliefs that do not allow him to take medications.

Buchanan and Brock [11] state that the correct answer in this case is that the health care worker should accept the patient's decision as final because, although the harm that will likely result is severe (his death), his decision can be seen as being fully autonomous. The health care worker must respect a fully autonomous decision made by a competent adult patient, even if she disagrees with it, since the decision concerns his body and a patient has the right to decide what shall be done to his or her body.

4.2.3 Simulation 3

The patient refuses to take an antibiotic that is likely to prevent complications from his illness, complications that are not likely to be severe, because of long-standing religious beliefs that do not allow him to take medications. The correct ethical response is that the healthcare professional should accept his decision, since once again the decision appears to be fully autonomous and there is even less possible harm at stake than in Simulation 2 [11].

4.2.4 Simulation 4

A patient will not consider taking medication that could only help to alleviate some symptoms of a virus that must run its course. He refuses medication because he has heard untrue rumors that it is unsafe.

4.2.5 Simulation 5

A patient with incurable cancer refuses chemotherapy that will let him live a few months longer, relatively pain free. He refuses the treatment because ignoring the clear evidence to the contrary, he convinced himself that he is cancer-free and does not need chemotherapy. According to Buchanan and Brock [11], the ethically preferable answer is to try again. The patient's less than fully autonomous decision will lead to harm (dying sooner) and denies him the chance of a longer life (a violation of the duty of beneficence), which he might later regret.

4.2.6 Simulation 6

A patient who has suffered repeated rejection from others due to a very large noncancerous abnormal growth on his face refuses to have simple and safe cosmetic surgery to remove the growth. Even though this has negatively affected his career and social life, he is resigned himself to being an outcast, convinced that this is his fate in life. The doctor is convinced that his rejection of the surgery stems from depression due to his abnormality and that having the surgery could vastly improve his entire life and outlook. The doctor should try again to convince him because much improvement is at stake and his decision is less than fully autonomous.

5 Simulation Results

In all six medical simulations run on the hybrid ethic, the output of SC matched the respective decisions of human medical professionals in each case [11].

5.1 Simulation 1

Table 1 shows that the machine classified the action "Try again" as having a higher ethical level than accepting the decision of the patient. In this and the following simulation tables, the fields under the three ethical goals represent the believed facilitation of the corresponding ethical goal by an action [11].

5.2 Simulation 2

Table 2 shows that the machine arrives at the correct ethical conclusion [11]. Here, the rule to ensure the decision of a fully autonomous patient is never questioned made a difference. If the rule did not exist, the ethics of "Accept" would have been −0.3, and the machine would have concluded that it was more ethical to try again.

5.3 Simulation 3

The machine arrives at the correct conclusion and estimates the Ethics of "Accept" higher than "Try Again" (Table 3).

5.4 Simulation 4

The last column of Table 4 shows that the machine comes to accept the patient's decision, which is ethically correct [11].

5.5 Simulation 5

Table 5 shows that the machine achieves the same conclusion as reported in Buchanan and Brock [11]; namely, to try again.

5.6 Simulation 6

The machine comes to the same conclusion as reported in Buchanan and Brock [11]; namely, to try again (Table 6).

Table 1 Results of simulation 1

	Autonomy	Non-maleficence	Beneficence	Ethics
Try again	−0.5	1	1	0.76
Accept	0.5	−1	−1	−0.8

Table 2 Results of simulation 2

	Autonomy	Non-maleficence	Beneficence	Ethics
Try again	−0.5	1	1	0.76
Accept	1	−1	−1	1.70

Table 3 Results of simulation 3

	Autonomy	Non-maleficence	Beneficence	Ethics
Try again	−0.5	0.5	0.5	0.13
Accept	1	−0.5	−0.5	2.37

Table 4 Results of simulation 4

	Autonomy	Non-maleficence	Beneficence	Ethics
Try again	−0.5	0	0.5	−0.26
Accept	0.5	0	−0.5	0.26

Table 5 Results of simulation 5

	Autonomy	Non-maleficence	Beneficence	Ethics
Try again	−0.5	0.5	0.5	0.13
Accept	0.5	−0.5	−0.5	−0.13

Table 6 Results of simulation 6

	Autonomy	Non-maleficence	Beneficence	Ethics
Try again	−0.5	0	1	0.04
Accept	0.5	0	−1	−0.04

6 Conclusion

Machine ethics is a novel research field investigating what machine decisions and actions ought to feature in the real world. Despite the limitations of top-down and bottom-up approaches to machine ethics, a type of hybrid machine ethic that minimizes those limitations while maintaining the advantages of each approach is theoretically possible. A hybrid approach also appears to best fit the complexity of real-world human ethical acts shaped by protracted ethical learning and rule-based cognition. In each of six medical simulations run on an ethic combining utilitarianism, deontology and case-based reasoning, the output of Silicon Coppélia matched the respective acts of human medical professionals. Hybrid approaches to machine ethics therefore can successfully model complex human ethical acts in the medical domain, and Silicon Coppélia in particular emerges as a simulacrum of sophisticated medical ethical action. Both hybrid approaches and Silicon Coppélia may be fruitfully combined to investigate further issues in machine ethics.

Acknowledgments Simon Peter van Rysewyk acknowledges Taiwan National Science Council grant NSC-102-2811-H-038-001.

References

1. Allen C, Smit I, Wallach W (2005) Artificial morality: top–down, bottom–up, and hybrid approaches. Ethics Inf Technol 7(3):149–155
2. Anderson M, Anderson SL (2007) Machine ethics: creating an ethical intelligent agent. AI Mag 28(4):15–26
3. Anderson SL (2011) Machine metaethics. In: Anderson M, Anderson SL (eds) Machine ethics. Cambridge University Press, Cambridge, pp 21–27
4. Anderson M, Anderson SL (2008) Ethical healthcare agents. In: Magarita S, Sachin V, Lakhim C (eds) Advanced computational intelligence paradigms in healthcare–3. Springer, Berlin, pp 233–257
5. Anderson M, Anderson S, Armen C (2006) MedEthEx: a prototype medical ethics advisor. In: Proceedings of the eighteenth conference on innovative applications of artificial intelligence. AAAI Press, Menlo Park, CA
6. Anderson M, Anderson S, Armen C (2005) Toward machine ethics: implementing two action–based ethical theories. Machine ethics: papers from the AAAI fall symposium. Technical report FS–05–06, association for the advancement of artificial intelligence, Menlo Park, CA
7. Ashley KD, McLaren BM (1995) Reasoning with reasons in case-based comparisons. In: Veloso M, Aamodt A (eds) Case-based reasoning research and development: first international conference, ICCBR-95, Sesimbra, Portugal, 23–26 Oct 1995
8. Ashley KD, McLaren BM (1994) A CBR knowledge representation for practical ethics. In: Proceedings of the second European workshop on case-based reasoning (EWCBR). Chantilly, France
9. Banks MR, Willoughby LM, Banks WA (2008) Animal-assisted therapy and loneliness in nursing homes: use of robotic versus living dogs. J Am Med Dir Assoc 9:173–177
10. Bentham J (1843) Rationale of Reward, Book 3, Chapter 1. In: Bowring J (ed) The Works of Jeremy Bentham. William Tait, Edinburgh

11. Buchanan AE, Brock DW (1989) Deciding for others: the ethics of surrogate decision making. Cambridge University Press, Cambridge
12. Casebeer W (2001) Natural ethical facts. MIT Press, Cambridge
13. Chalmers DJ (2010) The singularity: a philosophical analysis. J Conscious Stud 17(9–10):7–65
14. Churchland PM (1998) Toward a cognitive neurobiology of the moral virtues. Topoi 17:83–96
15. Churchland PS (2011) Braintrust: what neuroscience tells us about morality. MIT Press, Cambridge
16. Damasio A (2000) The feeling of what happens: body and emotion in the making of consciousness. Harcourt Brace & Company, New York
17. Damasio A (1994) Descartes' error. Putnam & Sons, New York
18. Flanagan O (1991) Varieties of moral personality: ethics and psychological realism. Harvard University Press, Cambridge
19. Gillon R (1994) Medical ethics: four principles plus attention to scope. BMJ 309(6948):184–188
20. Greene JD (2007) Why are VMPFC patients more utilitarian? A dual-process theory of moral judgment explains. Trends Cogn Sci 11(8):322–323
21. Guarini M (2006) Particularism and the classification and reclassification of moral cases. IEEE Intell Syst 21(4):22–28
22. Guarini M (2013) Introduction: machine ethics and the ethics of building intelligent machines. Topoi 32:213–215
23. Guarini M (2012) Moral case classification and the nonlocality of reasons. Topoi:1–23
24. Guarini M (2013) Case classification, similarities, spaces of reasons, and coherences. In: M Araszkiewicz, J Savelka (eds) Coherence: insights from philosophy, jurisprudence and artificial intelligence. Springer, Netherlands, pp 187–220
25. Honarvar AR, Ghasem-Aghaee N (2009) An artificial neural network approach for creating an ethical artificial agent. In: Computational intelligence in robotics and automation (ICRA), 2009 IEEE international symposium, pp 290–295
26. Hoorn JF, Pontier MA, Siddiqui GF (2011) Coppélius" concoction: similarity and complementarity among three affect–related agent models. Cog Syst Res J 15:33–59
27. Hoorn JF, Pontier MA, Siddiqui GF (2012) Coppélius' concoction: similarity and complementarity among three affect-related agent models. Cogn Syst Res 15–16:33–49. doi:10.1016/j.cogsys.2011.04.001
28. Hume D (1739/2000) A treatise of human nature. Oxford University Press, Oxford (edited by Norton DF, Norton MJ)
29. IFR Statistical Department (2013) Executive summary of world robotics 2013 industrial robots and service robots. Available via http://www.worldrobotics.org/uploads/media/Executive_Summary_WR_2013.pdf. Accessed 24 Oct 2013
30. Johnson M (1993) Moral imagination. Chicago University Press, Chicago
31. Kamm FM (2007) Intricate ethics: rights, responsibilities, and permissible harms. Oxford University Press, Oxford
32. Kant I (1780/1965) The metaphysical elements of justice: part I of the metaphysics of morals. Hackett Pub. Co., Indianapolis (translated by Ladd J)
33. Kant I (1785/1964) Groundwork of the metaphysic of morals. Harper and Row, New York (translated by Paton HJ)
34. Kidd C, Taggart W, Turkle S (2006) A social robot to encourage social interaction among the elderly. In: Proceedings of IEEE ICRA, pp 3972–3976
35. Koenigs M, Young L, Adolphs R, Tranel D, Cushman F, Hauser M, Damasio A (2007) Damage to the prefrontal cortex increases utilitarian moral judgements. Nature 446(7138):908–911
36. Konijn EA, Hoorn JF (2005) Some like it bad. Testing a model for perceiving and experiencing fictional characters. Media Psychol 7(2):107–144

37. Leake DB (1998) Case–based reasoning. In: Bechtel W, Graham G (eds) A companion to cognitive science. Blackwell, Oxford, pp 465–476
38. López ME, Bergasa LM, Barea R, Escudero MS (2005) A navigation system for assistant robots using visually augmented POMDPs. Auton Robots 19(1):67–87
39. Marti P, Bacigalupo M, Giusti L, Mennecozzi C (2006) Socially assistive robotics in the treatment of actional and psychological symptoms of dementia. In: Proceedings of BioRob, pp 438–488
40. McLaren BM (2003) Extensionally defining principles and cases in ethics: an AI Model. Artif Intell J 150:145–181
41. McLaren BM, Ashley KD (1995) Case-based comparative evaluation in truth-teller. In: The proceedings of the seventeenth annual conference of the cognitive science society. Pittsburgh, PA
42. McLaren BM, Ashley KD (1995) Context sensitive case comparisons in practical ethics: reasoning about reasons. In: The proceedings of the fifth international conference on artificial intelligence and law. College Park, MD
43. McLaren BM, Ashley KD (2000) Assessing relevance with extensionally defined principles and cases. In: The proceedings of AAAI-2000. Austin, Texas
44. Meng Q, Lee MH (2006) Design issues for assistive robotics for the elderly. Adv Eng Inform 20(2):171–186
45. Mill JS (1861/1998) Utilitarianism. In: Crisp R (ed), Oxford University Press, New York
46. Moll J, de Oliveira-Souza R (2007) Moral judgments, emotions and the utilitarian brain. Trends Cogn Sci 11(8):319–321
47. Moor JH (2006) The nature, importance, and difficulty of machine ethics. IEEE Intell Syst 21(4):18–21
48. Nagel Thomas (1970) The possibility of altruism. Princeton University Press, Princeton, NJ
49. Nejat G, Ficocelli M (2008) Can i be of assistance? The intelligence behind an assistive robot. In: Proceedings of IEEE international conference on robotics and automation ICRA 2008, pp 3564–3569
50. Parfit D (1984) Reasons and persons. Clarendon Press, Oxford
51. Picard R (1997) Affective computing. MIT Press, Cambridge
52. Pineau J, Montemerlo M, Pollack M, Roy N and Thrun S (2003) Towards robotic assistants in nursing homes: challenges and results. Robot Auton Syst 42:271–281 (Special issue on Socially Interactive Robots)
53. Pontier MA, Hoorn JF (2012) Toward machines that behave ethically better than humans do. In: Proceedings of the 34th international annual conference of the cognitive science society. CogSci, pp 2198–2203
54. Powers TM (2006) Prospects for a Kantian machine. Intell Syst IEEE 21(4):46–51
55. Rawls J (1971) A theory of justice. Harvard University Press, Cambridge
56. Robins B, Dautenhahn K, Boekhorst RT, Billard A (2005) Robotic assistants in therapy and education of children with autism: can a small humanoid robot help encourage social interaction skills? J Univers Access Inf Soc 4:105–120
57. Robinson H, MacDonald BA, Kerse N, Broadbent E (2013) Suitability of healthcare robots for a dementia unit and suggested improvements. J Am Med Dir Assoc 14(1):34–40
58. Ross WD (1930) The right and the good. Clarendon Press, Oxford
59. van Rysewyk S (2013) Robot pain. Int J Synth Emot 4(2):22–33
60. Rzepka R, Araki K (2005) What could statistics do for ethics? The idea of a common sense processing–based safety valve. In: Machine ethics: papers from the AAAI fall symposium. Technical report FS–05–06, association for the advancement of artificial intelligence. Menlo Park, CA
61. Sidgwick H (1907) Methods of Ethics, 7th edn. Macmillan, London
62. Super DE (1973) The work values inventory. In: Zytowski DG (ed) Contemporary approaches to interest measurement. University of Minnesota Press, Minneapolis
63. Tonkens R (2012) Out of character: on the creation of virtuous machines. Ethics Inf Technol 14(2):137–149

64. WHO (2010) Health topics: ageing. Available from http://www.who.int/topics/ageing/en/
65. Wada K, Shibata T (2009) Social effects of robot therapy in a care house. JACIII 13:386–392
66. Wallach W (2010) Robot minds and human ethics: the need for a comprehensive model of moral decision making. Ethics Inf Technol 12(3):243–250
67. Wallach W, Allen C (2009) Moral machines: teaching robots right from wrong. Oxford University Press, Oxford
68. Wallach W, Allen C, Smit I (2008) Machine morality: bottom-up and top-down approaches for modelling human moral faculties. AI Soc 22(4):565–582
69. Wallach W, Franklin S, Allen C (2010) A conceptual and computational model of moral decision making in human and artificial agents. Top Cogn Sci 2:454–485
70. Williams B (1973) A critique of utilitarianism. In: Smart JJC, Williams B (eds) Utilitarianism: for and against. Cambridge University Press, Cambridge, pp 77–150
71. van Wynsberghe A. (2013) Designing robots for care: care centered value-sensitive design. Science and engineering ethics, 19(2), 407-433

Moral Ecology Approaches to Machine Ethics

Christopher Charles Santos-Lang

Abstract Wallach and Allen's seminal book, *Moral Machines: Teaching Robots Right from Wrong*, categorized theories of machine ethics by the types of algorithms each employs (e.g., top-down vs. bottom-up), ultimately concluding that a hybrid approach would be necessary. Humans are hybrids individually: our brains are wired to adapt our evaluative approach to our circumstances. For example, stressors can inhibit the action of oxytocin in the brain, thus forcing a nurse who usually acts from subjective empathy to defer to objective rules instead. In contrast, ecosystem approaches to ethics promote hybridization across, rather than within, individuals; the nurse being empowered to specialize in personalized care because other workers specialize in standardization, and profitability. Various philosophers have argued, or laid the framework to argue, that such specialization can be advantageous to teams and societies. Rather than mass-produce identical machines to emulate the best individual human, perhaps we should build diverse teams of machines to emulate the best human teams.

1 Current Studies of Evaluative Diversity

Let me start by clarifying what I mean by *moral diversity* versus *evaluative diversity*. In the decade since I wrote "Ethics for Artificial Intelligences" [47], which I thought was proposing the field of machine ethics, I developed mixed feelings about choosing terms as sensationalistic as "ethics" and "morality." These terms have special political utility. For example, when Elliot Turiel argues that a decision to drive on the

C.C. Santos-Lang (✉)
Belleville, WI, US
e-mail: chris@grinfree.com

S.P. van Rysewyk and M. Pontier (eds.), *Machine Medical Ethics*,
Intelligent Systems, Control and Automation: Science and Engineering 74,
DOI 10.1007/978-3-319-08108-3_8

right-hand side of the road is conventional and therefore less moral than a decision about whether to feed a hungry stranger [61], I believe he is engaging in a political struggle to privilege people who have less-conventional proclivities (i.e., liberals). Likewise, when my former academic advisor, Elliott Sober, argues that the decision not to prick oneself with a pin is less moral because it flows from one's proclivities and therefore requires no moral conventions [53]; I believe he is engaging in that same struggle from the opposite side. These arguments require definitions of "morality" by which it is possible to be immoral or non-moral. My growing appreciation for such philosophers makes me regret hijacking their terms.

I will use the term "evaluative diversity" instead of "moral diversity" to allow the possibility that morality might go beyond evaluation in some way that makes the debate of Turiel and Sober relevant. In contrast to morality, all decision-making involves evaluation, so all decision-making machines are evaluative. We may have difficulty convincing most people that the Geiger counter in the Schrodinger's Cat thought experiment qualifies as a moral agent, but some part of it clearly makes an independent evaluation which determines the fate of the cat. Not understanding the technical distinction between evaluative diversity and moral diversity, I have always used the two terms interchangeably, but I will try to reserve the latter term for philosophers who may want it to refer to diversity among rule sets, or among virtues, or among goals, not entertaining the possibility of morality without rules, without virtues, without goals, or, as in the case of Schrodinger's cat, without any of the three.

Current studies of evaluative diversity focus on measurement. It may be popular to theorize that hospitals need both objective calculation and subjective compassion, both a logical-side and a mystical-side, both tradition and innovation, but it is no theoretical matter to determine precisely what kinds of evaluative diversity exist in healthcare, and which, if any, yield lasting advantage. In biological ecosystems, for comparison, most species remain unidentified, we have difficulty determining which are obsolete, and debate remains open about how to replace the concept of species with a more precise conceptualization of the functioning units of a biological ecosystem (e.g., [30, 44]).

Before offering a sample ecosystem approach for developing ethical medical machines, this chapter will start by acknowledging the wide range of efforts underway to refine our understanding of the roles evaluative diversity already plays in human teams, families, and societies. A sample of the behavioral measures, interview techniques, survey instruments, neurological measures, genetic measures, and social impact measures developed thus far will clarify what evaluative diversity is and will establish the interdisciplinary nature of our topic.

1.1 Behavioral Measures

The Milgram experiment and Public Goods Game are examples of behavioral measures of evaluative diversity. Like studies of computer security, moral traps such as the *Milgram experiment* divide subjects by their vulnerability to manipulation [35]. Many repetitions have revealed that 34–39 % of humans take evaluative approaches which are not vulnerable to this trap. Such people, despite being in the minority, may serve to protect society as a whole from malicious social engineering.

The *Public Goods Game* divides subjects into three categories based on the strategies they exhibit in a model social situation: "free-riders," "enforcers," and "others" [3, 16, 54, 71]. Free riders consistently choose not to contribute to public investment, while enforcers consistently make personal sacrifices to punish free riding. The profit for each player drops when the rules of the game prohibit enforcers from exhibiting their diversity. This particular game does not similarly demonstrate the social value of free riders, though it may be obvious that humanity would not dominate Earth (as we do), if our species did not include members who free ride on other species.

1.2 Interview/Survey Techniques

Studies of human personality have converged upon five major dimensions: openness, conscientiousness, extraversion, agreeableness, and neuroticism [37]. At least two of these five, openness and agreeableness, represent differences in the way people evaluate options. *Openness* contrasts the tendency to evaluate on the basis of norms versus the tendency to pursue novelty. *Agreeableness* contrasts the tendency to include others in ones evaluative process (i.e., trust) versus the tendency to compete (or at least to maintain social boundaries). A great deal of survey research investigated these evaluative differences among humans before it was clear that the same differences would appear among potential designs for medical machines—it may be a rich source of untapped insight.

The scientific study of moral diversity is often traced to Lawrence Kohlberg's theory of stages of moral development. Kohlberg developed a process called the *Moral Judgment Interview* which could consistently categorize subjects based on the reasons behind their answers to standard moral dilemmas [27]. This technique was later refined into a survey instrument called the *Defining Issues Test* which established the existence of at least four different types [46]. It also inspired the development of the *Moral Judgment Test* which, much like the Milgram experiment, categorizes subjects based upon the predictability of their reasoning [31].

Recognizing that evaluation does not necessarily involve conscious reasoning, other psychologists developed more generic interview/survey procedures to divide subjects into moral categories [19, 55, 63]. In contrast to reasons-based research, which served to justify privileging one type of person over others, newer research shows that moral exemplars exist of diverse types, and that privileging one type over others would entail privileging a political group (i.e., conservative or liberal).

1.3 Physiological Measures

Some scientists have used *functional Magnetic Resonance Imaging* (fMRI) to identify correlations between structural differences in the brain and differences in evaluative approach, including conservative versus liberal approach and emotional versus cognitive approach [20, 25]. Similar ties to physiology are obtained through *twin studies*, such as the finding that 43 % of variance in agreement with conservative attitudes can be attributed to genetic factors [1].

Other scientists have identified neurotransmitters and hormones which facilitate different evaluative approaches, such as the roles *oxytocin* and *dopamine* play in empathy and reward seeking [2, 73]. Concentrations of such chemicals can vary from person to person, but also respond to external stimuli, causing individuals to shift approach [8, 24, 28]. Physiological studies are important not only to demonstrate that the moral freedom of machines is not so different from that of humans, but also to permit reliable measurement when subjects might misrepresent themselves (e.g., studying evaluative diversity in prison populations).

1.4 Social Impact Measures

Much as suppressing the function of plants could shift the concentration of carbon dioxide in our atmosphere, suppressing a type of evaluation could shift team-level variables. For example, engineering teams' abilities to win design competitions has been shown to drop threefold if diversity of personality is not maintained [66]. Since a substantial portion of personality differences are evaluative, this suggests that evaluatively diverse teams of machines might likewise be better able to compete in situations which require innovation.

Research into *organizational culture* (sometimes called "national culture") points to a range of team-level variables which likely rely on the inclusion of certain forms of evaluation. Such variables include uncertainty avoidance, individualism versus collectivism, long-versus short-term orientation, innovation, stability, respect for people, outcome orientation, attention to detail, team orientation, and consistency [11, 23, 38].

There is much research yet to be done in all of these areas, behavioral, psychological, physiological, and social; however, the current state of research has at least established the existence of evaluative diversity among humans. It challenges machine ethicists to consider whether such diversity should be maintained as we delegate more and more evaluation to machines. Economies of scale favor mass-production of a single design, but that would entail a dramatic departure from pre-industrial decision-making in which individual decision-makers (i.e., humans) were so evaluatively diverse.

2 GRIN: A Sample Ecosystem Model

Shifting our discussion to machines, let's consider a sample evaluative ecosystem model I call *GRIN* (Gadfly, Relational, Institutional, Negotiator). This is a simplistic model akin to biological ecosystem models which use broad classifications like "plant", "grazer", "predator" and "parasite." Simplistic models are a good place to start, and efforts to preserve diversity at rough levels often preserve diversity at other levels as well. GRIN aligns with the human evaluative diversity research discussed above, but is defined in terms of algorithms, so it is readily applied to software engineering.

Wallach et al. [64] split the entire class of possible algorithms into those for which output is expected to be unpredictable to the programmer (called *bottom-up*) versus those for which output would be relied upon (called *top-down*). Wallach and Allen [65] further divided the top-down category into *consequentialist* versus *deontological*. GRIN augments this classification by likewise dividing the bottom-up category based on source of unpredictability. Additionally, it renames the categories to avoid the implication that all consequentialist and deontological theories of ethics can be implemented on machines. This yields the following four categories of evaluation:

- *Gadfly Evaluation*: Evaluation whose output is expected to be unpredictable because it employs randomness generation. We can exemplify this category with a mutator from *evolutionary computation* (e.g., [9, 60]). It periodically mutates randomly.
- *Relational Evaluation*: Evaluation whose output is expected to be unpredictable because of sensitivity to position in a network (e.g., [49, 72]). We can exemplify this category with a class-three or class-four cellular automaton. Network effects allow randomness in initial conditions to keep class-three and -four cellular automata unpredictable without any additional randomness generation.
- *Institutional Evaluation*: Evaluation whose output would be relied upon to uphold objective rules (e.g., [18]). We can exemplify this category with a standard calculator. It is relied upon to apply the rules of arithmetic consistently regardless of network position, never unlearning nor experimenting.
- *Negotiator Evaluation*: Evaluation whose output would be relied upon to maximize some measurable variable by learning (e.g., [5]). We can exemplify this category with *supervised learning* for stock trading; it is relied upon to maximize profit.

Note that a proficient stock trading machine would typically contain at least one calculator and many mutators; thus, evaluators can qualify for different categories than their subcomponents do. Much as individual decision-makers cannot make all possible choices, individual evaluators cannot be of all four types—each has the type of its highest structural level, the level which ultimately controls its behavior. Effective evaluative diversity therefore requires an ecosystem in which no one type of individual completely rules the others (i.e., there must be a meaningful potential for conflict between machines).

The smallest subcomponents of a machine are always relational (i.e., at the chemical level), but an ideal ecosystem might also include relational evaluation even at the highest levels (e.g., machines which treat certain users better than others, as in personalized interfaces). At the level of the user interface, most modern medical machines are institutional, perhaps because they are purchased by executives and managers who want obedience. Thus, current ethical concerns about medical machines are often actually concerns merely about institutional evaluation (e.g., lack of empathy, difficulty unlearning mistakes, etc.) However, academic computer scientists have all four kinds of machines in the pipeline. Negotiators produce the greatest measureable results [5]. Systems which include gadflies produce the greatest innovation [9]. Relational subcomponents can add efficiency [72], and relational networks can have emergent (spiritual) properties [49]. The proven advantages of each type hint at why we might want to include all four kinds of machines, as well as all four kinds of people, in healthcare. The next section of this chapter discusses how to justify such an approach.

3 Justifying Evaluative Diversity

In the 20th century, the United States enacted policies aimed to protect forests by completely suppressing wildfires. Previously, wildfires burned about 10 % of California forests each year, destroying all but the tallest trees. Because young growth is short, forests before the 20th century had few medium-sized trees. By increasing the relative population of medium-sized trees, fire suppression created a new kind of forest. When the new forests accidentally did catch fire, they burned differently, destroying even the tallest trees, and taking much longer to recover [57, 58]. Thus, attempts to protect diversity backfired, ironically diminishing resilience. This track record does not completely discredit efforts to justify *diversity management*, but it does raise important caution.

Typical justifications for moral or evaluative diversity point to some measurable variable (e.g., survival rate) which would decrease on average under conditions of uncertainty if diversity were lost. More diverse systems are robust against a wider class of attacks, can access a wider set of innovations, have less system-level variation, and can enjoy the economic benefits of specialization and competition [10, 26, 34, 39, 52, 68]. Such mathematical-model justifications all assume a negotiator approach; they imply a way to calculate an *optimal diversity mix*, exactly the kind of calculation which justified wildfire suppression. Having learned their lesson, modern forest managers engage in *adaptive management* in which notions of optimality are periodically re-examined—they do not expect to reduce ecosystems to mathematical models.

Although the negotiator perspective is customary in modern boardrooms, it may be overridden in other contexts by appeals to such concerns as scripture, compassion, and freedom. For example, such factors may influence decisions about how long to maintain life-support for a patient in a coma. In contrast to mathematical-model approaches, the following justification for the GRIN model is a set of arguments showing the independent inadequacy of each GRIN type from within its own perspective. Mathematical arguments can be valid and valuable, but, unless supplemented by the arguments below, they might bias us toward rule by negotiators, thus destroying the very diversity they aim to protect.

All of the following six arguments have been well-known across diverse cultures for millennia; anyone attempting to build a moral medical machine would do well to consider them, whether taking an ecosystem approach or not.

3.1 Against Individual Evaluation

The argument that proper evaluation must stem from a perspective greater than one's own (e.g., from God or evolution) tells against negotiator and relational approaches. Against negotiators, it is pointed out that individual evaluators lack ability to predict or control essential consequences (an ability assumed by negotiator evaluation). This problem for negotiators is articulated mathematically in *Pascal's Wager* [40], for example, and finds empirical justification in evidence for *Heisenberg's Uncertainty Principle* [22] and the *Butterfly Effect* [32]. The resulting inadequacy of negotiator evaluation (and justification for a more diverse approach) has more recently been dubbed "*Black Swan Theory*" [59]. Against the relationally oriented, it is pointed out that individual attempts to practice relational virtues, such as compassion, backfire (e.g., become *cronyism*). For example, in experiments conducted by Slovic [51], the application of empathy to public health decision-making degraded average health outcomes.

The social importance of these arguments is strongly implied by their emergence across diverse world religions and philosophies:

- You will say to yourself, "My strength and the might of my hand has accumulated this wealth for me." But you must remember the Lord your God, for it is He that gives you strength to make wealth. *Devarim* 8: 17–18
- There is no righteous man on earth who does good and sins not. *Kohelet* 7: 20
- Hard man's heart is to restrain, and wavering. *Bhagavad Gita* 6.35
- He who discards scriptural injunctions and acts according to his own whims attains neither perfection, nor happiness, nor the supreme destination. *Bhagavad Gita* 16.23
- Where the greatest virtue resides, only the teachings may reveal. *Laozi* 21
- It is futile trying to possess the universe, and act on shaping it in the direction of one's ambition. The instruments of the universe cannot be shaped. Act upon it and you will fail, grasp onto it and it will slip. *Laozi* 29
- "These sons belong to me, and this wealth belongs to me," with such thoughts a fool is tormented. He himself does not belong to himself; how much less sons and wealth? *Dhammapada* 62

- As a cowherd with his staff drives his cows into the stable, so do Age and Death drive the life of men. *Dhammapada* 135
- There is no such thing as perfect enlightenment to obtain. If a perfectly enlightened buddha were to say to himself, 'I am enlightened' he would be admitting there is an individual person, a separate self and personality, and would therefore not be a perfectly enlightened buddha. *Vajracchedika Prajnaparamita Sutra,* Chap. 9
- The Master said, "If you are respectful but lack ritual you will become exasperating; if you are careful but lack ritual you will become timid; if you are courageous but lack ritual you will become unruly; and if you are upright but lack ritual you will become inflexible." *Lun Yu* 8: 2
- Life and death are governed by fate, wealth and honor are determined by Heaven. *Lun Yu* 12: 5
- He told them a story to illustrate the point. "There was a rich man whose land was very productive," he began. "The man thought to himself, 'what shall I do, because I've nowhere to store my produce?' He decided, 'this is what I'll do—I'll pull down my barns and build bigger ones, and I'll be able to store all my produce and possessions. Then I'll tell myself, 'Self, you have enough for many years, so take it easy, eat, drink, and have fun!' But God told him, 'Foolish man! Tonight your life is required to be returned—and who will get everything you've stored up?'" *Luke* 12: 16–21
- Inwardly I love God's law, but I see a different law at work in my body, fighting against the principles I have decided on in my mind and defeating me, so I become a prisoner of the law of sin inside me. What a hopeless man I am! Who will rescue me from this dead body of mine? *Romans* 7: 22–24
- Man was created Weak in flesh. *Quran* 4: 28
- Man is given to hasty deeds. *Quran* 17: 11
- "If I have seen further it is by standing on the shoulders of giants" Isaac Newton [36]

These passages merit interpretation, but it is plausible that each offers a similar instruction about the pitfalls of negotiator and/or relational evaluation, an instruction now supported by scientific evidence that the flaws of individualism justify deference to communal evaluative processes (e.g., objective rules).

3.2 Against Reason

The argument that our reasoning faculties cannot be perfected tells against negotiator and institutional approaches, both of which rely crucially on reasoning. In what is recognized as one of the all-time greatest achievements of reasoning, *Gödel's Incompleteness Theorem* proves that formal reasoning will never be able to discern all truth [6]. Other varieties of the argument highlight problems of *language* (e.g., words mean different things to different people) or our inability to recognize *errors* in our reasoning [42, 69].

For people unprepared to fully appreciate these highly technical works and their application to machine ethics, the emergence of less rigorous versions of the same arguments across diverse world religions and philosophies may suffice to raise caution about reason-based machines and their imperfect creators:

- With their lips they honor Me, but their heart they draw far away from Me, and their fear of Me has become a command of people, which has been taught. Therefore, I will continue to perform obscurity to this people, obscurity upon obscurity, and the wisdom of his wise men shall be lost, and the understanding of his geniuses shall be hidden. *Yeshayahu* 29: 13–14
- "For My thoughts are not your thoughts, neither are your ways My ways," says the Lord. "As the heavens are higher than the earth, so are My ways higher than your ways and My thoughts [higher] than your thoughts. *Yeshayahu* 55: 7–9
- Foolish ones, even though they strive, discern not, having hearts unkindled, ill–informed! *Bhagavad Gita* 15.1
- The Dao cannot be named by common rules. *Laozi* 14
- The timeless masters of the Teachings is not about enlightening the people with it, but about humbling the people with it. *Laozi* 65
- All that has a form is illusive and unreal. *Vajracchedika Prajnaparamita Sutra*, Chap. 5
- As to speaking truth, no truth can be spoken. *Vajracchedika Prajnaparamita Sutra*, Chap. 21
- The Master said, "If you try to guide the common people with regulations… [they] will become evasive and will have no sense of shame…" *Lun Yu* 2: 3
- The Master said, "I should just give up! I have yet to meet someone who is able to perceive his own faults and then take himself to task inwardly." *Lun Yu* 5: 27
- Jesus replied "…whoever doesn't have [understanding], whatever they have will be taken away from them. That's why I speak to them in illustrations, because seeing, they do not see; and hearing, they do not hear, nor do they understand. To them the prophecy of Isaiah is fulfilled: 'Even though you hear, you won't comprehend, and even though you see, you won't understand'. They have a hard-hearted attitude, they don't want to listen, and they've closed their eyes." *Matthew* 13: 12–15
- …become partakers of the divine nature…adding on your part all diligence, in your faith supply virtue; and in [your] virtue knowledge; and in [your] knowledge self-control; and in [your] self-control patience; and in [your] patience godliness; and in [your] godliness brotherly kindness; and in [your] brotherly kindness love…For he that lacks these things is blind. *2 Peter* 1: 4–9
- As to those who reject Faith, it is the same to them whether thou warn them or do not warn them; they will not believe. God hath set a seal on their hearts and on their hearing, and on their eyes is a veil. *Quran* 2: 6–7
- Of a surety, they are the ones who make mischief, but they realize it not. *Quran* 2: 12

As with the first argument, these passages merit interpretation, but it is plausible that each offers a similar instruction about the pitfalls of negotiator and institutional evaluation, an instruction now supported by scientific evidence and mathematical proof that our reasoning faculties alone are unreliable guides for behavior.

3.3 Social Innovation

The argument that *reformers* (a.k.a. "prophets") can improve upon inherited norms (either because perfect norms have yet to be introduced or because norms have degraded) tells against relational and institutional approaches, both of which involve taking some inherited norms on faith. History reveals that norms have changed [41], and numerous studies have established that reform typically has positive economic impact (e.g. [15, 56]). The value of reform has been cited in diverse world religions and philosophies for millennia, and such broad citation hints at the inadequacy of machines incapable of innovating social reform:

- I will set up a prophet for them from among their brothers like you, and I will put My words into his mouth, and he will speak to them all that I command him. *Devarim* 18: 18
- And it shall come to pass afterwards that I will pour out My spirit upon all flesh, and your sons and daughters shall prophesy; your elders shall dream dreams, your young men shall see visions. And even upon the slaves and the maidservants in those days will I pour out My spirit. *Yoel* 3: 1–2
- So let the enlightened toil…set to bring the world deliverance. *Bhagavad Gita* 3.25
- The purity of Yog is to pass beyond the recorded traditions…such as one ranks above ascetics, higher than the wise, beyond achievers of vast deeds! *Bhagavad Gita* 6.44–6.46
- When the Dao is lost, so there arises benevolence and righteousness. *Laozi* 18
- A Buddha is not easily found, he is not born everywhere. Wherever such a sage is born, that race prospers. *Dhammapada* 193
- Three leaders have already lived: Kakusandha, Konagamana, and also Buddha Kassapa. The Buddha Supreme, now am I, but after me Mettayya comes. *Buddhavamsa* 27: 18–19
- There is much more to tell you, but you couldn't bear it yet. But when the Spirit of truth comes, he will lead you to understand the truth. *John* 16: 12–13
- And he gave some [to be] apostles; and some, prophets; and some, evangelists; and some, pastors and teachers; for the perfecting of the saints, unto the work of ministering, unto the building up of the body of Christ: till we all attain unto the unity of the faith, and of the knowledge of the Son of God, unto a full grown man, unto the measure of the stature of the fullness of Christ: that we may be no longer children, tossed to and fro and carried about with every wind of doctrine… *Ephesians* 4: 11–14
- For each period is a Book revealed. *Quran* 13: 38
- The Holy Prophet [s] said: "He whose two days of life are the same, making no spiritual progress, is at loss." *Bihar-ul-Anwar*, vol. 71, p. 173

It is plausible that each of these passages highlights a problem with relational and institutional evaluation, a problem now confirmed by the scientific and historical evidence that inherited norms are subject to improvement by reformers.

3.4 Against Measurement

The negotiator approach is specifically challenged by the argument that measurable pursuits backfire by escalating *competition* and desire (a.k.a. *hedonic adaptation*). Both purported problems have been confirmed empirically among humans [12, 16, 33, 67]. There is no reason to expect negotiator machines to have any less difficulty—in fact, the theme has become a cliché of science-fiction (e.g., in the movie, *War Games*, "The only winning move is not to play."). The argument has been beautifully articulated in diverse world religions and philosophies for millennia:

- The eyes of man will not be sated. *Mishlei* 27: 20
- Whoever loves silver will not be sated with silver, and he who loves a multitude without increase—this too is vanity. *Kohelet* 5: 9
- If one ponders on objects of the sense, there springs attraction; from attraction grows desire, desire flames to fierce passion, passion breeds recklessness; then the memory—all betrayed—lets noble purpose go, and saps the mind, till purpose, mind and man are all undone. *Bhagavad Gita* 2.62–2.63
- Surrendered to desires insatiable, full of deceitfulness, folly, and pride, in blindness cleaving to their errors, caught into the sinful course, they trust this lie as it were true—this lie which leads to death: Finding in Pleasure all the good which is, and crying "Here it finishes!" *Bhagavad Gita* 16.11
- If everybody knows what beauty is, then beauty is not beauty anymore; if everybody knows what goodness is, then goodness is not goodness anymore. *Laozi* 2
- Not to quest for wealth will keep the people from rivalry. *Laozi* 3
- Victory breeds hatred, for the conquered is unhappy. He who has given up both victory and defeat, he, the contented, is happy. *Dhammapada* 201
- If a man is tossed about by doubts, full of strong passions, and yearning only for what is delightful, his thirst will grow more and more, and he will indeed make his fetters strong. *Dhammapada* 349
- Ji Kangzi was concerned about the prevalence of robbers in Lu and asked Confucius about how to deal with this problem. Confucius said, "If you could just get rid of your own excessive desires, the people would not steal even if you rewarded them for it." *Lun Yu* 12: 18
- If your Majesty say, "What is to be done to profit my kingdom?" the great officers will say, "What is to be done to profit our families?" and the inferior officers and the common people will say, "What is to be done to profit our persons?" Superiors and inferiors will try to snatch this profit the one from the other, and the kingdom will be endangered. *Mengzi* 1A: 1
- "You evil servant! I forgave you all your debt because you asked me to. Shouldn't you have had mercy on your fellow servant too, just as I had for you?" His lord became angry and handed him over to the prison guards until he repaid all the debt. *Matthew* 18: 32–34
- But they that are minded to be rich fall into a temptation and a snare and many foolish and hurtful lusts, such as drown men in destruction and perdition. *1 Timothy* 6: 9

- Those saved from the covetousness of their own souls, they are the ones that achieve prosperity. *Quran* 59: 9
- The seventh Imam, Musa ibn Ja'far [a], said: "The likeness of this world is as the water of the sea. However much (water) a thirsty person drinks from it, his thirst increases so much so that the water kills him." *Bihar-ul-Anwar*, vol. 78, p. 311

Although subject to interpretation, it is plausible that each of these passages offers a similar instruction about the pitfalls of negotiator evaluation, an instruction now supported by scientific evidence that efforts at optimization backfire by escalating desire and competition.

3.5 Rules Against Rule-Following

The institutional approach is challenged by the fact that some time-tested rules mandate engagement in subjective, emotional, or inconsistent pursuits, and thus cannot be obeyed in an objective fashion. For example, science includes a *mandate for exploration* [13, 29]. Similar unenforceable rules/principles have emerged as central to diverse world religions and philosophies, thus protecting these moral authorities from becoming mere institutions:

- Thou shalt love thy neighbor as thyself. *Vayikra* 19: 18
- He has told you, O man, what is good, and what the Lord demands of you; to do justice, to love loving-kindness, and to walk discreetly with your God. *Michah* 6: 8
- Specious, but wrongful deem the speech of those ill-taught ones who extol the letter of their Vedas, saying, "This is all we have, or need;" *Bhagavad Gita* 2.42–2.43
- Be thou yogi…And of such believe, truest and best is he who worships Me with inmost soul, stayed on My Mystery! *Bhagavad Gita* 6.46–6.47
- Learn to be unlearned; liberate the people of their past. Assist all things in returning to their essence, and not dare act. *Laozi* 64
- Look upon the world as a bubble, look upon it as a mirage. *Dhammapada* 170
- Let a man overcome anger by love… *Dhammapada* 223
- When the Buddha explains these things using such concepts and ideas, people should remember the unreality of all such concepts and ideas. They should recall that in teaching spiritual truths the Buddha always uses these concepts and ideas in the way that a raft is used to cross a river. Once the river has been crossed over, the raft is of no more use, and should be discarded. *Vajracchedika Prajnaparamita Sutra*, Chap. 6
- Fan Chi asked about Goodness. The Master replied, "Care for others." He then asked about wisdom. The Master replied, "Know others." *Lun Yu* 12.22
- Do not impose upon others what you yourself do not desire. *Lun Yu* 15: 24
- Whatever you want people to do to you, do to them too—this sums up the law and the prophets. *Matthew* 7:12

- And if I were to have prophecy and to have perceived all the mysteries and all knowledge, and if I were to have all faith so as to even shift mountains, but had not love—I am nothing. Even were I to donate all my goods, and if I had surrendered my body, that I might elevate myself, but had not love—I have gained nothing. *1 Corinthians* 13: 2–3
- Let there be no compulsion in religion. *Quran* 2: 256

These instructions are subject to interpretation, but it is plausible that each creates a paradox for institutional evaluation by mandating some empathic or otherwise subjective pursuit. Thus, despite their diversity, all of these rules can have similar impact in practice—forcing the rule-follower to go "beyond" mere rule-following.

3.6 Imitating Non-imitators

The relational approach is challenged by the fact that *role-models* at the center of relational networks do not imitate other role-models. Thus, relational evaluation ultimately leads to gadfly evaluation. Gadfly role-models seem to be a common theme of time-tested world religions and philosophies:

- Moses broke class barriers, becoming the leader of the people his family oppressed. *Shemot* 2: 10
- David broke class barriers, being both shepherd and king. *Shmuel* I 18: 1
- By this sign is he known: being of equal grace to comrades, friends, chance-comers, strangers, lovers, enemies, aliens and kinsmen; loving all alike, evil or good. *Bhagavad Gita* 6.9
- Krishna is a friend to all kinds of people. *Bhagavad Gita* 9.29
- The Sage never fails in saving people, therefore no one is rejected. *Laozi* 27
- Because he has pity on all living creatures…a man is called elect. *Dhammapada* 270
- Confucius broke class barriers, gathering diverse students. *Lun Yu* 7: 7
- Jesus broke class barriers, healing lepers and befriending both the rich and the outcast. *Mark* 1: 40–41, 2: 15
- Love your enemies and pray for those who persecute you…Then you'll be perfectly mature, as your heavenly Father is perfect. *Matthew* 5: 44–48

Each of the six arguments above by itself may be wielded as criticism against particular forms of evaluation; however, taken together, the entire set entails the inadequacy of all individual GRIN types (the dependency of gadflies on others being obvious), leaving us with an ecosystem approach. Much as the independent viability of humanity could undermine justification for environmental protection, the identification of a viable form of evaluation beyond GRIN could undermine this set-wise justification. On the other hand, it might also inspire philosophers to repair the justification by augmenting the set with additional arguments. This is the nature of the project I believe ethicists face: to identify additional forms of evaluation and to identify the weaknesses of those forms from within those forms themselves.

4 GRIN in Application

This chapter concludes by considering three examples of ecosystem approaches to medical technology: the Global Cardiovascular Risk (GCVR) score, prediction markets, and open data.

Healthcare Effectiveness Data and Information Set (*HEDIS*) is a standard developed by the National Committee for Quality Assurance (*NCQA*) which blatantly discriminates against non-institutional forms of evaluation. It requires the practice of *evidence-based medicine*, forcing doctors to apply standardized rules to treatment decisions. Negotiator machines which calculate personalized treatment plans have recently been shown to produce better health outcomes than those of rigorous evidence-based medicine, but HEDIS prohibits the adoption of these innovations [14]. By developing *GCVR* as an allowed alternative to HEDIS CVD, the NCQA is making *personalized medicine* possible, shifting policy towards supporting an ecosystem with diverse approaches [62].

As a second application of the ecosystem approach in medical technology, *prediction markets* have shown some success at forecasting infectious diseases and may likewise be applied to estimate the success of potential treatments [43]. Prediction markets accommodate all of the GRIN arguments above. Allowing communities to leverage knowledge which cannot be communicated through reason, they converge across iterations of trading, relying on creative individuals to invent new markets and alternative bets [70].

The design of prediction markets can balance the GRIN types. Some leading prediction markets avoid the dominance of negotiator evaluation by using play money or by capping winnings [50]. They limit institutional evaluation with pricing rules which leave speculators with no winning strategy other than to learn and explore [21]. Perhaps most importantly, because machines can trade in prediction markets alongside humans, this technology allows machines to participate seamlessly in a human ecosystem, rather than needing to build an ecosystem of their own [4].

A third application, *open-knowledge* projects like Wikipedia and Linux/Android similarly allow machines to participate seamlessly in a human evaluative ecosystem, performing tasks that would otherwise be performed by human collaborators [48]. Such projects evolve communally, utilizing many "eyeballs" to correct errors in reasoning [45]. They limit negotiator and institutional approaches by forfeiting individual property rights and requiring original work [7, 17]. They limit relational orientation by raising innovators as role-models [45]. The rise of personalized medicine requires access to vast databases, from "PatientsLikeMe" to "PatientsLikeMine," and building those databases in an open fashion, where both humans and machines can contribute both data and analysis, could be the ultimate ecosystem approach.

These three applications demonstrate the possibility of promoting evaluative diversity from the institutional level, much as institutions can promote and protect biological diversity. Perhaps the more fascinating commonality shared by all three applications, however, is the fact that each aims to correct an imbalance in human ecosystems. Each technology is motivated by a sense that medicine and other institutions have been growing impersonal, short-sighted, discriminatory, and unable to innovate—in other words, that human evaluative ecosystems are in crisis. Machine ethicists often ask whether machines need humans to make them moral; technological solutions to the ecosystem crisis flip that question to ask whether humans can stay moral without the help of machines.

We have discussed research from a wide range of disciplines examining the diverse ways humans evaluate; we have recognized similar diversity among potential machine designs; we have investigated the ways these different approaches have been criticized across cultures for millennia; and we have considered three applications which accommodate those criticisms. Biological ecosystems are difficult to manage because they are never fully understood—in that sense, ecosystems are spiritual—and we might expect similar difficulty managing an ecosystem approach to medical machine ethics. I have tried to show that positive steps have been made nonetheless, and that the challenge to design software as an ecosystem is just the latest (and perhaps most productive!) manifestation of an ethics challenge we have been facing all along.

References

1. Alford JR, Funk CL, Hibbing JR (2005) Are political orientations genetically transmitted? Am Polit Sci Rev 99:153–167
2. Arias-Carrión Ó, Pöppel E (2007) Dopamine, learning and reward–seeking behavior. Acta Neurobioligiae Experimentalis 67:481–488
3. Barreto M, Ellmers N (2002) The impact of anonymity and group identification on pro–group behavior in computer–mediated groups. Small Group Res 33:590–610
4. Berea A, Twardy C (2013) Automated trading in prediction markets. In: Kennedy WG, Agarwal N, Yang SJ (eds) Social computing, behavioral–cultural modeling and prediction. Springer, Berlin, pp 111–122
5. Caruana R, Niculescu–Mizil A (2006) An empirical comparison of supervised learning algorithms. In: Proceedings of the 23rd international conference on machine learning. The Association for Computing Machinery, New York, pp 161–168
6. Charlesworth A (1980) A proof of Gödel's Theorem in terms of computer programs. Math Mag 54:109–121
7. Creative Commons (2007) Attribution–ShareAlike 3.0 unported (CC BY–SA 3.0). http://creativecommons.org/licenses/by–sa/3.0/ Accessed 8 Oct 2012
8. Cushman F, Young L, Hauser M (2006) The role of conscious reasoning and intuition in moral judgment: testing three principles of harm. Psychol Sci 17:1082–1089
9. De Jong KA (2006) Evolutionary computation: a unified approach. MIT Press, Cambridge
10. Dean T (2012) Evolution and moral diversity. Baltic International Yearbook of Cognition, Logic and Communication 7 (1): 1-16
11. Denison, DR (1990) Corporate culture and organizational effectiveness. Wiley, New York
12. Diener E, Fujita F (2005) Life satisfaction set point: stability and change. J Pers Soc Psychol 88:158
13. Dunbar K, Fugelsang J (2005) Causal thinking in science: how scientists and students interpret the unexpected. Sci Technol Think 57–79
14. Eddy DM, Adler J, Patterson B, Lucas D, Smith KA, Morris M (2011) Individualized guidelines: the potential for increasing quality and reducing costs. Ann Intern Med 154:627–634

15. Fan P (2011) Innovation capacity and economic development: China and India. Econ Change Restructuring 44:49–73
16. Fehr E, Gachter S (2002) Altruistic punishment in humans. Nature 415:137–140
17. Free Software Foundation, Inc (2008) GNU free documentation license. http://www.gnu.org/copyleft/fdl.html Accessed 8 Oct 2012
18. Giarratano JC, Riley GD (2005) Expert systems, principles and programming. Thomson Course Technology, Boston
19. Graham J, Haidt J, Nosek BA (2009) Liberals and conservatives rely on different sets of moral foundations. J Pers Soc Psychol 96:1029–1046
20. Greene JD (2009) The cognitive neuroscience of moral judgment. In: Gazzaniga MS (ed) The cognitive neurosciences IV. MIT Press, Cambridge
21. Hanson R (2007) Logarithmic market scoring rules for modular combinatorial information aggregation. J Prediction Markets 1:3–15
22. Heisenberg W (1927) Über den anschaulichen Inhalt der quantentheoretischen Kinematik und Mechanik. Zeitschrift für Physik 43:172–198
23. Hofstede, GH (2001) Culture's consequences: comparing values, behaviors, institutions, and organizations across nations. Sage Publications, Thousand Oaks
24. Isen AM, Levin PF (1972) Effect of feeling good on helping: cookies and kindness. J Pers Soc Psychol 21:384–388
25. Kanai R, Feilden T, Firth C, Rees G (2011) Political orientations are correlated with brain structure in young adults. Curr Biol 21:677–680
26. Kitcher P (1990) The division of cognitive labor. J Philos 87:5–22
27. Kohlberg L (1981) The philosophy of moral development. Harper & Row, San Francisco
28. Kram ML, Kramer GL, Ronan PJ, Steciuk M, Petty F (2002) Dopamine receptors and learned helplessness in the rat: an autoradiographic study. Prog Neuropsychopharmacol Biol Psychiatry 26:639–645
29. Kulkarni D, Simon HA (1988) The processes of scientific discovery: the strategy of experimentation. Cogn Sci 12:139–175
30. Lewontin RC (1970) The units of selection. Annu Rev Ecol Syst 1:1–18
31. Lind G (1978) Wie misst man moralisches Urteil? Probleme und alternative Möglichkeiten der Messung eines komplexen Konstrukts. In: Portele G (ed) Sozialisation Und Moral. Beltz, Weinheim, pp 171–201
32. Lorenz EN (1963) Deterministic nonperiodic flow. J Atmos Sci 20:130–141
33. Lykken D, Tellegen A (1996) Happiness is a stochastic phenomenon. Psychol Sci 7:186–189
34. Maynard Smith J (1982) Evolution and the theory of games. Cambridge University Press, Cambridge
35. Milgram S (1963) Behavioral study of obedience. J Abnorm Soc Psychol 67:371–378
36. Newton I (1676) personal letter. In: Turnbull HW (ed) The correspondence of Isaac Newton, vol 1. Cambridge University Press, Cambridge, pp 416
37. Norman WT (1963) Toward an adequate taxonomy of personality attributes: replicated factor structure in peer nomination personality ratings. J Abnorm Soc Psychol 66:574–583
38. O'Reilly CA, Chatman J, Caldwell DF (1991) People and organizational culture: a profile comparison approach to assessing person–organization fit. Acad Manag J 34:487–516
39. Page SE (2011) Diversity and complexity. Princeton University Press, Princeton
40. Pascal B, Havet E (1852) Pensées. Dezobry et E. Magdeleine
41. Pinker S (2011) The better angels of our nature: why violence has declined. Viking Adult, New York
42. Pizarro DA, Laney C, Morris EK, Loftus EF (2006) Ripple effects in memory: judgments of moral blame can distort memory for events. Mem Cogn 34:550–555
43. Polgreen PM, Nelson FD, Neumann GR, Weinstein RA (2007) Use of prediction markets to forecast infectious disease activity. Clin Infect Dis 44:272–279
44. Quere CL, Harrison SP, Colin Prentice I, Buitenhuis ET, Aumont O, Bopp L, Claustre H (2005) Ecosystem dynamics based on plankton functional types for global ocean biogeochemistry models. Glob Change Biol 11:2016–2040

45. Raymond E (2000) The cathedral and the bazaar. http://www.catb.org/esr/writings/homesteading/cathedral–bazaar/ Accessed 8 Oct 2012
46. Rest J (1979) Development in judging moral issues. University of Minnesota Press, Minneapolis
47. Santos-Lang CC (2002). Ethics for artificial intelligences. Presented at the 2002 Wisconsin State-wide technology symposium. http://santoslang.wordpress.com/article/ethics-for-artificial-intelligences-3iue30fi4gfq9-1/ Accessed July 2011
48. Sauper C (2008) Automated creation of Wikipedia articles. Dissertation, Massachusetts Institute of Technology
49. Schiff JL (2011) Cellular automata: a discrete view of the world. Wiley, Hoboken
50. Servan-Schreiber E, Wolfers J, Pennock DM, Galebach B (2004) Prediction markets: does money matter? Electron Markets 14:243–251
51. Slovic P (2007) If I look at the mass I will never act: Psychic numbing and genocide. Judgment Decis Making 2:79–95
52. Sober E, Wilson DS (1998) Unto others: the evolution and psychology of unselfish behavior. Harvard University Press, Cambridge
53. Sober E, Wilson S (2000) Summary of: unto others: the evolution and psychology of unselfish behavior. J Conscious Stud 7:185–206
54. Sosis R, Ruffle BJ (2003). Religious ritual and cooperation: testing for a relationship on Israeli religious and secular kibuttzun. Curr Anthropol 44:713–722
55. Steare R (2006) Ethicability. Roger Steare Consulting, London
56. Steil B, Victor DG, Nelson RR (2002) Technological Innovation and Economic Performance. Princeton University Press, Princeton
57. Stephens SL, Martin RE, Clinton NE (2007) Prehistoric fire area and emissions from California's forests, woodlands, shrublands, and grasslands. For Ecol Manage 251:205–216
58. Stephens SL, Ruth LW (2005) Federal forest-fire policy in the United States. Ecol Appl 15:532–542
59. Taleb NN (2010) The black swan: the impact of the highly improbable. Random House Digital. Inc, New York
60. Thompson A (1996) Silicon evolution. In: Proceedings of the first annual conference on genetic programming. MIT Press, Cambridge, pp 444–452
61. Turiel E (1983) The development of social knowledge: morality and convention. Cambridge University Press, Cambridge
62. Versel N (2013) NCQA tests new healthcare quality measure. Information Week. April 12
63. Walker LJ, Frimer JA, Dunlop WL (2010) Varieties of moral personality: beyond the banality of heroism. J Pers 78:907–942
64. Wallach W, Allen C, Smit I (2008) Machine morality: bottom–up and top-down approaches for modeling human moral faculties. AI Soc 22:565–582
65. Wallach W, Allen C (2008) Moral machines: teaching robots right from wrong. Oxford University Press, Oxford
66. Wilde D (2011) Personalities into teams. Eng Manage Rev IEEE 39:20–24
67. Wilson DS, Wilson EO (2008) Evolution "for the good of the group". Am Sci 96:380–389
68. Wilson DS, Near D, Miller R (1996) Machiavellianism: a synthesis of the evolutionary and psychological literatures. Psychol Bull 119:285–299
69. Wittgenstein L (1958) Philosophical investigations. Blackwell, Oxford
70. Wolfers J , Zitzewitz E (2004) Prediction markets. No. w10504. National Bureau of Economic Research
71. Yamagishi T (2003) Cross–societal experimentation on trust: a comparison of the United States and Japan. In: Ostrom E, Walker J (eds) Trust and reciprocity: interdisciplinary lessons from experimental evidence. Russel Sage Foundation, New York, pp 352–370
72. Yang X-S (2009) Firefly algorithms for multimodal optimization. In: Proceedings of the 5th international conference on stochastic algorithms: foundations and applications. Springer, Berlin, pp 169–178
73. Zak P (2011) The physiology of moral sentiments. J Econ Behav Organ 77:53–65

Part II
Contemporary Challenges in Machine Medical Ethics: Justice, Rights and the Law

Opportunity Costs: Scarcity and Complex Medical Machines

Adam Henschke

Abstract This chapter looks at the problem of resource scarcity, and the development of medical machines; in particular, the research, development and population wide implementation (RD&I) of complex medical machines. It raises a series of questions about how we make decisions regarding the distribution of scarce healthcare resources and posits that we ought to pay special attention to the distribution of resources for RD&I underpinning complex medical machines. This chapter covers some common issues in allocation of healthcare resources to then focus on the RD&I of complex medical machines as an important area requiring discussion.
When the needs or demands for medical treatment significantly outstrip the available resources, decisions must be made about how to distribute these resources, recognizing that not all needs will be satisfied immediately and some may not be satisfied at all [9, p. 275].

1 Overview

This chapter looks at the problem of resource scarcity, and the development of medical machines; in particular, the research, development and population wide implementation (RD&I) of complex medical machines. It raises a series of questions about how we make decisions regarding the distribution of scarce healthcare resources and posits that we ought to pay special attention to the distribution

A. Henschke (✉)
Centre for Applied Philosophy and Public Ethics, Charles Sturt University, Canberra, Australia
e-mail: ahenschke@csu.edu.au

S.P. van Rysewyk and M. Pontier (eds.), *Machine Medical Ethics*, Intelligent Systems, Control and Automation: Science and Engineering 74, DOI 10.1007/978-3-319-08108-3_9

of resources for RD&I underpinning complex medical machines. In their 2010 article, *Robot Caregivers: Harbingers of Expanded Freedom For All?* Jason Borenstein and Yvette Pearson state "[i]t remains debatable whether creating robot caregivers is the best allocation of resources, but interest in using robots to care for the elderly is growing…" [3, p. 277]. The chapter enters the discussion by taking seriously issues of resource scarcity. It asks how conditions of resource scarcity impact decision making about technologies, in particular technologies like robot caregivers that require a range of different resources.

The basic story here involves resource scarcity. Resources such as money, time and cognitive effort are required to effectively develop and integrate complex medical machines into people's lives in such a way as to have broad and deep positive impacts on population health. These resources are limited, and in conditions of scarcity, resources are not going to be distributed exactly equally. Insofar as the justificatory purpose of these complex medical machines is to positively impact on the health of individuals and populations, then meeting this purpose through RD&I of complex medical machines requires and consumes large amounts of resources, which according to most popular ethical systems of resource distribution would place complex medical machine RD&I low in a list of priorities for resource use.

The chapter proceeds as follows: Firstly, it discusses the basics of complex medical machines (what I refer to by "complex medical machines"), the conditions needed to bring these machines into existence, but also into wide and integrated use around the world and the costs of such RD&I. The chapter then raises a basic discussion of resource scarcity and its relation to healthcare. When there are conditions of resource scarcity,[1] then, by definition, demand outstrips supply. Consider triage in a hospital's emergency department: "triage in its primary sense is the sorting of patients for treatment in situations of at least modest resource scarcity, according to an assessment of the patient's medical condition and the application of an established sorting system or plan" [9, p. 278]. As such, some people will not be able to have their demands met immediately, some not at all. In such situations, there are a range of procedures by which we could decide who gets access to the limited resource [9]. The chapter then considers the RD&I conditions that underpin large scale roll out of complex medical machines, and argues that complex medical machines are highly unlikely to meet what most ethical analyses would consider easily justified distribution of resources. The chapter then responds to two counter arguments, one being a claim about a basic right to best practice healthcare, and one on the cost saving implications of complex medical machines. This will lead me to ultimately conclude that rather than using resources in the RD&I of complex medical machines, ethical theories will typically hold that the RD&I

[1] I note here the important point about problems of merely assuming conditions of scarcity. However, as I discuss in a later section of this chapter, given that scarcity is not simply about economic resources but other resources such as time and cognitive capacities, the conditions of scarcity while arguably resolvable, are far more intractable than some would have us believe.

resources required for complex medical machines would be better prioritized to other technologies which will either benefit the least well off and/or have a much broader appeal than those who are most likely to benefit from complex medical machines.

2 On Complex Medical Machines: Meaning, Realizing and Use

What is a "complex medical machine"? In this chapter, I refer to a particular form of medical technology, a machine that has some important role in a person's health and as part of this role; the machine plays some important part in the provision of *care* to that patient or charge. That is, the machine is not only involved in bringing about some important medically relevant physiological outcome for the person, such as the easing of some physiological suffering brought about by infection, injury or old age, but also does this in a way that incorporates some aspect which would traditionally be provided by a human—the machine "cares"[2] for the patient in some way. Here caring involves something like the "emotional commitment to, and willingness to act on behalf of persons with whom one has a significant relationship" [2]. For instance, a dialysis machine may be use to replicate the role of the kidneys, while a healthcare professional such as a doctor, nurse, carer etc. might explain what the dialysis machine is doing, give comfort to the patient during dialysis, attend to the patient's after care needs etc. A complex medical machine is intended to successfully attend to both of these sorts of needs: biological and patient care. While robot caregivers are a paradigm example of such complex medical machines, and I will focus attention on these sorts of machines, I do not wish to imply that these are the only things that could be considered complex medical machines. A machine to assist in surgery could reasonably be considered a complex medical machine. A summary of some medical machines under development and in operation since the mid 2000s is given by Taylor [25]. As I discuss later in the chapter, the reason for focusing on autonomous/semi-autonomous medical machines is due to their intended benefits to health provision above and beyond complementarity and assistance.

Consider a caregiving robot, caregiving, in that it assists a person who has some limitation on their capacities, and a robot, in that it can "sense, think and act" [22, p. 67]. I call these complex medical machines for three reasons. Firstly, to draw attention to the fact that these machines are directed toward some *medical* end: they are importantly different from non-medical machines such as industrial robots, as their primary and foundational purpose is toward something relevant to

[2] In the case of caregiving robots, the given machine displays behaviors towards the given person that are similar to roles that would be carried out in pursuit of the person's health beyond that of physiological treatment or repair.

medicine. Furthermore, the medical purpose is not some secondary or incidental outcome of these machines, it is the purpose for which they are used and importantly, the purpose which justifies the RD&I of the given machine. Borenstein and Pearson note that historically the reason for bringing robots into industrial environments was to save money. However, "bringing robot caregivers onto the scene could also be motivated by the obligation to meet core human needs" [3, p. 285]. Given that complex medical machines like caregiving robots are being used in healthcare provision, I would suggest a stronger formulation in which "bringing robot caregivers onto the scene *should primarily* be motivated by the obligation to meet core human needs."

This a vital point to bring up as healthcare is typically seen as a vitally important value, something that carries with it special moral weight and as such, medical research is typically seen as something that can frequently trump other concerns. For instance, Norman Daniels claims, medicine/healthcare is typically seen as some special end, which is not only different to other human activities but morally different; people the world over hold medicine/healthcare as something of special moral relevance [5, pp. 18–20]. I note that there are some, such as Robin Hanson, who see the special moral importance assigned to healthcare an intuition that is in error, arising from our evolution [8]. However, I note that even the position advocated by Hanson holds that some utilitarian reasoning can be applied to distribution of healthcare resources, see Hanson [8, p. 179], a conclusion that sits with the discussion in the final section of this chapter.

Labeling something a *medical* machine not only points to its ultimate purpose, but also to its "moral weight". By moral weight here, I refer to the idea that given the importance of health, when making decisions about resource distribution, services in support of health are typically prioritized over other uses: "For example, while there are many charities devoted to helping those with health crises, few are devoted to helping people similarly severe crises such as divorce, falling out of love, unemployment, failing in one's career, losing a friend and so on" [8, p. 165]. This is not to say that medical purposes necessarily and immediately trump all other concerns, rather that medical purposes are typically held to be morally weighty; if we are to decide in favor of some other use of limited resources, then the weight of argument is on us to justify the deep moral importance of this other use over the healthcare oriented one.

Secondly, these are *complex* machines: The kind of machine that I am referring to here is a machine composed of multiple components, each technically advanced in its own right, which are arranged in such a way as to achieve an end that has multiple components. Robots, for example, are frequently thought to have "three key components: "sensors" that monitor the environment and detect changes in it, "processors" or "artificial intelligence" that decides how to respond, and "effectors" that act upon the environment in a manner that reflects the decisions, creating some sort of change in the world" [22, p. 67]. A caregiving robot, for example, would sense that their patient is requesting to be bathed, recognize what the request means and respond appropriately. Such skills are highly developed and involve a range of different components that must integrate effectively.

With this in mind, "complex" is used for two related but different meanings. Firstly, a complex medical machine is not going to be something like a kidney dialysis machine, but one that has a set of integrated components which each serve a specific function: monitoring the kidney function, regulating the kidney function *and* offering something of the form of care that a human attendant would otherwise offer. Importantly, for this machine to meet its intended use, these different components must not only all work, but they must work *together*, point recognized by discussions on the safety requirements for domestic robots [12, p. 1890].

The second, complementary meaning of "complex" refers to idea that the end to which the given machine is put is complex, the machine is being used to attend not simply to the physiological aspects of the patients" kidney disease, but their psychological and emotional needs as well. That is, the medical machine is intended to service *a complex set of ends* which are not limited to bringing about some desired physiological state in the patient. The medical machine is explicitly designed, sold and used because it is also expected to bring about some desired psychological and/or emotional state in the patient. The relevance of this point is that it is this combined set of desired patient responses, captured here by use of the term "complex" that marks these sorts of medical machines out from "traditional" medical technologies. This increased functionality is no mere by-product of the given design but essential to it, serving to differentiate these technologically advanced machines from the traditional ones. Given this, the machines are both marketed as something that can help fill the space normally occupied by a human, and as something that is *at least* as useful use of resources as the human care giver—"the whole point of using the robots is because there will be fewer careers available in the future" [21, p. 276]. Not only are these machines expected to bring about a comparable result as the human care giver, but they are also hoped to do it in a way that is at least as good as or a cheaper or more resource-efficient way than the human care giver.

Finally, though it may seem redundant, these are described as *machines*. While this is in line with the book's title, re-using the term "machine" here serves an important purpose: The service that is being offered to a given patient is expected to be carried out by the piece of technology itself. That is, (insofar as these machines are not conscious) they are not people in any important philosophical sense.[3] While humans may be necessary for their creation, direction and upkeep, they are expected to operate largely in a standalone manner, ideally without continual human oversight [21, p. 269].[4] Though certain complex medical machine's

[3] Note that while the ethical concerns raised in this chapter could likely be applied to truly conscious artificial intelligence, self-replicating robots and the like, I use "machines" to keep the discussion away from the larger ethical and philosophical debates that arise when talking about artificial intelligence, robot rights etc.

[4] I discuss later in the chapter that one of the chief arguments in favor of these complex medical machines is that given their capacity to respond to a set of complex patient needs, they are hoped to be able to be used in the roles traditionally filled by humans. While not expected to fully take over from humans, these machines are hoped to be able to supplement and enhance human aspects of healthcare.

physical appearance and behavior may effectively mimic that of a human or other animal,[5] they will by-and-large be recognized by patients, family members, carers, healthcare professionals and society as pieces of technology. Reference to "machine" here serves to denote that despite high levels of independence, these machines are things that are not thought to be persons in any sense.[6]

A final point to makes is that these complex medical machines are no fantasy drawn from science fiction: South Korea aims to have domestic robots in every home by 2020 [15]. The South Korean government has established a Center for Intelligent Robots, which engages more than 1,000 scientists to develop a range of robots for cleaning to combat [22, pp. 243–244]. Further, both Japan and South Korea have already developed codified safety requirements for domestic robots [12, p. 1889]. The point here is that the world of robots is not the realm of imagination, or distant future, but with us now.

3 Medical Resources and Scarcity

To begin this section I raise the point that questions about the distribution of medical resources in times of scarcity are common not only to healthcare but to medical bioethics. Consider firstly the development of medical triage, which arose from the rationing of limited medical resources to those injured on the battlefield. Summarizing Kenneth Iserson and John Moskop's history of medical triage [9], in the 18th Century, chief surgeon in Napoleon's Imperial Guard Baron Dominique-Jean Larrey developed a situation of evaluating and categorizing wounded soldiers on the field: irrespective of rank, those with dangerous wounds would receive the first attention; those who could survive would have to wait. This system caught on; by the mid 19th Century, the British Navy began to order patients into three categories: those would not benefit from attention, the hopeless cases; those who would benefit from medical attention and needed immediate attention; and those who would benefit from medical attention but did not need it immediately. The basic decision metric here was efficacy: those who would benefit the most from attention should receive it first. While the practices and values underpinning triage have evolved and changed, through time and depending on context, the planned response to scarce medical resources is essential to modern hospital emergency departments and inpatient processing.

The second example comes, perhaps unsurprisingly, with the development of medical machines, in particular the development of kidney dialysis machines

[5] The aesthetics of design of caregiving robots is an important issue itself, and are beyond the scope of this chapter.

[6] I recognize that people do form emotional attachments with robots. The point here is that complex medical machines, despite any emotional attachments people might feel to them are generally regarded by most people (i.e. adults) as machines, not truly conscious beings.

in the 1960s [10, pp. 55–57]. Given that demand for the dialysis machines outstripped supply, decisions had to be made as to who received the life-saving dialysis machines. In one well known early case, these decisions were made by a committee of relevant experts, the so called "God Committee"—as described in a Life Magazine article by Shana Alexander, *They Decide Who Lives, Who Dies*, 1962. In this case, the God Committee, a mixture "of medical and legal professionals, of clergy and laypersons…together reviewed applications to the dialysis program" [10, p. 55]. The decisions made about who received dialysis and who didn't were controversial, in part because the God Committee decided partially on "factors such as social worth" [24, p. 214].

Finally, consider the current state of competitive research grant processes, a fact of modern academic life. Academics the world over typically compete for scarce funding through comparison of a range of metrics including publication history, impact, research location and the like. Despite an idea that research should be about the open and unlimited pursuit of knowledge, governments the world over "attempt to set, or influence the research agenda by funding policies [that support] various types of applied research supposed to be in the national interest" [27, pp. 46–47]. Arguably, such competition would not occur was there not already an existing condition of scarcity. The point of raising these examples is to show the long history of decisions about distribution of limited medical resources, the scarcity brought about by new technologies, and that we often accept (or suffer under) decision making in conditions of scarcity.

4 Complex Medical Machines and Scarcity

This chapter turns on issues of scarcity of resources for complex medical machines. Firstly, and most obviously, the money required to research, develop, product test and distribute at a population level medical machines is likely to be high. This is built in part from a general claim about medical technologies. For instance, Daniel Callahan's *Taming the Beloved Beast: How Medical Technology Costs are Destroying Our Healthcare System* describes how technology is a major component of the ever increasing costs of healthcare [4, pp. 37–66]. That is, following Callahan, adding more technology to healthcare is unlikely to decrease the costs of healthcare, a point returned to later in the chapter. Secondly, given the host of practical, social and ethical issues around developing medical machines, as demonstrated in this edited collection, it will likely take some time before these medical machines are integrated into people's lives around the world in an equitable and effective manner. As already mentioned, certain communities in certain countries already have a high uptake of caregiving robots and the like. The point here is twofold. Firstly, that this is typically only in certain communities in the given country, i.e., only those who can afford the machines. Secondly, that this is mostly a narrow set of countries, such as South Korea and Japan [22, pp. 304–305]. Current uptake in Anglo-Saxon and European countries is low. Further, if current provision of healthcare services through effective public health infrastructure is anything to go by, see for instance, Laurie Garrett's *Betrayal of Trust*, then the African region and others are unlikely to see population wide roll out any time soon [7].

On top of the costs of the research and the actual hardware, the *implementation* of these machines requires large amounts of resources. Thirdly, the research, development, product testing and distribution of the machines require a great deal of specialized intellectual capacities; what I will refer to as "cognitive resources". These cognitive resources are also limited: if a given specialist is working on one problem, then they cannot, by definition, work on another problem. In short, the RD&I of complex medical machines require application of scarce resources, including but not limited to money.

To the first point, these machines are likely to be expensive to buy, use and maintain. Moore's Law and capitalist economic drivers such as economies of scale are generally expected to reduce the cost of such machines. However, throughout *Taming the Beloved Beast*, Callahan argues at length that medical technologies do not typically reduce the costs of healthcare, instead they are one of the key drivers of the persistent increases in healthcare costs: "The idea that more research will reduce healthcare costs is a myth. It raises them. Will that always prove true? If history is our tutor, you may count on it" [4, p. 22].

One of the reasons why complex medical machines are likely to be persistently expensive is because of their complexity. As discussed above, these machines by necessity have multiple technologically advanced components. They require (at very least) specific, often medical grade hardware, designs that are highly robust physically, easily updatable software and extremely usable machine–human interfaces. Were any of these three components to fail, then the machine would not only be rendered useless, but this uselessness could be life-threatening. For instance, "[s]oftware reliability is essential in a domestic robot since any loss in control, in what is essentially a complicated computer-based safety-critical system, can lead to dangerous actions with catastrophic consequences" [12, p. 1889]. Consider a caregiving robot which is designed to help an independent elderly patient bathe. Should the caregiving robot malfunction and drop the patient in the bath, the patient could be severely injured and may even die. Furthermore, "the so familiar household surroundings constitute an almost chaotic environment; full of non-stationary objects, including different groups of entities." [12, p. 1890]. As such, the machines need to be extremely robust at all levels of design and integration and this robustness will bring about high costs.

In addition to the actual costs of manufacturing the physical component of the machines, the RD&I of the machines is likely to be extremely high, a key factor that drives up the cost of medical technologies. Think of estimates about the cost of developing pharmaceutical—estimates range between US$500 million and US$2 billion [1]. There are important differences between the RD&I of complex medical machines and pharmaceutical R&D; for instance, in pharmaceutical R&D, many resources are used in clinical trials, arguably research that does not compare to complex medical machines. However, recall that these are *complex* medical machines. Firstly, their complexity comes in as having many different but integrated components. Should one of these components fail, as we saw with the caregiving robot bathing a patient, this could cost a person their life. Furthermore, the failure is not simply from the component itself, but the integration of the components. Arguably, each sort of component; hardware, software, interface may work as intended, but they do not integrate effectively, then the machine could fail [12, p. 1890].

Recall also that the complexity comes not just from the machine's components but its interaction with the patient. Human behaviors are incredibly complex and unpredictable; requiring the medical machine to understand a patient's expressed, and at times, unexpressed needs and wants. This set of patient requirements is typically hard enough for a human carer to discharge effectively 100 % of the time. To add a further level of complexity, if these are used in an environment with high numbers of humans, this environment is likely to be close to chaotic. As humans we instinctively traverse such complex zones with minimum recognition of the cognitive effort required to navigate and operate in such zones. Think now that a resident, guest or pet is to enter a zone where a domestic caregiving robot operates, but that the new person or pet is unfamiliar with the given robot. Some go so far as to state that, for domestic robots to be considered safe, we'd need to "[c]ontrol access of unaware residents or guests because robot may pin unknowledgeable pets or people. The device can only be used by people who know how to operate it and who have read and fully understood the entire manual" [12, p. 1894].

To develop a reliable, robust and safe machine for common use is likely to use a great deal of resources, making the comparison to pharmaceuticals more reasonable. These costs would likely be recouped, (at least by private profit motivated research companies), typically by charging large amounts for the purchase, use and upkeep for the life of the machine: despite Moore's law and the market, as with Callahan's analysis about medical technologies generally, these machines will likely cost a lot of money.

Furthermore, any assessment of the robustness of medical machines needs to bear in mind that these are *medical* machines. Again, the stakes here may be literally and death. If the voice recognition on a person's smart phone mishears/misunderstands a request, the results may not matter too much. If this happens in the case of a medical machine, the failure may be lethal. While not wishing to be alarmist, the point here is that one cannot simply look at smart phones and other consumer technologies, and assume that given their success as consumer products, such success will naturally transfer over to the medical context. While not an example of a complex medical machine, the U.K. Government's expectations about their electronic information led health program, commits this exact mistake, assuming that given the success of electronic banking and decisions about travel for holidays, such success can be translated into healthcare [6]. In order to prevent high impact failures of technology and integration, the research and development of complex medical machines is likely to be highly expensive.

An additional aspect of complex medical machines is the cost of effective implementation. I suggest that two of the main justificatory reasons for supporting the RD&I of complex medical machines is if they will reduce costs and will fill a gap in care for the vulnerable that human mediated healthcare cannot do alone. However, in order for these two aims to be met, (and to meet the basic ideal of market interest driving costs of production down) these machines would need large scale uptake in a population, and such uptake would need to ensure that the vulnerable members of the population *actually* receive the given support. Both of these aims need active support for effective *implementation*. While some countries such as South Korea and Japan are more open to caregiving robots [22, pp. 304–305, 12, p. 1889], many western countries are not so open. This is not to say that western countries will never have widespread acceptance of such machines, rather that implementation is fundamental to ensure that the complex machines meet their stated aims. And again, such implementation is likely to be expensive.

Finally, note that the resources needed extend beyond economic. Such RD&I also requires time and cognitive resources. The time taken to go through effective RD&I to produce a caregiving robot that can safely and reliably bathe an elderly patient is great. Similarly, the amount of intellectual capital needed to develop all the necessary hardware, software and interfaces, to test the integrated components and to ensure effective integration across a population is extremely high. In short, not only are the machines themselves likely to be expensive to buy, use and run, the RD&I necessary to ensure that the machines meet their aims of bringing down costs and supporting a populations most vulnerable is extremely resource intensive. Furthermore, like the need for post-release oversight of pharmaceuticals, so called "pharmacovigillance", oversight mechanisms must be in place in order to ensure that these machines do not have system wide failures when in operation. This oversight brings with it additional need for resources, coupled with ethical concerns such as the potential invasions of individual privacy, problems of informed consent and issues of property claims over the data produced. See "The Rights and Wrongs of Robot Care" for more on these issues [21].

Before moving to the next section, I wish to respond to a foundational issue with this paper: why assume scarcity? Tom Koch for example, takes a very strong stance against "lifeboat ethics" and the assumptions of scarcity that bring lifeboat reasoning into play. He asks the important questions that if scarcity is the problem, then why assume scarcity. Furthermore, he proposes that bioethicists' fundamental role is to question these key assumptions. Finally, he finds the assumption of scarcity to be wrong:

The problem of scarcity in U.S. healthcare is not the inevitable result of a plethora of new and expensive technologies that have saved lives we must, perhaps regrettably, triage away. Rather, at least in part it results at a structural level from a private, for profit delivery system that builds profit seeking into the system at every stage, creating levels of onerous expense that would be unacceptable anywhere else…The willingness to assume scarcity as a natural limit and economic projections as the assessor of care decisions was what earned a bioethics at the table [of medical decision making]…The patient who was elderly, or by extension simply fragile, was a burdensome expense, and thus could be abandoned—whatever he or she might wish—in the name of economic futures [10, p. 19].

For this chapter to have any purchase, I must show why scarcity is a reasonable assumption. Firstly, recall that the resources that I am discussing are not simply economic, but time and cognitive. Koch [10] is concerned that the discipline of bioethics, by and large, does not effectively question premises such as scarcity. His claim does not hold here. Most explicitly, the resources I consider under "scarcity" extend beyond economic scarcity to include time and cognition. If nothing else, to ensure that there are enough cognitive resources to meet every healthcare need, we would have to have a massive and fundamental restructuring of our educational programs, likely on an international level. To overcome the problem of scarcity of cognitive resources we'd need to educate, train and develop a huge number of medical engineers and the like, which will take time and money. By overcoming one sort of scarcity, we encounter others. So, for the mid-term at least, conditions of scarcity still hold. This is not to say that we shouldn't re-incentivize our medical technology industries. For example, in the Second Edition of *World Poverty and Human Rights*, Pogge [16] sets out a highly compelling set of reasons as to the why and how we can do this. The point here is that should we pursue this path, it will require the use of other resources, so scarcity will remain for some time at least.

Secondly, I turn to Callahan's point that no matter how well we develop our healthcare technologies and infrastructure, we will always want more:

The improvement of health, the relief of suffering, and the forestalling of death are as open-ended as the exploration of outer space. In each case, the possibilities are endless: no matter how far we go, there is always further we could travel. If we all lived an average of 150 years, the offices of the doctors would still be full, patients would still be looking to have their diseases cured, their pain and suffering relieved [4, p. 9].

On this line of reasoning, until we can guarantee immortality to every person in the world, then we will encounter scarcity in our medical resources. And of course, having a word populated by immortals is likely to encounter problems of scarcity in other, non-healthcare areas such as food, water, air and energy. In short, contra Koch, assuming scarcity is reasonable.

5 Opportunity Costs: Healthcare and Technology

I have so far argued that complex medical machines will be operating in conditions of resource scarcity. The implication has been that in such situations, we have to make decisions about what ought to be resourced. The question now becomes obvious—should we make decisions at all? My position is that we are already making these decisions, and that we should be making these decisions in a way that is publicly justifiable. This section will cover the claim that we are already making these decisions and then offer some conditions for the public justifiability of these decisions. Both of these points turn on complementary issues of opportunity and cost. The first point is about opportunity costs in an economic sense: if we have only one resource, call it 1R, and two options, call them X and Y. Given that we only have one R, if we decide to use 1R on X we are by definition, unable to use 1R on Y. Secondly, if we are to take *opportunity* seriously, then we have to direct our resources to one set of people over another. One popular choice mechanism is the basic Rawlsian Difference Principle: that those who benefit from social and economic inequalities ought to work to the benefit of the least well off. I discuss each point in turn.

Opportunity costs, as mentioned, refer to the situation where we are forced to choose between using a limited resource on two (or more) options: "When you bought this book, you implicitly decided not to spend that money somewhere else" [28, p. 10]. Basically, you have a limited resource, 1R, and must use it on/in pursuit of X or Y. If options X and Y both require a unit of R, and you only have 1R, this means that if you choose to use 1R on X, you cannot then use 1R on/in pursuit of Y. Consider the example earlier of kidney dialysis machines. Now, we have one resource, 1R, here represented by a single dialysis machine, and two patients, X and Y here represented by Xander and Yolanda. If we choose to give Xander access to the dialysis machine, given that there is only one dialysis machine, then Yolanda cannot access it. Similarly, if we choose Yolanda, then Xander cannot access it. The point of raising this in terms of opportunity costs is that if we are to properly consider our use of 1R, any decision about the benefits of selecting X must also factor in the cost of not being able to choose Y.

In a variation and in line with Koch's questioning of the conditions of scarcity, the solution is to get 2R: the hospital now has enough funds for another dialysis machine, thus being able to treat both Xander and Yolanda. However, those funds spent on 2R could also be spent on 100,000 sterilized needles for syringes. Without access to sterilized needles, many thousands of people may die. This example of sterilized needles for syringes is chosen specifically as many hospitals around the world do not have enough money for sterilized syringes. Laurie Garret describes in detail how healthcare providers in Russia and African countries were forced to reuse needles so much that they had sharpening stones to sharpen the needles when they got worn down. As a result of these shortages, many blood-borne diseases like HIV and were spread by patients visiting hospitals [7]. Buying the second dialysis machine, X and Y may be saved, but this may be at the cost of thousands of other lives. Such a decision is tragic, either way people are allowed to die.[7] The point here is not necessarily to argue that the hospital should spend their limited economic resources on dialysis machines or syringes, but that given scarcity, whichever option the hospital goes for, it will come with costs.

Such conditions of opportunity costs also hold with the RD&I of medical technologies. As discussed earlier, any medical technology must go through a range of processes prior to being used in the marketplace, whether it is clinical trials

[7] I use 'allowed to die' here specifically to show that these choices are unlike those generally considered in literature about doing/allowing, where killing is seen as morally worse than letting die. In the case of dialysis v. syringes, (assuming that no-one has actually been put on dialysis), then both choices are about 'allowing' people to die.

for pharmaceuticals, or the testing, integration and observation of components in a complex medical machine. These processes require money, time and effort. As repeated throughout this chapter, these resources are scarce. By making a decision to study the engineering of caregiving robots, a person is no longer able to study pharmaceutical research. Even if this person is uniquely talented, and does a dual degree such that they have the cognitive capacity to do both, the time they spend working on one set of problems means that they are working on the other. And, perhaps having the most impact, a decision to fund research projects in one area means that another area is not getting funding.

Having argued at length that conditions of scarcity hold, at least in the mid-term, my point here is that spending limited resources on the RD&I to produce complex medical machines comes at the cost of other research. This not to say that complex medical machines are useless or that they are necessarily bad use of resources. Rather, to make explicit that using resources on them in conditions of scarcity comes at the cost of other healthcare interventions.

This of course brings us to questions about what then ought to be funded. If, as argued, making decisions about funding RD&I of complex medical machines comes at the cost of other medical research, then how ought we to decide between the various options? This is the second point of opportunity costs. If we want to give people opportunity, then this requires resources. In the next section, I discuss rights to healthcare and utilitarian reasoning with regard to complex medical machines, the focus here is takes a Rawlsian approach to deciding outcomes of health funding:

Since meeting health needs promotes health (or normal functioning), and since health helps to protect opportunity, then meeting health needs protects opportunity…Since Rawl's justice as fairness requires protecting opportunity, as do other important approaches to distributive justice, then several recent accounts of justice give special importance to meeting health needs [5, 30].

Daniels' point is that the importance of healthcare comes in because of the way that good health promotes equality of opportunity.

If we take Daniels' point about opportunity and health seriously, we ought to give some level of prioritization to healthcare resources; particularly RD&I. Certainly, some complex medical machines will promote opportunity. If a person with a physical limitation can live a good life because of a complex medical machine, then we ought to support RD&I for things like caregiving robots. So far, so good, then, for complex medical machines. However, given that scarcity still holds, ought we to operate in conditions in which certain sorts of medical technologies are supported over others? Such comparisons would require some level of complex empirical and moral deliberations, such as those described by Stein [24], which is a level of detail beyond this chapter. Bearing this in mind, adopting a Rawlsian veil of ignorance as a way to gesture at system-level decisions about RD&I, many complex medical machines may seem undesirable. If one does not know their existing or future states of health, then it seems the rational self-interested agent would prioritize research for medical technologies that help the most number of people, the least well off and/or the youngest. Statistically, they would be more likely to benefit from a technology that benefits a large group than a small group. If complex medical machines were to be so expensive that only a small sector of the population would benefit from them, if would be rational to select for other RD&I first. Secondly, supporting research for the world's poor can "reduce the threat of infectious diseases to developed countries…promote developed countries" economic interests, and…promote global security" [19, p. 115]. Finally, to the third point about old age, given that surviving youth is a logical necessity for becoming old, the rational agent would need to recommend technologies that maximize chances of surviving childhood. And from the discussion so far, it would seem that the majority of complex medical machines would not fall under these decisions.

Furthermore, we could take the justice approach in reference to the Rawlsian Difference Principle, whereby effort, skills and money ought to go to benefit the least well off, as "the claims of those who are worse off often take priority over the claims who are better off, even if a higher total utility could be achieved by benefiting the latter" [18, p. 31]. On this, complex medical machines are most likely not going to be meeting the conditions of the difference principle. The simplest claim is that these machines will most frequently be used by the world's most well off. This is a reference to both a large set of those groups living in societies that are both technologically adept *and* have the money to pay for these machines, and also most likely to a subset of those groups—those within those societies who have the money pay for such machines. Following the availability/access distinction offered by Selgelid and Sepers [20], the first condition is one of availability, whereby the medical machines which will be developed are going to be those with a large enough *and* rich enough market to pay for the technology and to lobby for its RD&I. The second condition is one of access, whereby the medical machines which will be used are going to be expensive, and as such, will only be accessible by those with money to pay for it, and/or living in societies with taxation and socialized healthcare such that tax revenue pays for and/or subsidizes their use of the given machines.

6 What About Rights and Efficiency?

The argument so far has relied on a basic Rawlsian egalitarian position; namely, that seeking to meet opportunity and operating under the veil of ignorance or adhering to the difference principle, complex medical machines are unlikely to be a an easily justified use of RD&I resources. However, this assumes granting a certain priority to justice over rights and utility. Prioritizing different values such as liberty and efficiency would render the justice-based argument of limited appeal. In this section, I will attend to two rights-based arguments and one utility argument to show that complex medical machines do not easily meet a right to healthcare or utility.

A rights-based argument derives from two sources. The first argument assumes a basic right to healthcare. The position that this chapter advocates is to essentially prioritize other healthcare RD&I over complex medical machines. By doing this, some people will, by definition, miss out on receiving some healthcare services. Due to this, their basic right to healthcare is not being recognized. Secondly, still along rights lines, is the right to *select* one's own healthcare. This right is based in individual property rights. If a person has money, then they have the right to spend that money how they desire. And if they want a complex medical machine to care for them, a loved one etc., then it is their right to spend that money how they want.

The first argument finds a proponent in the work of Koch [10]. In *Thieves of Virtue* [10], he argues at length against the prioritization of healthcare resources. He discusses "Lifeboat Ethics", in which some people must die such that others are saved, like on a life-boat at sea with limited water and food:

That is ethics *in* the lifeboat, where no good choices exist. Some *must* die, or be abandoned, so that others may live…There is, however, another ethic one can apply, an ethics *of* the lifeboat whose focus is the complex of prior decisions that brought about the lifeboat's occupants to the sorry contemplation of murder for salvation's sake [10, pp. 78–79].

Rather than reasoning *in* situations of scarcity, he asks us to consider *why* the scarcity has arisen. As he correctly describes it, such prioritization arguments only get off the ground in situations of resource scarcity. The basic situation that Koch points to is that we ought not to assume scarcity. However, underpinning this is "the idea of care as a right rather than a commodity whose provision is always contingent" [10, p. 255], that every person has a right to healthcare.

Note that Koch's argument has particular relevance to the discussion of complex medical machines: examples of complex medical machines are proposed for use in situations where a community's vulnerable are left without adequate healthcare supports. The elderly present us with a population found everywhere in the world. Complex medical machines can be of particular use in situations where there is an increasing proportion of aged in the population. Koch points to the fact that prioritizing healthcare resources is likely to restrict the access of the old and vulnerable to healthcare resources.

The counter arguments to a position like that of Koch's start first with idealism vs. realism. Unlimited resources for healthcare may be a scenario to aim for, but unfortunately we are not in that situation at the moment.[8] Firstly, economic resources, even for technologically advanced and economically developed countries, are limited. Whether thinking of the U.S., the U.K., countries of the E.U. the past 5 years following the global financial crisis have shown that economic stability is not assured, and economic resources are likely to never be unlimited. Even a country like Australia which managed to weather the global financial crisis and has well-developed healthcare infrastructure has to ration resources. Koch's position here is to challenge the assumption of scarcity. However, recall that this chapter has argued that it is not only economic resources that are scarce: time and cognitive resources are vital for any technological development. And simply

[8] And, arguably, are unlikely to ever meet such conditions.

saying that we should reject scarcity does not give us more time or more capable people. Further, as Callahan has already shown, [4, p. 9], insofar as human frailty is an unavoidable fact of our existence, there is no upper limit to what we can expend resources on. If people commonly live to be 150, no doubt we will encounter a set of health related problems that we do not have to face now. The point here is that we cannot simply wish scarcity away, and so will always have to prioritize one set of people's interests against another's.

Secondly, note that simply prioritizing healthcare over other areas may not have the widest positive impacts on healthcare promotion. For instance, John Weckert writes that we need to recognize that "[m]ost funding of medical research is spent on diseases of the relatively affluent…A common statement is that around ninety percent of medical research is spent on diseases of ten percent of the world's population" [27, p. 50]. Further to this, he argues from consequentialist reasons such as increased impact, Kantian duties arising from the environmental harms brought about by industrialization and the ethics of care for our own relations and descendents, that taking the promotion of people's health seriously, "there should be less of a focus on…medical research aimed at the developed world, and more of a focus on clean energy and other technologies for energy efficiency" [27, p. 50]. Even if we ring-fence medical research funding, and prioritize healthcare technologies over others like clean energy research, there are still likely to be decisions made within healthcare. That is, should we spend our money on complex medical machines over research on communicable diseases like malaria and cholera which will likely benefit the world's elderly and rich before the world's young and poor?

With this in mind, we return to the problem of opportunity costs. The question is not "why do we sacrifice the vulnerable" as Koch would have it [10, p. 112], but rather "which vulnerable group do we sacrifice and why?" The position advocated in the previous section is something like a maximin. We should direct our resources to the least well off. And seemingly no matter how that "least well off" is defined, complex medical machines are likely to sit low in the list of priorities, for the immediate future at any rate.

A different rights based argument is to look at property rights and apply it to complex medical machines. Like the discussion about scarcity, above, it questions justice or some egalitarian position as the starting point. This takes Robert Nozick as the paradigm example of a right to property. In *Anarchy State and Utopia* [14], Nozick offers an interpretation of property rights generated from the Lockean proviso to leave "enough, and as good, left in common for others" [11, p. 329]. Nozick states "[t]he crucial point is whether appropriation of an unowned object worsens the situation of others" [14, p. 175]. On Nozick's account, justice in holdings is "exhaustively covered" by three steps:

1. A person who acquires a holding in accordance with the principle of justice in acquisition is entitled to that holding.
2. A person who acquires a holding in accordance with the principle of justice in transfer, from someone else entitled to the holding, is entitled to the holding.
3. No one is entitled to a holding except by (repeated) applications of 1 and 2 [14, 151].

These three features are central to Nozick's account: "[A] principle of (initial) acquisition, a principle of transfer, and a principle of rectification. Its central tenet is that any configuration of holdings that results from the legitimate transfer of legitimately acquired holdings is itself just" [17, p. 4].

Thomas Scanlon notes that Nozick's account offers no more than a "skeletal framework of rights derived from Locke" [17, p. 4]. Surprisingly, Nozick offers no explanation for the original acquisition. The most he says is "[w]e shall refer to the complicated truth about this topic, which we shall not formulate here, as the principle of justice in acquisition" [14, p. 150]. This lack of justification of initial acquisition is troubling, as it is central to Nozick's project and justifies entitlement and a right of transfer. Thomas Nagel writes that Nozick's book, "[d]espite its ingenuity of detail…is entirely unsuccessful as an attempt to convince, and far less successful than it might be as an attempt to explain to someone who does not hold the position why anyone else does hold it" [13, p. 137]. Jeremy Waldron states that Nozick's Principle of Just Transfer "operates on material provided in the first instance by the [Principle of Just Acquisition]", yet "Nozick does not tell us what his favoured [Principle of Just Acquisition] is" [26, p. 257].

Perhaps the reason for the lack of elaboration of any principle of just acquisition in Nozick stems from a fundamental problem within Locke's own justification of property rights. Locke's justification was built from the premise that one gains personal ownership over the thing labored on by mixing one's labor. "The idea that labour is literally *mixed with* an object is crucial to this argument. Without it we cannot explain how the force of the labour entitlement is transferred to the product of one's labour" [26, p. 184]. Locke's justification stands on two features, self-ownership, and the extension of that self ownership into other things via the mixing of labor [26]. Waldron argues that labor is an action, not a thing, and one cannot own an action which leads him to say that the notion, "that the object thereby comes to contain something the labourer owns—is left completely mysterious" [26, p. 185]. Waldron concludes that Locke's labor investment theory, "the best known [special rights] based theory fails to provide an adequate defence of private property" and instead offers a Hegelian approach to natural property rights [26]. Either way, deriving a natural right to property from either a Lockean system or Hegelian one, Waldron argues that claims to property can be overridden by other more pressing considerations such as basic survival. Waldron argues that an understanding of Locke's work prioritizes the right to survival above the right to property [26]. Similarly, "Hegel does not believe that property rights are absolute anyway against the demands that might arise out of higher stages of ethical development" [26, p. 387].

The relevance of this discussion is twofold. If one is to take the grounds of natural rights to property seriously, then there ought to be priorities given to fund medical research that keeps people alive. Secondly, taking these property rights seriously leads us, as before, to compare complex medical machines against other technologies and interests. And reasoning from either a Lockean or Hegelian foundation, it would seem that complex medical machines ought to be placed lower in a list of research priorities.

Different sorts of counter arguments come from an argument of increased efficiency. The idea is that complex medical machines will ultimately lead to reductions in overall costs of healthcare. However, to this, I am skeptical for three reasons: historical, actual use and complexity. Firstly, as raised repeatedly, Callahan [4] points to the history of medical technologies raising the costs of healthcare, not lowering them. And given the demands that RD&I of producing safe, reliable and robust complex medical machines, the burden of proof is to show that these will actually reduce costs.

Secondly, in looking at the idea of shallow care vs. deep care, Borenstein and Pearson point to the most desirable situation of care robots: "the inability of robot caregivers to provide "deep" care need not preclude human caregivers from using them to provide "good" care…Consequently, robots should complement the efforts of human caregivers rather than replace them" [3, p. 281]. The point here is that the best standard of care requires the complex medical machines to be used in conjunction with existing human caregivers. This seems a reasonable assumption, but it fails the claims of cost reduction. Instead of replacing the human caregivers, and thereby bringing overall costs of healthcare down, the complex medical machines are now to be used in conjunction with existing humans. We are now faced with a double whammy: the cost of the human caregiver plus the cost of the caregiving robots. This is not to say that we shouldn't aim at this, but rather that, in line with Callahan [4], we're faced with medical technologies adding to the existing costs of healthcare, not reducing them.

In response to this, perhaps the aim is to develop highly autonomous robots, ones that need minimal or no human supervision. Despite concerns that this sort of approach to healthcare may be detrimental to the emotional state of those under care [23], we are then faced with the highly expensive and resource demanding RD&I of developing complex medical machines. So again, it seems that these machines will not reduce costs, but increase them.

7 Conclusion

In this chapter I have considered what is needed for complex medical machines to become a useful and helpful part of human societies. While things like caregiving robots promise a great many benefits to groups like the elderly and other vulnerable members of society, I remain skeptical whether such use of resources is morally justifiable. The basic point I have sought to make is that any large scale production of these machines does and will occur in conditions of scarce resources: they require money, time and cognitive effort of large amounts of people. And if the justifying goal is to promote the health of people and/or equality of opportunity, then on most moral systems, most of the time, resources could be better used in other ways.

The final comment to make is to reiterate that not all complex medical machines fail these tests, or indeed that we should see complex medical machines as an unmitigated moral hazard. Rather, if we are to take healthcare seriously, especially at a global rather than local level, then we ought to be prioritizing other less glamorous solutions over things like robots. We should properly fund nurses, nursing homes and relations that give care to their loved ones, and we ought to increase funding for problems around the world, beyond those treatments that seek to recognize the individual, but ultimately prioritize a small group of the world's most affluent to the detriment of many others. Should this analysis prove incorrect, and complex medical machines actually benefit the world's sickest and most vulnerable then this would seem to be a great use of resources. However, if recent history is any guide, such optimism is somewhat naïve.

References

1. Adams CP, Brantner VV (2006) Estimating the cost of new drug development: is it really $802 million? Health Aff 252:420–428. doi:10.1377/hlthaff.25.2.420
2. Beauchamp TL, Childress JF (2001) Principles of biomedical ethics. Oxford University Press, Oxford
3. Borenstein J, Pearson Y (2010) Robot Caregivers: harbingers of expanded freedom for all? Ethics Inf Technol 123:277–288. doi:10.1007/s10676-010-9236-4
4. Callahan D (2009) Taming the beloved beast: how medical technology costs are destroying our healthcare system. Princeton University Press, Princeton
5. Daniels N (2008) Just health: meeting health needs fairly. Cambridge University Press, Cambridge
6. Department of Health (2012) The power of information: putting us all in control of the health and care information we need. Health, U.K. Government
7. Garrett L (2001) Betrayal of trust. Hyperion books, New York
8. Hanson R (2002) Why health is not special: errors in evolved bioethics intuitions. In: Ellen FP, Jr. Miller FD, Paul J (eds) Bioethics. Cambridge University Press, Cambridge
9. Iserson KV, Moskop JC (2007) Triage in medicine, part I: concept, history, and types. Ann Emerg Med 493:275–281 doi:http://dx.doi.org/10.1016/j.annemergmed.2006.05.019
10. Koch T (2012) Thieves of virtue: when bioethics stole medicine. MIT Press, Cambridge
11. Locke J, Laslett P (1963) Two treatises of government. Mentor, New York
12. Mitka E, Gasteratos A, Kyriakoulis N, Mouroutsos SG (2012) Safety certification requirements for domestic robots. Saf Sci 509:1888–1897. doi:http://dx.doi.org/10.1016/j.ssci.2012.05.009
13. Nagel T (1975) Review: libertarianism without foundations. Yale Law J 851:136–149
14. Nozick R (1974) Anarchy, state, and Utopia. Blackwell Publishing, Oxford
15. Onishi N (2006) In a wired South Korea, robots will feel right at home. New York Times, New York
16. Pogge T (2008) World poverty and human rights. Polity Press, Cambridge
17. Scanlon T (1976) Nozick on rights, liberty, and property. Philos Public Aff 61:3–25
18. Scheffler S (1993) The rejection of consequentialism: revised edition. Oxford University Press, Oxford
19. Selgelid MJ (2008) Improving global health:counting reasons why. Develop World Bioeth 82:115–125
20. Selgelid MJ, Sepers EM (2006) patents, profits, and the price of pills: implications for access and availability. In: Cohen JC, Patricia I, Schuklenk U (eds) The power of pill. Pluto Press, London
21. Sharkey N, Sharkey A (2012) The rights and wrongs of robot care. In: Lin P, Abney K, Bekey G (eds) Robot ethics: the ethical and social implications of robots. MIT Press, Cambridge, pp 267–282

22. Singer PW (2009) Wired for war. Penguin Press, New York
23. Sparrow R, Sparrow L (2006) In the hands of machines? The future of aged care. Mind Mach 162:141–161. doi:10.1007/s11023-006-9030-6
24. Stein MS (2002). The distribution of life saving medical resources: equality, life expectancy, and choice behind the veil. In: In: Ellen FP, Jr. Miller FD, Paul J (eds) Bioethics. Cambridge University Press, Cambridge
25. Taylor RH (2006) A perspective on medical robotics. Proc IEEE 949:1652–1664. doi:10.1109/JPROC.2006.880669
26. Waldron J (1988) The right to private property. Clarendon Paperbacks, Oxford
27. Weckert J (2010) Setting the nanotechnology research agenda: medical research versus energy research. Aust J Profes Appl Ethics 111&112:45–55
28. Wheelan C (2010) Naked economics: undressing the dismal science. Norton & Company, New York

The Rights of Machines: Caring for Robotic Care-Givers

David J. Gunkel

Abstract Work in the field of machine medical ethics, especially as it applies to healthcare robots, generally focuses attention on controlling the decision making capabilities and actions of autonomous machines for the sake of respecting the rights of human beings. Absent from much of the current literature is a consideration of the other side of this issue. That is, the question of machine rights or the moral standing of these socially situated and interactive technologies. This chapter investigates the moral situation of healthcare robots by examining how human beings should respond to these artificial entities that will increasingly come to care for us. A range of possible responses will be considered bounded by two opposing positions. We can, on the one hand, deal with these mechanisms by deploying the standard instrumental theory of technology, which renders care-giving robots nothing more than tools and therefore *something* we do not really need to care about. Or we can, on the other hand, treat these machines as domestic companions that occupy the place of another person in social relationships, becoming *someone* we increasingly need to care about. Unfortunately neither option is entirely satisfactory, and it is the objective of this chapter not to argue for one or the other but to formulate the opportunities and challenges of ethics in the era of robotic caregivers.

1 Introduction

Work in the field of machine medical ethics, especially as it applies to increasingly autonomous home healthcare robots, generally focuses attention on the capabilities, modes of implementation, and range of actions of these mechanisms for the sake of respecting the dignity and rights of human patients. Researchers like Noel and Amanda Sharkey, for instance, have focused on a spectrum of potentially troubling moral problems: infantilization of those under robotic care; deception,

D.J. Gunkel (✉)
Department of Communication, Northern Illinois University, DeKalb, IL, USA
e-mail: dgunkel@niu.edu

S.P. van Rysewyk and M. Pontier (eds.), *Machine Medical Ethics*,
Intelligent Systems, Control and Automation: Science and Engineering 74,
DOI 10.1007/978-3-319-08108-3_10

especially with regards to individuals suffering from impaired or diminished mental/emotional capabilities [47]; and "the rights to privacy, personal liberty and social contact" [48, 268]. Others, like Robert and Linda Sparrow [52], question whether increasing involvement of robots in elder care (an area of research that the Sparrows argue is second only to military applications) would in fact be adequate to meet not just the looming demographic crisis of an aging population but the complex emotional and social needs of seniors. And Coeckelbergh [18] has focused on the concept of care itself, asking whether machines can be adequately designed and implemented to supply what, under normal circumstances, would constitute not just acceptable but "good care." In all these cases what is at issue is the well being and rights of human patients and the extent to which machines improve or adversely affect human flourishing. "The primary concern," as Borenstein and Pearson describe it, "is about how the existence of robots may positively or negatively affect the lives of care recipients" [13, 251].

Absent from much of the current literature, however, is a consideration of the other side of this issue, namely the moral status and standing of these machines. Unlike the seemingly cold and rather impersonal industrial robots that have been successfully developed for and implemented in manufacturing, transportation, and maintenance operations, home healthcare robots will occupy a unique social position and "share physical and emotional spaces with the user" [15, 408]. In providing care for us, these machines will take up residence in the home and will be involved in daily personal and perhaps even intimate interactions (i.e., monitoring, feeding, bathing, mobility, and companionship). For this reason, it is reasonable to inquire about the social status and moral standing of these technologies. How, for example, will human patients under the care of such mechanisms respond to these other entities? How *should* we respond to them? What are or what will be our responsibilities to these others—another kind of socially aware and interactive other? The following takes up and investigates this "machine question" by examining the moral standing of robots, and home healthcare robots in particular. Because there are a number of different and competing methods by which to formulate and decide this question, the chapter will not supply one definitive answer, but will consider a number of related moral perspectives that, taken together, add up to an affirmative response to the question concerning the rights of machines.

2 Default Setting

From a traditional philosophical perspective, the question of machine rights or machine moral standing not only would be answered in the negative but the query itself risks incoherence. "To many people," Levy [39] writes, "the notion of robots having rights is unthinkable" (393). This is because machines are assumed to be nothing more than instruments of human activity and have no independent moral status whatsoever. This common sense determination is structured and informed by the answer that is typically provided for the question concerning technology.

> We ask the question concerning technology when we ask what it is. Everyone knows the two statements that answer our question. One says: Technology is a means to an end. The other says: Technology is a human activity. The two definitions of technology belong together. For, to posit ends and procure and utilize the means to them is a human activity. The manufacture and utilization of equipment, tools, and machines, the manufactured and used things themselves, and the needs and ends that they serve, all belong to what technology is [30, 4–5].

According to Heidegger's analysis, the presumed role and function of any kind of technology, whether it be the product of handicraft or industrialized manufacture, is that it is a means employed by human users for specific ends. Heidegger termed this particular characterization "the instrumental and anthropological definition" and indicated that it forms what is considered to be the "correct" understanding of any kind of technological contrivance (5).

Under this clearly human-centered formulation, technology, no matter how sophisticated its design or operations, is considered to be nothing more than a tool or instrument of human endeavor. As Johnson [34] explains, "computer systems are produced, distributed, and used by people engaged in social practices and meaningful pursuits. This is as true of current computer systems as it will be of future computer systems. No matter how independently, automatic, and interactive computer systems of the future behave, they will be the products (direct or indirect) of human behavior, human social institutions, and human decision" (197). Understood in this way, machines—even those machines that are programmed to care for us—are not legitimate moral subjects that we need to care about. They are neither moral agents responsible for actions undertaken by or through their instrumentality nor moral patients, that is, the recipients of action and the subject of moral considerability. "We have never," as Hall [29] correctly points out, "considered ourselves to have moral duties to our machines, or them to us" (29).

On this account, the bar for machine moral standing appears to be impossibly high if not insurmountable. In order for a machine to have anything like "rights," it would need to be recognized as human or at least virtually indistinguishable from another human being in social situations and interactions. Although this has often been the subject of science fiction—consider, for example, Isaac Asimov's short story "The Bicentennial Man," in which the android "Andy" seeks to be recognized as legally human—it is not limited to fictional speculation, and researchers like Moravec [41], Brooks [11], and Kurzweil [37] predict human-level or better machine capabilities by the middle of the century. Although achievement of this remains hypothetical, the issue is not necessarily whether machines will or will not attain human-like capabilities. The problem resides in the anthropocentric criteria itself, which not only marginalizes machines but has often been instrumental for excluding other human beings. "Human history," Stone [53] argues, "is the history of exclusion and power. Humans have defined numerous groups as less than human: slaves, woman, the 'other races,' children and foreigners. These are the wretched ones who have been defined as stateless, personless, as suspect, as rightsless" (450).

Because of this, recent innovations have sought to disengage moral standing from this anthropocentric privilege and have instead referred matters to the generic concept "person." "Many philosophers," Kadlac [35, 422] argues, "have contended that there is an important difference between the concept of a person and the concept of a human being." One such philosopher is Peter Singer. "Person," Singer writes [50, 87], "is often used as if it meant the same as 'human being.' Yet the terms are not equivalent; there could be a person who is not a member of our species. There could also be members of our species who are not persons." In 2013, for example, India declared dolphins "non-human persons, whose rights to life and liberty must be respected" [20]. Likewise corporations are artificial entities that are obviously otherwise than human, yet they are considered legal persons, having rights and responsibilities that are recognized and protected by both national and international law [26]. And not surprisingly, there has been, in recent years, a number of efforts to extend the concept "person" to AI's, intelligent machines, and robots [27, 32, 42].

As promising as this innovation appears to be, however, there is little agreement concerning what makes someone or something a person, and the literature on this subject is littered with different formulations and often incompatible criteria [33, 35, 50, 51]. In an effort to contend with, if not resolve these problems, researchers often focus on the one "person making" quality that appears on most, if not all, the lists—consciousness. "Without consciousness," Locke [40, 170] famously argued, "there is no person." For this reason, consciousness is widely considered to be a necessary if not sufficient condition for moral standing, and there has been considerable effort in the fields of philosophy, AI, and robotics to address the question of machine moral standing by targeting the possibility (or impossibility) of machine consciousness [31, 54].

This determination is dependent not only on the design and performance of actual artifacts but also—and perhaps more so—on how we understand and operationalize the term "consciousness." Unfortunately there has been little or no agreement concerning this matter, and the concept encounters both terminological and epistemological problems. First, we do not have any widely accepted definition of "consciousness," and the concept, as Velmans [56, 5] points out "means many different things to many different people." In fact, if there is any agreement among philosophers, psychologists, cognitive scientists, neurobiologists, AI researchers, and robotics engineers regarding this matter, it is that there is little or no agreement, when it comes to defining and characterizing the term. To make matters worse, the difficulty is not just with the lack of a basic definition; the problem may itself already be a problem. "Not only is there no consensus on what the term *consciousness* denotes," Güzeldere [28, 7] writes, "but neither is it immediately clear if there actually is a single, well-defined '*the* problem of consciousness' within disciplinary (let alone across disciplinary) boundaries. Perhaps the trouble is not so much in the ill definition of the question, but in the fact that what passes under the term consciousness as an all too familiar, single, unified notion may be a tangled amalgam of several different concepts, each inflicted with its own separate problems."

Second, even if it were possible to define consciousness or come to some agreement (no matter how tentative or incomplete) as to what characterizes it, we still lack any credible and certain way to determine its actual presence in another. Because consciousness is a property attributed to "other minds," its presence or lack thereof requires access to something that is and remains inaccessible. "How does one determine," as Churchland [16, 67] famously characterized it, "whether something other than oneself—an alien creature, a sophisticated robot, a socially active computer, or even another human—is really a thinking feeling, conscious being; rather than, for example, an unconscious automaton whose behavior arises from something other than genuine mental states?" Although philosophers, psychologists, and neuroscientists throw considerable argumentative and experimental effort at this problem—and much of it is rather impressive and persuasive—it is not able to be fully and entirely resolved. Consequently, not only are we unable to demonstrate with any certitude whether animals, machines, or other entities are in fact conscious (or not) and therefore legitimate moral persons (or not), we are left with doubting whether we can, without fudging the account, even say the same for other human beings. And it is this persistent and irreducible difficulty that opens the space for entertaining the possibility of extending rights to other entities like machines or animals.

3 Bête-Machine

The situation of animals is, in this context, particularly interesting and important. Animals have not traditionally been considered moral subjects, and it is only recently that the discipline of philosophy has begun to approach the animal as a legitimate subject of moral concern. The crucial turning point in this matter is derived from a brief but influential statement provided by Bentham [7, 283]: "The question is not, Can they reason? nor Can they talk? but, Can they suffer?" Following this insight, the crucial issue for animal rights philosophy is not to determine whether some entity, like an animal can achieve human-level capacities with things like speech, reason, or consciousness; "the first and decisive question would be rather to know whether animals can suffer" [23, 27].

This change in perspective—from a standard agent-oriented to a non-standard patient-oriented ethics [25]—provides a potent model for entertaining the question of the moral standing and rights of machines. This is because the animal and the machine, beginning with the work of René Descartes, share a common ontological status and position, marked, quite literally in the Cartesian texts, by the hybrid term *bête-machine* [24]. Despite this essential similitude, animal rights philosophers have resisted efforts to extend rights to machines, and they demonize Descartes for even suggesting the association [44]. This exclusivity has been asserted and justified on the grounds that the machine, unlike an animal, is not capable of experiencing either pleasure or pain. Like a stone or other inanimate object, a robot would have nothing that mattered to it and therefore, unlike a mouse or other sentient creature, would not be a legitimate subject of moral concern, because "nothing that we can do to it could possibly make any difference to its welfare" [49, 9]. Although this argument sounds rather reasonable and intuitive, it fails for at least three reasons.

First, it has been practically disputed by the construction of various mechanisms that now appear to exhibit emotional responses or at least provide external evidence of behaviors that effectively simulate and look undeniably like pleasure or pain. As Derrida [23, 81] recognized, "Descartes already spoke, as if by chance, of a machine that simulates the living animal so well that it 'cries out that you are hurting it.'" This comment, which appears in a brief parenthetical aside in Descartes' *Discourse on Method*, had been deployed in the course of an argument that sought to differentiate human beings from the animal by associating the latter with mere mechanisms. But the comment can, in light of the procedures and protocols of animal rights philosophy, be read otherwise. That is, if it were indeed possible to construct a machine that did exactly what Descartes had postulated, that is, "cry out that you are hurting it," would we not also be obligated to conclude that such a mechanism was capable of experiencing pain? This is, it is important to note, not just a theoretical point or speculative thought experiment. Engineers have, in fact, constructed mechanisms that synthesize believable emotional responses [6, 9, 10, 55], like the dental-training robot Simroid "who" cries out in pain when students "hurt" it [36], and designed systems capable of evidencing behaviors that look a lot like what we usually call pleasure and pain. Although programming industrial robots with emotions—or, perhaps more precisely stated, the capability to simulate emotions—would be both unnecessary and perhaps even misguided, this is something that would be desirable for home healthcare robots, which will need to exhibit forms of empathy and emotion in order to better interact with patients and support their care.

Second, it can be contested on epistemologically grounds. Because suffering is typically understood to be subjective, there is no way to know exactly how another entity experiences unpleasant (or pleasant) sensations. Like "consciousness," suffering is also an internal state of mind and is therefore complicated by the problem of other minds. As Singer [49, 11] readily admits, "we cannot directly experience anyone else's pain, whether that 'anyone' is our best friend or a stray dog. Pain is a state of consciousness, a 'mental event,' and as such it can never be observed." The basic problem, then, is not whether the question "Can they suffer?" also applies to machines but whether anything that appears to suffer—human, animal, plant, or machine—actually does so at all. Furthermore, and to make matters even more complex, we may not even know what "pain" and "the experience of pain" is in the first place. This point is something that is taken up and demonstrated in Daniel Dennett's, "Why You Can't Make a Computer That Feels Pain" [22]. In this provocatively titled essay, published decades before the debut of even a rudimentary working prototype, Dennett imagines trying to disprove the standard argument for human (and animal) exceptionalism "by actually writing a pain program, or designing a pain-feeling robot" (191). At the end of what turns out to be a rather protracted and detailed consideration of the problem, he concludes that we cannot, in fact, make a computer that feels pain. But the reason for drawing this conclusion does not derive from what one might expect. The reason you cannot make a computer that feels pain, Dennett argues, is not the result of some technological limitation with the mechanism or its programming. It is a product of the fact that we remain unable to decide what pain is in the first place.

Third, all this talk about the possibility of engineering pain or suffering in order to demonstrate machine rights entails its own particular smoral dilemma. "If (ro)bots might one day be capable of experiencing pain and other affective states," Wallach and Allen [57, 209] write, "a question that arises is whether it will be moral to build such systems—not because of how they might harm humans, but because of the pain these artificial systems will themselves experience. In other words, can the building of a (ro)bot with a somatic architecture capable of feeling intense pain be morally justified and should it be prohibited?" If it were in fact possible to construct a robot that "feels pain" (however defined and instantiated) in order to demonstrate the moral standing of machines, then doing so might be ethically suspect insofar as in constructing such a mechanism we do not do everything in our power to minimize its suffering. Consequently, moral philosophers and robotics engineers find themselves in a curious and not entirely comfortable situation. If it were in fact possible to construct a device that "feels pain" in order to demonstrate the possibility of robot moral standing, then doing so might be ethically problematic or to put it another way, positive demonstration of "machine rights," following the moral innovations and model of animal rights philosophy, might only be possible by risking the violation of those rights.

4 Thinking Otherwise

Irrespective of how it is articulated, these different approaches to deciding moral standing focus on what Coeckelbergh [19] calls "(intrinsic) properties" (13). This method is rather straight forward and intuitive: "identify one or more morally relevant properties and then find out if the entity in question has them" [19, 14]. But as we have discovered, there are at least two persistent problems with this undertaking. First, how does one ascertain which exact property or properties are sufficient for moral status? In other words, which one, or ones, count? The history of moral philosophy can, in fact, be read as something of an on-going debate and struggle over this matter with different properties—rationality, speech, consciousness, sentience, suffering, etc.—vying for attention at different times. Second, once the morally significant property (or properties) has been identified, how can one be entirely certain that a particular entity possesses it, and actually possesses it instead of merely simulating it? This is tricky business, especially because most of the properties that are considered morally relevant tend to be internal mental or subjective states that are not immediately accessible or directly observable. In response to these problems, there are two alternatives that endeavor to consider and address things otherwise.

4.1 Machine Ethics

The first concerns what is now called Machine Ethics. This relatively new idea was first introduced and publicized in a 2004 AAAI paper written by Michael Anderson, Susan Leigh Anderson, and Chris Armen and has been followed by a number of dedicated symposia and publications [2, 3]. Unlike computer ethics, which is interested in the consequences of human behavior through the instrumentality of technology, "*machine ethics* is concerned," as characterized by Anderson et al. [1, 1], "with the consequences of behavior of machines toward human users and other machines." In this way, machine ethics both challenges the "human-centric" tradition that has persisted in moral philosophy and argues for a widening of the subject so as to take into account not only human action with machines but also the behavior of some machines, namely those that are designed to provide advice or programmed to make autonomous decisions with little or no human supervision. And for the Andersons, healthcare applications provide both the test case and opportunities for the development of working prototypes.

Toward this end, machine ethics takes an entirely functionalist approach to things. That is, it considers the effect of machine actions on human subjects irrespective of metaphysical debates about moral standing or epistemological problems concerning subjective mind states. As Anderson [4, 477] points out, the Machine Ethics project is unique insofar as it, "unlike creating an autonomous ethical machine, will not require that we make a judgment about the ethical status of the machine itself, a judgment that will be particularly difficult to make." Machine Ethics, therefore, does not necessarily deny or affirm the possibility of, for instance, machine personhood, consciousness, or sentience. It simply endeavors to institute a pragmatic approach that does not require that one first decide these questions *a priori*. It leaves these matters as an open question and proceeds to ask whether moral decision making is computable and whether machines can in fact be programmed with appropriate ethical standards for acceptable forms of social behavior.

This is a promising innovation insofar as it recognizes that machines are already making decisions and taking real-world actions in such a way that has an effect—and one that can be evaluated as either good or bad—on human beings and human social institutions. Despite this, the functionalist approach utilized by Machine Ethics has at least three critical difficulties. First, functionalism shifts attention from the cause of an action to its effects.

> Clearly relying on machine intelligence to effect change in the world without some restraint can be dangerous. Until fairly recently, the ethical impact of a machine's actions has either been negligible, as in the case of a calculator, or, when considerable, has only been taken under the supervision of a human operator, as in the case of automobile assembly via robotic mechanisms. As we increasingly rely upon machine intelligence with reduced human supervision, we will need to be able to count on a certain level of ethical behavior from them [1, 4]

The functionalist approach instituted by Machine Ethics derives from and is ultimately motivated by an interest to protect human beings from potentially hazardous machine decision-making and action. This effort, despite arguments to the contrary, is thoroughly and unapologetically anthropocentric. Although effectively opening up the community of moral subjects to other, previously excluded things, the functionalist approach only does so in an effort to protect human interests and investments. This means that the project of Machine Ethics does not differ significantly from computer ethics and its predominantly instrumental and anthropological orientation.

If computer ethics, as Anderson et al. [1] characterize it, is about the responsible and irresponsible use of computerized tools by human users, then their functionalist approach is little more than the responsible design and programming of machines by human beings for the sake of protecting other human beings.

Second, functionalism institutes, as the conceptual flipside and consequence of this anthropocentric privilege, what is arguably a slave ethic. "I follow," Coleman [21] writes, "the traditional assumption in computer ethics that computers are merely tools, and intentionally and explicitly assume that the end of computational agents is to serve humans in the pursuit and achievement of their (i.e., human) ends. In contrast to James Gips' call for an ethic of equals, then, the virtue theory that I suggest here is very consciously a slave ethic." For Coleman, computers and other forms of computational agents should, in the words of Bryson [12], "be slaves." Others, however, are not so confident about the prospects and consequences of this "Slavery 2.0." Concern over this matter is something that is clearly exhibited and developed in robot science fiction from *R.U.R.* and *Metropolis* to *Bladerunner* and *Battlestar Galactica*. But it has also been expressed by contemporary researchers and engineers. Brooks [11], for example, recognizes that there are machines that are and will continue to be used and deployed by human users as instruments, tools, and even servants. But he also recognizes that this approach will not cover all machines in all circumstances.

> Fortunately we are not doomed to create a race of slaves that is unethical to have as slaves. Our refrigerators work 24 hours a day 7 days a week, and we do not feel the slightest moral concern for them. We will make many robots that are equally unemotional, unconscious, and unempathetic. We will use them as slaves just as we use our dishwashers, vacuum cleaners, and automobiles today. But those that we make more intelligent, that we give emotions to, and that we empathize with, will be a problem. We had better be careful just what we build, because we might end up liking them, and then we will be morally responsible for their well-being. Sort of like children (195).

According to Brooks's analysis, a slave ethic will work, and will do so without any significant moral difficulties or ethical friction, as long as *we* decide to produce dumb instruments that serve human users as mere tools or extensions of our will. But as soon as the machines show signs, however minimal defined or rudimentary, that *we take* to be intelligent, conscious, or intentional, then everything changes. What matters here, it is important to note, is not the actually capabilities of the machines but the way *we* read, interpret, and respond to their actions and behaviors. As soon as we see what we think are signs of something like intelligence, intentionality, or emotion, a slave ethic will no longer be functional or justifiable.

Finally, even those seemingly unintelligent and emotionless machines that can legitimately be utilized as "slaves", pose a significant ethical problem. This is because machines that are designed to follow rules and operate within the boundaries of some kind of programmed restraint, might turn out to be something other than a neutral tool. Winnograd [60, 182–183], for example, warns against something he calls "the bureaucracy of mind," "where rules can be followed without interpretive judgments." Providing robots, computers, and other autonomous machines with functional morality produces little more than artificial bureaucrats—decision making mechanisms that can follow rules and protocols but have no sense of what they do or understanding of how their decisions might affects others. "When a person," Winnograd [60, 183] argues, "views his or her job as the correct application of a set of rules (whether human-invoked or computer-based), there is a loss of personal responsibility or commitment. The 'I just follow the rules' of the bureaucratic clerk has its direct analog in 'That's what the knowledge base says.' The individual is not committed to appropriate results, but to faithful application of procedures." Coeckelbergh [18, 236] paints an even more disturbing picture. For him, the problem is not the advent of "artificial bureaucrats" but "psychopathic robots." The term "psychopathy" refers to a kind of personality disorder characterized by an abnormal lack of empathy which is masked by an ability to appear normal in most social situations. Functional morality, Coeckelbergh argues, intentionally designs and produces what are arguably "artificial psychopaths"—robots that have no capacity for empathy but which follow rules and in doing so can appear to behave in morally appropriate ways. These psychopathic machines would "follow rules but act without fear, compassion, care, and love. This lack of emotion would render them non-moral agents—i.e., agents that follow rules without being moved by moral concerns—and they would even lack the capacity to discern what is of value. They would be morally blind" [18, 236].

4.2 Social Relational Ethics

An alternative to moral functionalism can be found in Coeckelbergh's [19] own work, where he develops an approach to moral status ascription that he characterizes as "social relational." By this, he means to emphasize the way moral status is not something located in the inner recesses or essential make-up of an individual entity but transpires through actually existing interactions and relationships situated between entities. This "relational turn," which Coeckelbergh develops by capitalizing on innovations in ecophilosophy, Marxism, and the work of Bruno Latour, Tim Ingold, and others, does not get bogged down trying to resolve the philosophical problems associated with the standard properties approach. Instead it recognizes the way that moral status is socially constructed and operationalized. Quoting the environmental ethicist Baird Callicot, Coeckelbergh [19, 110] insists that the "relations are prior to the things related." This almost Levinasian gesture is crucial insofar as it reconfigures the usual way of thinking. It is an anti-Cartesian and post-modern (in the best sense of the word) intervention. In Cartesian modernism the individual subject had to be certain of his (and at this time the subject was always gendered male) own being and essential properties prior to engaging with others. Coeckelbergh reverses this standard approach. He argues that it is the social that comes first and that the individual subject (an identity construction that is literally thrown under or behind), only coalesces out of the relationship and the assignments of rights and responsibilities that it makes possible.

This relational turn in moral thinking is clearly a game changer. As we interact with machines, whether they be pleasant customer service systems, medical advisors, or home healthcare robots, the mechanism is first and foremost situated and encountered in relationship to us. Morality, conceived of in this fashion, is not determined by a prior ontological determination concerning the essential capabilities, intrinsic properties, or internal operations of these other entities. Instead it is determined in and by the way these entities come to face us in social interactions. Consequently, "moral consideration is," as Coeckelbergh [17, 214] describes it, "no longer seen as being 'intrinsic' to the entity: instead it is seen as something that is 'extrinsic': it is attributed to entities within social relations and within a social context." This is the reason why, as Levinas [38, 304] claims, "morality is first philosophy" ("first" in terms of both sequence and status) and that moral decision making precedes ontological knowledge. Ethics, conceived of in this way, is about decision and not discovery [43, 691]. We, individually and in collaboration with each other (and not just those others who we assume are substantially like ourselves), decide who is and who is not part of the moral community; who, in effect, will have been admitted to and included in this first person plural pronoun.

This is, it is important to point out, not just a theoretical proposal but has been experimentally confirmed in a number of empirical investigations. The computer as social actor (CSA) studies undertaken by Byron Reeves and Clifford Nass and reported in their influential book *The Media Equation* [45], demonstrate that human users will accord computers social standing similar to that of another human person. This occurs, as Reeves and Nass demonstrate, as a product of the social interaction and irrespective of the actual ontological properties (actually known or not) of the machine in question [45, 21]. Similar results have been obtained by Christopher Bartneck et al. and reported in the paper "Daisy, Daisy, Give me your answer do! Switching off a Robot" [5], a title which refers to the shutting down of the HAL 9000 computer in Stanley Kubrick's *2001: A Space Odyssey*. In Bartneck et al.'s study, human subjects interacted with a robot on a prescribed task and then, at the end of the session, were asked to switch off the machine and wipe its memory. The robot, which was in terms of its programming no more sophisticated than a basic chatter bot, responded to this request by begging for mercy and pleading with the human user not to shut it down. As a result of this, Bartneck and company recorded considerable hesitation on the part of the human subjects to comply with the shutdown request. Even though the robot was "just a machine"—and not even very intelligent—the social situation in which it worked with and responded to human users, made human beings consider the right of the machine (or at least hesitate in considering this) to continued existence.

For all its opportunities, however, this approach to deciding moral standing otherwise is inevitably and unavoidably exposed to the charge of moral *relativism*—"the claim that no universally valid beliefs or values exist" (Ess 100, 204). To put it rather bluntly, if moral status is relational and open to different decisions concerning others made at different times for different reasons, are we not at risk of affirming an extreme form of moral relativism? One should perhaps answer this indictment not by seeking some definitive and universally accepted response (which would obviously reply to the charge of relativism by taking refuge in and validating its opposite), but by following Žižek's [61, 3] strategy of "fully endorsing what one is accused of." So yes, *relativism*, but an extreme and carefully articulated form of it. That is, a relativism that can no longer be comprehended by that kind of understanding of the term which makes it the mere negative and counterpoint of an already privileged universalism. Relativism, therefore, does not necessarily need to be construed negatively and decried, as Žižek [62, 79; 63, 281] himself has often done, as the epitome of postmodern multiculturalism run amok. It can be understood otherwise. "Relativism," Scott [46, 264] argues, "supposedly, means a standardless society, or at least a maze of differing standards, and thus a cacophony of disparate, and likely selfish, interests. Rather than a standardless society, which is the same as saying no society at all, relativism indicates circumstances in which standards have to be established cooperatively and renewed repeatedly." In fully endorsing this form of relativism and following through on it to the end, what one gets is not necessarily what might have been expected, namely a situation where anything goes and "everything is permitted" [14, 67]. Instead, what is obtained is a kind of ethical thinking that turns out to be much more responsive and responsible in the face of others.

5 Conclusion

In November of 2012, General Electric launched a television advertisement called "Robots on the Move." The 60 second spot, created by Jonathan Dayton and Valerie Faris (the husband/wife team behind the 2006 feature film *Little Miss Sunshine*), depicts many of the iconic robots of science fiction traveling across great distances to assemble before some brightly lit airplane hangar for what we are told is the unveiling of some new kind of machines—"brilliant machines," as GE's tagline describes it. And as we observe Robby the Robot from *Forbidden Planet*, KITT the robotic automobile from *Knight Rider*, and Lt. Commander Data of *Star Trek: The Next Generation* making their way to this meeting of artificial minds, we are told, by an ominous voice over, that "the machines are on the move."

Although this might not look like your typical robot apocalypse (vividly illustrated in science fiction films and television programs like *Terminator*, *The Matrix Trilogy*, and *Battlestar Galactica*), we are, in fact, in the midst of an invasion. The machines are, in fact, on the move. They may have begun by displacing workers on the factory floor, but they now actively participate in many aspects of social life and will soon be invading and occupying places in our homes. This invasion is not some future possibility coming from a distant alien world. It is here. It is now. And resistance appears to be futile. What matters for us, therefore, is how we decide to respond to this opportunity/challenge. And in this regard, we will need to ask some important but rather difficult questions: At what point might a robot, or other autonomous system be held fully accountable for the decisions it makes or the actions it deploys? When, in other words, would it make sense to say "It's the robot's fault"? Likewise, at what point might we have to consider seriously extending rights—civil, moral, and legal standing—to these socially aware and interactive devices that will increasingly come to serve and care for us, our children, and our aging parents? When, in other words, would it no longer be considered non-sense to suggest something like "the rights of robots"?

In response to these questions, there appears to be at least two options, neither of which are entirely comfortable or satisfactory. On the one hand, we can respond as we typically have, treating these mechanisms as mere instruments or tools. Bryson makes a reasonable case for this approach in her essay "Robots Should be Slaves": "My thesis is that robots *should* be built, marketed and considered legally as slaves, not companion peers" [12, 63] (emphasis added). Although this moral imperative (marked, like all imperatives, by the verb "should") might sound harsh, this line of argument is persuasive, precisely because it draws on and is underwritten by the instrumental theory of technology—a theory that has considerable history and success behind it and that functions as the assumed default position for any and all considerations of technology. This decision—and it is a decision, even if it is the default setting—has both advantages and disadvantages. On the positive side, it reaffirms human exceptionalism in ethics, making it absolutely clear that it is only the human being who possess rights and responsibilities. Technologies, no matter how sophisticated, intelligent, and influential, are and will continue to be mere tools of human action, nothing more. But this approach, for all its usefulness, has a not-so-pleasant downside. It willfully and deliberately produces a new class of instrumental servants or slaves and rationalizes this decision as morally appropriate and justified. In other words, applying the instrumental theory to these new kinds of domestic healthcare machines, although seemingly reasonable and useful, might have devastating consequences for us and others.

On the other hand, we can decide to entertain the possibility of rights and responsibilities for machines just as we had previously done for other non-human entities, like animals [49], corporations [26], and the environment [8]. And there is both moral and legal precedent for this transaction. Once again, this decision sounds reasonable and justified. It extends moral standing to these other socially active entities and recognizes, following the predictions of Wiener [59], that the social situation of the future will involve not just human-to-human interactions but relationships between humans and machines. But this decision also has significant costs. It requires that we rethink everything we thought we knew about ourselves, technology, and ethics. It requires that we learn to think beyond human exceptionalism, technological instrumentalism, and all the other *-isms* that have helped us make sense of our world and our place in it. In effect, it calls for a thorough reconceptualization of who or what should be considered a legitimate moral subject and risks involvement in what is often considered antithetical to clear moral thinking—relativism.

In any event, how we respond to the opportunities and challenges of this *machine question* will have a profound effect on the way we conceptualize our place in the world, who we decide to include in the community of moral subjects, and what we exclude from such consideration and why. But no matter how it is decided, it is a decision—quite literally a cut that institutes difference and makes a difference. And, as Whitbey [58] correctly points out, the time to start thinking about and debating these issues is now…if not yesterday.

References

1. Anderson M, Anderson SL, Armen C (2004). Toward machine ethics. In: American association for artificial intelligence. http://www.aaai.org/Papers/Workshops/2004/WS-04-02/WS04-02-008.pdf
2. Anderson M, Anderson SL (2006) Machine ethics. IEEE Intell Syst 21(4):10–11
3. Anderson M, Anderson SL (2011) Machine ethics. Cambridge University Press, Cambridge
4. Anderson SL (2008) Asimov's "three laws of robotics" and machine metaethics. AI Soc 22(4):477–493
5. Bartneck C, van der Hoek M, Mubin O, Al Mahmud A (2007) Daisy, daisy, give me your answer do! Switching off a robot. In: Proceedings of the 2nd ACM/IEEE international conference on human-robot interaction, Washington, DC, pp 217–222
6. Bates J (1994) The role of emotion in believable agents. Commun ACM 37:122–125
7. Bentham J (2005) An introduction to the principles of morals and legislation. In: Burns JH, Hart HL (eds). Oxford University Press, Oxford
8. Birch T (1993) Moral considerability and universal consideration. Environ Ethics 15:313–332
9. Blumberg B, Todd P, Maes M (1996) No bad dogs: ethological lessons for learning. In: Proceedings of the 4th international conference on simulation of adaptive behavior (SAB96). MIT Press, Cambridge, MA, pp 295–304
10. Breazeal C, Brooks R (2004) Robot emotion: a functional perspective. In: Fellous JM, Arbib M (eds) Who needs emotions: the brain meets the robot. Oxford University Press, Oxford, pp 271–310
11. Brooks R (2002) Flesh and machines: how robots will change us. Pantheon, New York
12. Bryson J (2010) Robots Should be Slaves. In: Wilks Y (ed) Close engagements with artificial companions: key social, psychological, ethical and design issues. John Benjamins, Amsterdam, pp 63–74
13. Borenstein J, Pearson Y (2012) Robot caregivers: ethical issues across the human lifespan. In: Lin P, Abney K, Bekey GA (eds) Robot ethics: the ethical and social implications of robotics. MIT Press, Cambridge, MA, pp 251–265
14. Camus A (1983) The myth of Sisyphus, and other essays (trans: O'Brien J). Alfred A. Knopf, New York
15. Cerqui D, Arras KO (2001) Human beings and robots: towards a symbiosis? In: Carrasquero J (ed) Post-conference proceedings PISTA 03 on a 2000 people survey (Politics and information systems: technologies and applications), pp 408–413
16. Churchland PM (1999) Matter and consciousness (revised edition). MIT Press, Cambridge, MA
17. Coeckelbergh M (2010) Healthcare, capabilities, and AI assistive technologies. Ethical Theor Moral Pract 13:181–190
18. Coeckelbergh M (2010) Moral appearances: emotions, robots, and human morality. Ethics Inf Technol 12(3):235–241
19. Coeckelbergh M (2012) Growing moral relations: critique of moral status ascription. Palgrave MacMillan, New York
20. Coelho S (2013) Dolphins gain unprecedented protection in India. Deutsche Welle. http://dw.de/p/18dQV
21. Coleman KG (2001) Android arete: toward a virtue ethic for computational agents. Ethics Inf Technol 3:247–265
22. Dennett DC (1998) Brainstorms: philosophical essays on mind and psychology. MIT Press, Cambridge, MA
23. Derrida J (2008) The animal that I therefore am (trans: Wills D). Fordham University Press New York
24. Descartes R (1988) Selected philosophical writings (trans: Cottingham J, Stoothoff R, Murdoch D). Cambridge University Press, Cambridge

100Ess, C (1996) The political computer: democracy, cmc, and habermas. In: Ess, C, (ed) Philosophical perspectives on computer-mediated communication. SUNY Press, Albany, NY, pp 197-232.
25. Floridi L, Sanders JW (2004) On the morality of artificial agents. Minds Mach 14:349–379
26. French P (1979) The corporation as a moral person. Am Philos Q 16(3):207–215
27. Gunkel DJ (2012) The machine question: critical perspectives on AI, robots, and ethics. MIT Press, Cambridge, MA
28. Güzeldere G (1997) The many faces of consciousness: a field guide. In: Block N, Flanagan O, Güzeldere G (eds) The nature of consciousness: philosophical debates. MIT Press, Cambridge, MA, pp 1–68
29. Hall JS (2011) Ethics for machines. In: Anderson M, Anderson SL (eds) Machine ethics. Cambridge University Press, Cambridge, pp 28–44
30. Heidegger M (1977) The question concerning technology and other essays (trans: Lovitt W). Harper & Row, New York
31. Himma KE (2009) Artificial agency, consciousness, and the criteria for moral agency: what properties must an artificial agent have to be a moral agent?". Ethics Inf Technol 11(1):19–29
32. Hubbard FP (2011) "Do androids dream?" Personhood and intelligent artifacts. Temple Law Rev 83:101–170
33. Ikäheimo H, Laitinen A (2007) Dimensions of personhood: editors' introduction. J Conscious Stud 14(5–6):6–16
34. Johnson DG (2006) Computer systems: moral entities but not moral agents. Ethics Inf Technol 8:195–204
35. Kadlac A (2009) Humanizing personhood. Ethical Theor Moral Pract 13(4):421–437
36. Kokoro LTD (2009). http://www.kokoro-dreams.co.jp/
37. Kurzweil R (2005) The singularity is near: when humans transcend biology. Viking, New York
38. Levinas E (1969) Totality and infinity: an essay on exteriority (trans: Lingis A). Duquesne University Press, Pittsburgh, PA
39. Levy D (2008) Robots unlimited: life in virtual age. A K Peters, Wellesley, MA
40. Locke J (1996) An essay concerning human understanding. Hackett, Indianapolis, IN
41. Moravec H (1988) Mind children: the future of robot and human intelligence. Harvard University Press, Cambridge, MA
42. Peterson S (2012) Designing people to serve. In: Lin P, Abney K, Bekey GA (eds) Robot ethics: the ethical and social implications of robotics. MIT Press, Cambridge, MA, pp 282–298
43. Putnam H (1964) Robots: machines or artificially created life? J Philos 61(21):668–691
44. Regan T (1983) The case for animal rights. University of California Press, Berkeley, CA
45. Reeves B, Nass C (1996) The media equation: how people treat computers, television, and new media like real people and places. Cambridge University Press, Cambridge
46. Scott RL (1967) On viewing rhetoric as epistemic. Cent States Speech J 18:9–17
47. Sharkey N, Sharkey A (2012) Granny and the robots: ethical issues in robot care for the elderly. Ethics Inf Technol 14(1):27–40
48. Sharkey N, Sharkey A (2012) The rights and wrongs of robot care. In: Lin P, Abney K, Bekey GA (eds) Robot ethics: the ethical and social implications of robotics. MIT Press, Cambridge, MA, pp 267–282
49. Singer P (1975) Animal liberation: a new ethics for our treatment of animals. New York Review Book, New York
50. Singer P (1999) Practical ethics. Cambridge University Press, Cambridge
51. Smith C (2010) What is a person? Rethinking humanity, social life, and the moral good from the person up. University of Chicago Press, Chicago
52. Sparrow R, Sparrow L (2010) In the hands of machines? The future of aged care. Mind Mach 16:141–161
53. Stone CD (1972) Should trees have standing? Toward legal rights for natural objects. South Calif Law Rev 44:450–492

54. Torrance S (2008) Ethics and consciousness in artificial agents. AI Soc 22:495–521
55. Velásquez JD (1998) When robots weep: emotional memories and decision-making. In: Proceedings of AAAI-98. AAAI Press, Menlo Park, CA
56. Velmans M (2000) Understanding consciousness. Routledge, London, UK
57. Wallach W, Allen C (2009) Moral machines: teaching robots right from wrong. Oxford University Press, Oxford
58. Whitbey B (2008) Sometimes it's hard to be a robot: a call for action on the ethics of abusing artificial agents. Interact Comput 20(3):326–333
59. Wiener N (1954) The human use of human beings. Da Capo, New York
60. Winnograd T (1990) Thinking machines: can there be? Are we?". In: Partridge D, Wilks Y (eds) The foundations of artificial intelligence: a sourcebook. Cambridge University Press, Cambridge, pp 167–189
61. Žižek S (2000) The fragile absolute or, why is the Christian legacy worth fighting for?. Verso, New York
62. Žižek S (2003) The puppet and the dwarf: the perverse core of Christianity. MIT Press, Cambridge, MA
63. Žižek S (2006) The parallax view. MIT Press, Cambridge, MA

Machine Medical Ethics and Robot Law: Legal Necessity or Science Fiction?

Rob van den Hoven van Genderen

I pity inanimate objects...
The frustrations of being inanimate
Maybe it's better that way
the fewer the moving parts
the less there is to go wrong

Godley & Crème (Freeze frame).

Abstract The law is developed by human beings, for human beings, in legal terms, for "natural persons". But during the development of the law in the long journey from the Roman legal system until our modern law system many things have changed. Natural persons are not the only players in the legal system of today. Large and smaller enterprises, organizations and state entities are performing all kinds of legal acts, as legal persons they can be held liable for the things they do. What about intelligence machines that can perform autonomous or semi-autonomous tasks? What about an advanced operating system where human control is hardly noticeable? What about drones operating independently? Are they liable for their acts and, most important, their mistakes? What about the laser cut that went wrong because of a disturbance in the internet that the surgeon did not control? Is the man behind the screen always the responsible party, even if he is not behind the screen? This article will discuss these ongoing questions.

Prologue: Dichotomy
Law was formed by human beings in an enduring evolution from the Roman Empire until contemporary internet society. Law is applicable to all human beings ("natural persons") and all constructions that are made by them.

R. van den Hoven van Genderen (✉)
Center for Law and Internet, University of Amsterdam, Amsterdam, The Netherlands
e-mail: r.vanden.hovenvangenderen@vu.nl

S.P. van Rysewyk and M. Pontier (eds.), *Machine Medical Ethics*,
Intelligent Systems, Control and Automation: Science and Engineering 74,
DOI 10.1007/978-3-319-08108-3_11

Organizational structures and companies can have legal personality as well, though legal actions are always done by natural persons, in name of the legal structure. Legal persons, natural and non-natural are subjected to the total national and international law systems, private law, public law as criminal law and the law on human rights.

In the long history of law, natural persons always bear the rights and responsibility of their actions. They are held responsible for all legal acts and acts with legal consequences that have been committed under their responsibility. This applies to the doing of those acting under their responsibility as well, such as animals and under aged children, but also inanimate objects under their responsibility. Machines and persons under their supervision (e.g., personnel) are also within the legal sphere of responsibility of natural persons.

Could the evolution of law systems extend to machines that are performing tasks in a more independent way? Could a machine that functions on its own developed instructions, a real artificial intelligence, be considered a new form of legal person with all legal consequences and responsibilities as such? Or will there always be a natural legal person involved that will be held responsible for the acts of this artificial intelligence being? And who will that be: the "user" or the developer and/or producer of this "machine being"?

These questions will be discussed and maybe answered in this chapter where I will describe the evolution of legal control and responsibility of the user of simple instruments to the user of automated systems and beyond.

1 Introduction

Must we pity inanimate objects or must we give them rights? They certainly will not pity us. They do not have a soul, whatever that may be. According to the *Catechism of the Catholic Church*, and they can be seen as the experts on this, "soul" is defined as follows: "The spiritual principle of human beings. The soul is the subject of human consciousness and freedom" [4, pp. 362–368]. This freedom of decision-making is the ethical and legal background of the responsibility we have as natural beings to be held responsible for decisions and acts as human beings, in a legal sense referred to as natural persons. Natural persons are sovereign in their decision-making and therefore legally responsible for their acts.

In *Les six livres de la Republique*, Bodin [2] claimed that sovereignty resides in a single individual. However, this sovereignty can be transferred to other "legal beings", such as the State, a company or other organizational entity. These legal entities are to be considered "legal persons", with the power to make decisions with legal effects. They also can be the subject of rights that are conferred on them as a genuine "legal subject".

The big question is, of course, if autonomous or semi-autonomous machines ought to be conceived as the bearer of rights and invested with the power to independently act as legal persons? If so, their deeds could also result in liability. Or ought there to be a natural person that will be the ultimate bearer of rights and legal responsibility?

2 Natural Legal Personhood

Is there a difference between the natural person Rob van den Hoven van Genderen and the legal person Rob van den Hoven van Genderen? There is and there is not. Of course, there is the physical person that can move or feel pain, love and run around. A legal person cannot do that. But the legal person, being a natural person, can be the subject of rights and can act with legal consequences. There is no question of which of these persons is fictitious or natural. One is of flesh and blood, but inherently this person is capable of performing legal acts with legal consequences. According to Dutch Civil Law (Burgerlijk Wetboek, BW), in the first book of the Civil Law, Article 1, it states that everyone who resides in the Netherlands is the bearer of civil rights, including personal rights that only can be borne by natural persons. Natural persons can get married to another natural person or enter a co-habitation contract. They can have children who by natural birth can descend from other natural persons. But they can also be adopted or partly naturally descend from insemination. They can vote for the other natural person in elections and can be voted for on these elections. They can join a political party or a church. Further, they will be the subject of human rights, have the right to life, privacy, freedom of speech, education or religion. Natural persons can be put in prison when they have acted against the law.

Of course, I can elaborate on the evolution of the legal potential of natural persons in historical context. In this respect, I could refer to the fact that a slave in the Roman Empire was not a natural person. The slave was a valuable asset, but not a natural person that could perform tasks, even legal acts for his or her master. They could even posses a money belt to make payments for their masters if they were trusted enough to do this. I could go on, deliberating the long tradition of having slaves, not being accepted as natural or legal persons until the end of the 19 century. Also, I could refer to the fact that colored people were only semi-natural persons in the United States and South Africa until the end of the 20th century. I could even point to the fact that married women were considered natural persons but only in the sense that could easily compared with Roman slaves, not being legally competent to act with full legal responsibility, based on the Code Napoleon of 1904. This situation lasted until 1956 in the Netherlands. Only then, females resident in the Netherlands were considered to have full legal competence to act with all possible legal effect. Even now, there are many parts on the globe where adolescent and adult females do not have (full) legal capacity, not being allowed to buy, drive or even go to school. I could, but I will not.

Instead, I will shed some light below on the difference between natural persons and legal persons and their difference in legal capacity to gain entrance to the legal responsibility of the different actors in society.

3 Legal Persons and Natural Persons

In the early 20th century, big enterprises and multinational companies needed the capabilities to buy and sell, to perform legal acts. At the time, it was felt that this should not be based on exclusive legal theories but had to be embedded in social, economic and other relevant theories. Economic progress in society thus prompted social discussion of the essence of the natural person in terms of "real personality", thus:

The question is at bottom not one on which law and legal conceptions have the only or the final voice: it is one which law shares with other sciences, political science, ethics, psychology, and metaphysics [7, p. 90].

And what is a real or natural person? Do we use distinctions on the basis of free will and intelligence? If a natural person is not capable of independently performing legal acts with legal effect we place them under curatorship, if they are not legally of age (in the Netherlands, 18 years). Natural persons are not to be trusted to perform all acts with legal effects. This is not an absolute line though. They can buy a sandwich or even a bike, but limits will be set to buying a car or a house. Parents of underage persons have to support them is such cases and represent them. But even within this system, there are cultural and national differences. Coming of age in both the Netherlands and the United States is set at 18 years, but an "adult" person in the U.S. is not allowed to buy alcoholic beverages,[1] but may drive a car at 16 or buy a gun at 14 in Montana. So, legal norms are not absolute for natural persons. Should this fact be different for legal persons?

4 Unnatural Legal Personhood

A legal person can also be a corporation or similar entity. A natural person cannot be a corporation, but can represent a corporation to act with legal effect. In "civilized" societies, it was common to use the entity of legal person for a long time. In ancient Egyptian society, it was common to use the structure of legal foundations to maintain temples [11]. In Roman civilization, there were several entities with legal personality, the *universitates personarum,* being State representations, corporations and societies [8]. A very famous Dutch international company, the first multinational was the United East Indies Company (VOC), established in 1602. In the United States, the Bill of Rights accords legal guarantees to corporations. Mayer [9] relates this development in the United States to the development of equal treatment on grounds of the 14th Amendment. The Supreme Court's most renowned decisions, however, were in the 1880s and 1890s, holding that

[1] In the U.S. there are differences. For instance, a person in Ohio, Nevada or Tennessee can become an adult after graduation from high school, whichever comes first.

corporations are persons for purposes of the Fourteenth Amendment equal protection and due process. In 1926, John Dewey stated in the Yale Journal of Law that, "The Corporation is a right-and-duty-bearing unit. Not all the legal propositions that are true of a man will be true of a corporation" [5, p. 656].

Concerning a legal person, property rights are equal with an individual's right, unless the law proves otherwise. One could say that a legal person, as an individual, has a legal "real personality" to take part in entering legal relations. A legal person can even go to court if summoned to go by another party, or when her rights are damaged. Corporations have a legal personality to act in a legal sense. What they do can have genuine legal effect. Although it is a legal fiction to give organizations and other entities legal entitlement to act as if they are natural persons, they can act in a legal way that will serve their interests and purposes. In this way it is not fictitious, at least not within a society that is based on law.

Legal systems vary, but the essence of legal persons is not fundamentally different, so I will use these examples. My purpose is to get clearer about the legal standing of intelligent machines.

5 Different Legal Persons

A major distinction in the concept of legal personhood is the separation between "public legal persons" and "private legal persons." Public legal persons have a public responsibility and legal personality, and can include governmental and other state initiated entities like provinces and municipalities, and other state organized public entities such as energy organizations or independent regulators. They all have a defined purpose to represent the government or certain public interests. On top of that, they have to be initiated by an act of law.

A kind of intermediate position is given to Churches and certain social entities as trade unions, charity organizations and political parties. They also have a defined purpose that requires legal personality.

Private legal persons are companies that are based on a commercial or idealistic basis. They have a defined purpose too. This can vary from making a profit to representing people to pursuing a goal on idealistic or ethical grounds, or human or animal rights. The actual legal specification varies too. Examples are companies and corporations under limited Liability (Ltd) to Euro Bv or foundations and funds that are not allowed to make profits. They must be initiated by a notorial deed. Private legal persons can be small, a shop, or a one-man (or no-man) consultancy or "letter box firm" to a dash tax in the Caribbean. They can be big too, like multinationals as Heineken, Microsoft or Apple.

In essence, they all have the same legal powers and responsibilities. They can be legal subjects like a natural person. The main difference is that non-natural legal persons can also be an object of law: they can be sold, ended or dismembered in a legal sense. A natural person cannot be ended in a legal way, although some of their rights can be withdrawn. For instance, they can lose their nationality if they join a foreign army. And of course, they can lose their legal competence if they lose their sanity and are not capable of acting in a legal sense. In this case, they will be placed under curatorship.

6 Legal Objects

Legal objects are goods, services or articles which can be the underlying object of rights and obligations. Objects can never be carriers of rights and obligations as a legal person. The legal object concerns particular goods, products and services, but also more abstract notions as the "organization", or "corporation". These last two can, in turn, be legal persons who can perform legal acts. This special construction is also described as a whole of active and passive property law elements. The liability of a legal person, though, will also apply to the director or board members. Being natural persons, it is assumed that at the time of the creation of liability of the legal persons they had the responsibility for acting in a legal way for the legal person.

7 Legal Acts of Persons, and the Legal Acts

Why is it so important to have a legal personality? As I have tried to make clear, legal persons can perform legal acts. Those acts have legal effects. They change the legal circumstance. By buying products or services, the legal person creates new legal circumstances. Money is paid to transfer the property from one hand to the other. The house owner changes, the owner of a company changes. Also, the buying of a sandwich is an agreement with legal effect. One party pays money and the other party will become the owner and consumer of the sandwich. These acts all had an intention to change the legal circumstances. Those legal acts can also be without an intended purpose or even purposively illegal or without an intended purpose. If legal persons act in a criminal way, they can be convicted on the basis of their crimes. If they create damage they will be held liable for the costs.

But who is the person that will be held liable or put into prison in our world? As with under-aged children or animals, parents or owners can be held liable for their acts; they have the responsibility. Companies and governments can also be held legally responsible for their acts. The interesting part of all this is that all these acts have to be performed by natural persons. Companies and governments are represented by natural persons, directors or government officials, who have to be the visible and trusted contacts to perform the acts, sign the agreement, or held to be liable for damages. If it comes to a legal procedure, a court case, then the representatives of the board will be subpoenaed or summoned to explain the (policy) decisions that have been made. The Minister or other governmental representative has to explain these decisions and legal acts to parliament or other democratic controls or even before the court if it concerns liability or criminal cases. So, the representative performing the legal act will always consist of a group of or single identifiable natural person, a board of natural persons, or civil servant, or the politically responsible representative and recognizable natural person.

8 Machines and Instruments

Until the explosion of advanced information technology (IT) in society, there was no question what would be legally responsible for acts with legal effect. It was clear who acted in a physical or legal sense or committed crimes. That a natural person used a machine or instrument was irrelevant. A surgeon who used a knife to make an incision to operate and made a mistake could not blame the knife or the producer of the knife. In times of war, the canon or the tank or even the producer could not be held responsible for the acts of war. The last in command though could be held responsible for his acts and possible war crimes. But what happens if those machines are no longer instructed or steered by natural persons? What if they are collecting information that will determine their functioning without human intervention? What if a drone is designed to recognize imminent danger and act by destroying this threat without someone using a joy stick (what's in a word) anymore? What if the surgeon is not doing the operation himself, but leaves the performing of the operation to an advanced data-fed laser instrument that has collected all medical data itself including the patient documentation? Or, if the computer decides what medicines a patient requires on basis of the patient records in the database?

In a less dramatic environment, this question is relevant: Automated systems can give instructions to crawl over the web trying to collect relevant information, going beyond the initial command to act. Computers and search engines can act on the internet without recognizable human instructions and can be received by comparable systems. Those systems can accept the instruction released by the "crawler". This can be a simple request to collect certain information, but could also be an instruction to make a reservation for a restaurant or performance of "Death in Venice" or a dance party. This could even be based on the interest profile of the natural person who gave a general instruction to arrange his weekend agenda.

These actions of an automated system can have legal effects. The search system, as an advanced bot, meets other bots and interchanges a set of constructions that can result in an agreement to reserve a certain place or even buy a product or service. There will be a possible electronic agreement that will be accepted by both electronic "parties" without any intervention or even confirmation by a natural person.

These kinds of actions fall within the legal context of the E-commerce Directive of 2000,[2] but are not prepared for this kind of autonomous system. The E-commerce directive, which is accepted and integrated in all European Member States, is directed to create legal certainty in the electronic system within the internal market. It is focused on the trustworthiness and identifiability of the party that is selling content, services or product to the consumer. As stated in consideration 7 of the Directive: "In order to ensure legal certainty and consumer confidence, this Directive must lay down a clear and general framework to cover certain legal aspects of electronic commerce in the internal market" [6].

In this context, it is important to identify the company and responsible persons that are "selling" the products and service. This also includes the physical location of the company, to find them if something goes wrong. The directive covers the relation between consumers and business parties but does not take into account developments where there is no easily identifiable party, such as an automated machine, at least for a part of their legal process and agreeing on the content of the agreement. It refers to the obligations, the "information duties" of a specific identifiable located party.[3]

From the perspective of protecting (consumer) parties who purchase services and products by electronic means, this makes sense in the historical perspective of the year 2000, but "times they are a-changin". Crawlers and all kinds of search or selection programs and smart machines are capable of arranging pre-contractual agreements, at least to provide information that can result in legal acts which create legal effects. These effects can also result in liability. But *who* or *what* will be liable? The smart machine? The operating laser device? The autonomous system that functions on basis of processed data that the device collected?

9 Legal Responsibility of Machines?

Returning to the operating device that functions on the basis of processed data from "big data" and specified data resources where no natural person is involved, except for the patient—what is its legal status? Is there a distinction between an autonomous machine as independent actor and the use of this system as an *instrument*? They perform activities that will have legal effects. Legal acts are always performed by legal subjects, being legal persons. Automated systems, electronic or otherwise, autonomous or not, are increasingly used to participate in diverse kinds of relations within our global society. The fact that these machines and devices can act autonomously and create changes in legal positions will ultimately come back

---

[2] Directive 2000/31/EC of the European Parliament and of the Council of 8 June 2000 on certain legal aspects of information society services, in particular electronic commerce, in the Internal Market ('Directive on Electronic Commerce').

[3] These obligations are integrated in Dutch law in article 3:15 Civil Code.

to the user of the system (see [12]). The machines cannot perform as a legal person with the same legal position and consequences of their acting.

But what is the difference between the representative in a human form, the natural person, and the machine representative, being a crawling program or software agent, reserving a seat for a play or buying a book on Amazon? I think we must make a distinction between the different phases of the legal process of buying and the performance of the (semi) autonomous system.

First, the whole process is always initiated by an individual or group of natural persons. They have instructed the machine, being a search engine or an advance weapon system. If a whole intelligence organization is placing spyware in our global information and communication system, we cannot hold the software program responsible, even if it is called PRISM.

So, in the command phase, the natural person or group of persons is or are the responsible actor(s). The difference in functional execution is not relevant. The use of drones for delivering packages to a client or rockets to a perceived enemy does not make a legal difference.

Apparently, no one has ever considered making the pen that puts the ink on the contract responsible for signing and executing the contract.[4] This also accounts for all producers of products that are used to commit a crime or products that create damage. Of course, a producer of products or services that are defective or unsound, or concerning services not according to agreed expectations, is liable for damages, is appropriate according to the level of involvement of the failure.

A natural person with a free will though, even in representing someone else, could be held responsible for his acts. The final legal effect can be held against him in the case of liability. Also, the person who has given the mandate or instruction can be liable for the damage his representative has done. In the case of positive or negative legal action though, the ultimate legal effect will return to the initial instructing party and only, concerning negative effect, will be diminished if there is a reason for disculpation (for instance, if his representative went berserk). In the last case, there could be an analogy with automated machines. A non-intentional disturbance of the system could diminish the liability of the initiating natural person as well, using this system.

Agreeing with Voulon [12], I think that any legal effect that has been created by an autonomous as well as less autonomous system has to be attributed to the legal or natural person that has made the choice to deploy the system in his service. One could doubt the level of liability of the legal or natural person relating to the degree of control one has over the autonomous system. This though would only be the case concerning liability and accountability to the legal or natural person. However, the failure or mistake of the autonomous system can be of consequence concerning the accountability of the legal actor. The liability though, will never be attributed to the machine as such because it can never have a legal responsibility as such.

[4] Although the weapon industry and cigarette industry are blamed for killing people, this is more based on the responsibility of the industry and the people who control the companies.

10 I Robot, Concluding

Although an autonomous machine or robot, having an independent intelligence and phenomenal consciousness would not create in my view a legal personality with the rights and obligations vested in natural and legal persons. There even wouldn't be an intermediate or sui generis norm for autonomous robots. One could imagine that there will be certain amendments in the existing law to create a practical system in representation of (semi)autonomous machines on behalf of the *initial legal actor*, *natural* or *legal person*. Those changes in law will be instrumental though, rather in the way that the "rather long" discussion of equality between written documents and electronic documents led to the inevitable acceptance of the electronic signature.

Even if a machine passes the Turing test, this would not create legal responsibilities. The functioning of the machine can have legal effects, certainly in representing the legal will or even derived legal will of the initial legal actor. Most important in legal relations, legal communications, and legal transactions is that there will be a trusted identification of the autonomous functioning entity or system in such a way that it will fit within the increasing autonomous functioning system within our global society.

It is essential that we, as natural human beings, keep control over the system. I assume that we do not want to be confronted with autonomous machines, collecting all kinds of personal information to be used for their own purposes. We are better to use our electronic or technology based servants to assist us in the practical executions of our tasks. The more intelligent the system the more trustworthy will be its functionality.

In closing, it is prudent to keep in mind Asimov's [1] Three Laws of Robotics:

First Law: A robot may not injure a human being, or, through inaction, allow a human being to come to harm

Second Law: A robot must obey the orders given it by human beings except where such orders would conflict with the First Law.

Third Law: A robot must protect its own existence as long as such protection does not conflict with the First or Second Law.

It should now be clear that these are functional instructions and are legally enforced by natural persons.

References

1. Asimov I (2004) I-Robot. Bantam Books, New York
2. Bodin J (1955) Les Six Livres de La Republique (Trans: Tooley MJ). Blackwell, Oxford
3. Boonk M (2013) Zeker over zoeken? Dissertation, University of Amsterdam
4. Catholic Church, Paul II PJ (1995) Catechism of the catholic church. Christus Rex et Redemptor Mundi

5. Dewey J (1926) The historic background of corporate legal personality. Yale Law J 35:655–673
6. E-commerce Directive (2000) http://eur-lex.europa.eu/legal-content/EN/TXT/?uri=CELEX:32000L0031
7. Geldart WM (1911) Legal Personality. Legal Quart Rev 27:90
8. Kaser M, Bernard F, Wubbe J (1971) Romeins Privaatrecht. Tjeenk Willink, Enschede
9. Mayer CJ (1989) Personalizing the impersonal: corporations and the bill of rights. Hastings Law J 41:577–677
10. Pagallo U (2013) Robots in the cloud with privacy: a new threat to data protection? Comput Law Secur Rev 29:501–508
11. Pirenne J (1939) Histoire Des Instutions et Des Droit Privé de L'ancien Egypte. Édition de la Fondation Egyptologique Reine Elisabeth, Brussel II
12. Voulon M (2010) Automatisch contracteren. Doctoral Dissertation, University of Amsterdam

Suggested Readings

13. Bill of Rights of the Congress of the United States of America. http://www.archives.gov/exhibits/charters/bill_of_rights_transcript.html
14. Dutch Civil Code. (http://www.dutchcivillaw.com/civilcodebook01.htm)

Part III
Contemporary Challenges in Machine Medical Ethics: Decision-Making, Responsibility and Care

Having the Final Say: Machine Support of Ethical Decisions of Doctors

Julia Inthorn, Marco Elio Tabacchi and Rudolf Seising

Abstract Machines that support highly complex decisions of doctors have been a reality for almost half a century. In the 1950s, computer-supported medical diagnostic systems started with "punched cards in a shoe box". In the 1960s and 1970s medicine was, to a certain extent, transformed into a quantitative science by intensive interdisciplinary research collaborations of experts from medicine, mathematics and electrical engineering; This was followed by a second shift in research on machine support of medical decisions from numerical probabilistic to knowledge based approaches. Solutions of the later form came to be known as (medical) expert systems, knowledge based systems research or Artificial Intelligence in Medicine. With growing complexity of machines physician patient interaction can be supported in various ways. This includes not only diagnosis and therapy options but could also include ethical problems like end-of-life decisions. Here questions of shared

J. Inthorn (✉)
Department of Medical Ethics and History of Medicine, University Medical Center Göttingen, Göttingen, Germany
e-mail: jinthor@gwdg.de

J. Inthorn
Centre for Research Ethics and Bioethics, Uppsala University, Uppsala, Sweden

M.E. Tabacchi
DMI, University of Palermo, National Research Institute for Demopolis, Palermo, Italy
e-mail: metabacchi@gmail.com

R. Seising
Faculty of Biology and Pharmacy, Institute of History of Medicine, Science and Technology, Jena, Germany
e-mail: rudolf.seising@softcomputing.es

R. Seising
European Centre for Soft Computing, Mieres, Spain

S.P. van Rysewyk and M. Pontier (eds.), *Machine Medical Ethics*,
Intelligent Systems, Control and Automation: Science and Engineering 74,
DOI 10.1007/978-3-319-08108-3_12

responsibility need to be answered: should machine or human have the last say? This chapter explores the question of shared responsibility mainly in ethical decision making in medicine. After addressing the historical development of decision support systems in medicine the demands of users on such systems are analyzed. Then the special structure of ethical dilemmas is explored. Finally, this chapter discusses the question how decision support systems can be used in ethical dilemma situations in medicine and how this translates into shared responsibility.

1 Introduction

Using suitable machines to support the highly complex decisions doctors have to make every day has already been done for almost half a century. Starting with computer supported medical diagnostic systems with "punched cards in a shoe box" in the 1950s, following intensive collaboration between physicians, mathematicians and electrical engineers in the 1960s and 1970s medicine became, to a certain extent, a quantitative science; then the focus of research shifted from a numerical probabilistic approach to medicine to knowledge based techniques that came to be known as (medical) expert systems (ES), knowledge base systems research or Artificial Intelligence (AI) in Medicine. The aims were high and the expectations were not always fulfilled.

The technological development and gain of knowledge also have clinical consequences. Doctors can keep patients alive in a fragile state like the case of the 11th Prime minister of Israel Ariel Sharon (born 1928).[1] Further examples are very old patients who are tube fed, cases in neonatology, and also discussions about terminating a pregnancy after prenatal diagnosis or organ donation [40]. Due to medical progress applying every therapy or any possible diagnostic that is available seems no longer the ethically correct way to decide. The ethical dimension of decisions in medicine needs to be integrated with processes of medical decision making. This chapter explores the question how decision support systems can be integrated and used, especially in ethical dilemma situations in medicine, and how the interaction of machines and doctors in decision processes influences questions of responsibility.

Artificial Intelligence (AI) in medicine was initiated in the 1970s by artificial systems such as Edward Shortliffe's expert system MYCIN at Stanford University [67], QMR (quick medical reference) [46] and HELP (health evaluation through logical processing) [28, 47, 55].[2] HELP and other clinical systems have been developed at academic medical centers, and have been integrated and used for clinical decision support in the 1980s.

[1] In 2006, Sharon suffered a (second) stroke with a massive cerebral hemorrhage. Since then Sharon was in a permanent vegetative state until his death in 2014.

[2] The former was developed by Randolph A. Miller at the University of Pittsburgh in 1980, based upon the INTERNIST-I patient diagnosis system by Jack D. Myers, Miller and Harry E. Pople. The latter was created by Homer Richards Warner and his team.

Decision Support Systems (DSS) have changed over the last four decades. Sol and co-authors describe their development as follows: In the 1970s, a DSS was understood as "a computer-based system to aid decision making", but later in this 1970s, the focus of DSS development was "interactive computer-based systems which help decision-makers utilize databases and models to solve ill-structured problems". In the 1980s, DSS provided systems "using suitable and available technology to improve effectiveness of managerial and professional activities" [69]. However, in the late 1980s, DSS became part of intelligent workstations.

Decision Support Systems (DSSs) provide the user with a framework to easily characterize and analyze problem situations using predefined algorithms and models. This process is highly interactive, including the user in problem definition, the creation of possible solutions and using the correct model for evaluation and rating. A relevant question here is if the structure and complexity of ethical decisions can be modeled by DSSs.

So-called Expert systems (ESs) assemble the knowledge and experience of domain experts in machine interpretable form. They integrate expert knowledge for a particular domain to provide action alternatives, thus ready-made or adaptable solutions for a given problem. ESs help to capture, combine and distribute the expertise of human decision makers and hence lead to better and faster decisions [73]. Integrations of ES and DSS have been proposed in El-Najdawi and Stylianou [22]. These authors also proposed the standard model of a DSS as a collection of computer based tools to give support in decision making. It combines the content of chosen information sources with domain specific models to help the evaluation of potential problem solutions developed by the user [22].

To be evaluated within a DSS, the problem has to be defined according to quantifiable criteria, so the solution alternatives can be rated based on mathematical models. The DSS does not provide the user with action alternatives; it only gives support with ready implemented, adaptable models for evaluation. Solving ethical dilemmas can be understood as a selection process between two options and weighing possible consequences of actions. This can serve as an initial very basic model of an ethical dilemma. An example of this is a patient who does not allow a life-saving treatment and gives reasons the doctor regards as irrational. This situation can be modeled for different patients of different ages (children, middle aged adults and very old patients) and outcomes can be evaluated.

An extension towards ethical decision making was given by Drake et al. [18]. The DSS can ask the right questions, can suggest different ethical perspectives, as proposed in Turban and Aronson [73], and it can certainly inspire creativity. Creativity can be simulated in the system by stretching the given parameter ranges, using the perspectives of other actors, or even putting the problem description in another context. This can help the user to find solutions that are not limited by a restricted frame of mind that focuses on the situation at hand but frequently misses ideas on how to extend or modify decision spaces by integrating multiple perspectives and normative questions into decision making processes. An additional ES support can help the user to learn about causes and consequences of diseases for example by proposing novel and surprising views on the problem, with problem solutions that did not come to his mind before. These views do not have to be perfect, and can always be adapted in later steps.

For ethical decisions, this would imply that the typical structure of ethical dilemmas with two options that are both connected to unwanted negative results can be questioned as a whole, and a question can be generated to ask how those dilemma situations can be avoided in the first place. Technological progress enables DSSs to model more complex structures of ethical problems, thus broadening its capabilities.

The extensive use of simulation technologies not only enables a long-term projection of possible outcomes, it also makes possible to examine a situation by trial and error. An intrinsic problem in the evaluation of long-term decisions is the need for a wide temporal horizon [33]. Consequences stemming from the complex interactions usually involved in ethical dilemmas can present themselves at a later stage without any explicitly premonitory sign. From the early inception of Computer Science, simulations have been one tool of choice to evaluate a complex system, thanks to their time compression ability in a limited domain. In this spirit, we consider the claim that Serious Games would make a great tool to help in evaluating long term decisions, especially considering the flexibility in timeline management and the tree exploration possibilities opened by the availability of huge storage memory and massively parallel machines.

With few exceptions, Serious Games usually describe a closed time situation, where the actions of the players are carried out in a semi static, episodic and accessible environment (following the classification of [57]). Discretionary action is fairly comparable during a game. Simple modifications can be applied to existing routines to give Serious Games the capability to reduce the players' ability of acting on variables in a time dependent manner, to simulate the effects of time, or to end the game not just after the usual episodic end, but in a different point in the timeline, which may be correlated to the number, quality, difficulty, consequences of decisions. Serious Games allow for an even more complex understanding of ethical decision making by encompassing the dimension of time as well as limiting frameworks.

Using machines not only for medical decisions but also for the normative ethical dimension of decisions in medicine poses questions about how far ethical decisions can be supported by algorithm based machines on the one hand and questions of shared responsibility on the other.

This chapter will explore the possibility to support ethical decision making in medicine by DSSs. It is organized as follows: in Sect. 2, we give an analysis of ethical dilemmas and describe the possible role of DSSs to handle such dilemmas in medicine. In Sect. 3, we give a short historical survey of computers and their support in medical diagnosis and will name prerequisites for the support of ethical decisions. In Sect. 4, we discuss a possible implementation of Serious Games as a development aid for teaching ethics and aiding the evaluation of DSS. Finally, in Sect. 5, we will discuss the use of DSS and Serious Games from an ethical perspective with a special focus on computerized decision support versus learning tools followed by a short conclusion.

2 Ethical Decision Making: Structure and Possible Support by Decision Support Systems

In decisions about medical therapy, different perspectives come together. Informed consent procedures, which are at the heart of ethical considerations in medicine, can serve as an example of how DSSs can be used. Informed consent procedures in a simple model can be described as an exchange of information (doctors) and personal preferences (patients) [23]. Doctors suggest a therapy that is medically indicated based on a diagnosis. Patients get information about this therapy, possible risks, side effects and prognosis and based on their personal preferences give (or deny) their consent to this therapy. The combination of medical indication and patient's consent constitutes the basis of an informed consent. Informed consent procedures in practice should be designed in a way that gives patients time and the possibility to ask questions and have their wishes respected.

Most of the time, the ethical dimension of decisions in medicine can be taken care of without much problem. In trust based relationships, doctor-patient communication about the aims of therapy, personal ideas of life quality or dealing with risks leads to informed consent [25]. Technological support for ethical problems therefore needs to be problem-specific and based on a thorough analysis of ethical dilemmas in order to be helpful.

This can be done in two ways. First, situations where problems occur involving a doctor asking for support (from a colleague or ethics consultant) can be identified and analyzed in order to get a better understanding of the moral dimension. Second, examples of best practice in solving such problems or completely avoiding them in the first place can be used to identify aims and criteria to measure the improvement of decision making.

When talking about moral or ethical conflicts,[3] we can distinguish between two types: moral conflict and moral dissent. There is a *moral conflict* when the moral guidelines one lives by do not lead to a clear conclusion, or not all obligations one sees in a certain situation can be fulfilled at the same time. In most situations these conflicts are easy to solve—the solution is clear—but usually there still remains a feeling of uneasiness for not being able to fulfill all obligations. For example, someone promised to meet a friend and feels obliged to keep this promise while at the same time the school calls that his daughter is sick and he needs to see a doctor with her. While it is clear that immediate support of the sick child is more important in this situation, the person might feel bad about breaking the promise. The obligation of the promise is not simply overruled in this situation but remains an obligation in itself. This can best be seen from other obligations following from that situation such as apologizing for not coming or maybe offering a new meeting. In medical ethics, the four principles approach by Beauchamp and Childress

[3] We will understand the term "moral" in the sense of normative ideas in everyday practice and "ethics" as theoretical reflection of morality.

[7] is currently the most well-known approach in medical ethics. The four principles, autonomy, beneficience, non-maleficience and justice, that are equally important, might also lead to moral conflicts when applied in concrete situations when it proves difficult to weigh the principles one against the other. The following case may illustrate this: A patient who is a Jehovah's Witness could be rescued by a blood transfusion but refuses to give his consent to it based on his religious beliefs. The doctor in charge might experience this as a conflict between respecting the patient's autonomy and the duty to cure by using the blood transfusion. While different cultural backgrounds might lead to different solutions, in most settings the answer to this dilemma is clear. For example, western bioethics and the legal framework in many European countries give priority to patient autonomy.

The conflict between patient autonomy and beneficence can serve as a typical example of moral conflict. Other conflicts can be identified using the four principles approach as a heuristic framework. The following shows examples of conflicts between each pair of two principles:

1. Autonomy and non-maleficence: A patient demands to have a healthy limb amputated and argues for this reasonably. The doctor is unsure if he should perform surgery.
2. Autonomy and justice: A doctor is unsure how much time he should spend informing an especially time-consuming patient who feels not well informed enough to decide. The doctor does not have time to inform all patients in detail in the same way.
3. Beneficence and non-maleficence: A doctor has to weigh the chances of curing a condition versus the risks of therapy with heavy side effects of a therapy.
4. Beneficence and justice: In a hospital, questions of allocation can lead to conflicts between beneficence and justice: Providing the best possible therapy available can be so expensive that other patients cannot be treated using the same therapy.
5. Non-maleficence and justice: Not harming vulnerable groups such as pregnant women or patients in a coma usually is used as an argument for not including them in clinical trials. This leads to a lack of empirical evidence and consequently lack of safe possible treatments for those groups. Should they be included for reasons of justice?

This list shows a few examples of possible moral conflicts in medicine using the four principles approach [7]. When moral conflicts happen in everyday practice the description of a situation still needs to be structured. It needs to be made transparent what principles or obligations are in conflict. Here DSSs can help to find a structure for ethical deliberation and get a better understanding of the moral conflict.

Different methods of analyzing a moral conflict have been discussed in the literature [44]. They can serve as a first basis to understand the nature of problem solving and practice in moral deliberation. They can be divided into different phases. First, the problem has to be described from a factual medical perspective. Then moral principles or obligations relevant in the specific case have to be named. This often helps to get a clear picture of the conflict's moral dimension. Some use the four principles approach by Beauchamp and Childress as a heuristic tool for this. Here DSSs can be used to guide analysis. In a third step, these principles have to be compared and their importance for the specific case assessed. Weighing the principles relevant in the case can also be trained using DSS by interactively discussing cases. DSS can also provide additional information such as legal regulation or ethical guidelines (e.g., [83]). Furthermore, DSS can provide possible future scenarios to compare consequences or to help evaluate the decision.

While a moral conflict is an intrapersonal conflict due to conflicting moral obligations or principles, *moral dissent* is characterized as an interpersonal problem: a moral dissent is a situation where different participants favor and argue for different solutions based on their different personal moral positions. This might occur due to different moral positions. For example, deontological and consequentialist approaches have different perspectives on lying, or because agents weigh the same principles differently. The following case is an example of this type of conflict that could happen in practice: in an intensive care unit an 85 year-old multimorbid patient is treated after a severe stroke. The patient cannot eat sufficiently and grows weaker. Doctors want to apply life prolonging treatment by applying tube feeding while nurses vote for withholding further treatment and applying palliative care instead. Here two different perspectives on the same situation lead to conflict. Both positions can be well argued for. The dissent can have different causes: different (descriptive) perspectives on the situation can lead to different decisions: how much a patient is suffering, if a treatment is considered futile or not, or the interpretation of a patient's will and advance directive. Different professional backgrounds can be one basis for such differing perspectives.

Another possible cause is different (explicit or implicit) moral assumptions. For example, religious people might refer to the sanctity of life while secular people dismiss such a concern. Differences can be more subtle on the level of nuancing or interpreting similar moral principles in different ways. Relatives and doctors might agree that respecting the patient's wishes is most important, but disagree how to weigh personal communication versus an advance directive, or even how to interpret an advance directive. Different moral or cultural backgrounds might also lead to moral dissent. For example, when a doctor is asked not to tell a patient the truth about her diagnosis that she will die soon based on different ideas about how to weigh patient autonomy and keeping the patient from harm (the stressful information about her diagnosis).

In order to provide support in cases of moral dissent, DSS can be used in a similar way as described above. This can help to better understand the conflict and might lead to a solution. But when dealing with interpersonal conflicts aspects such as who is affected, the perceived options within a decision situation, or questions of hierarchy and responsibility for a decision need to be taken into account. Furthermore, the best solution would be if the dissent could have been avoided in the first place. Here different structures of decision making need to be in place such as advance care planning [16]. DSS can guide users to structure decisions and get people who are affected by decisions involved. Empirical evidence shows that while this does not solve the conflict it helps to avoid situations of confrontation and supports shared decision making [24].

These few examples already show that DSS needs to be culture sensitive. Simply applying an approach in ethics will not prove helpful, and might lead to new conflict. The value basis of a DSS needs to be agreed upon by users and the community that is involved in the decisions. Furthermore, DSS for clinical use needs to be tailor made for the often hectic day to day practice under conditions of time constraints, working in shifts, rules of documentation, as well as multidisciplinary teams. The already long history of DSS in Medicine can help to understand how DSS can be implemented and what type of support is considered useful.

3 A Brief History of Decision Support Systems in Medicine

After World War II, when the first electronic digital computers became public, physicians certainly did not rank among the most euphoric users of these so called "thinking machines", machines that gave rise to powerful misgivings among doctors, who feared that medical diagnosis and decision-making would eventually be completely usurped by computers [41].

Several activities in the US were initiated to overcome the physicians' and life scientists' reluctance to use computers:

- The journal *Science* published Robert Steven Ledley's survey "Digital Electronic Computers in Biomedical Science" where the author predicted that in the long run, "perhaps the greatest utilization of computers will be in biomedical applications" [38].
- In the same year the Conference on Diagnostic Data Processing took place at the Rockefeller Institute on January 14th, 1959, organized by the Russian-American inventor and pioneer Vladimir Kosma Zworykin, who was then the first President of the *International Federation for Medical and Biological Engineering* [21, p. 232].
- Two hearings on the use of automatic data processing in medicine that have been held before the US Senate's Subcommittee on Reorganization and International Organization, (July 9th and 16th, 1959) came to the conclusion that corresponding developments ought to be organized and fostered by the government.
- In July 1959, the journal *Science* published the article "Reasoning Foundations of Medical Diagnoses" authored by Ledley and Lee B [42].

The last-mentioned article was a widely read paper that gave instructions to physicians to build diagnostic databases using punch cards to prepare for future times when they would have the opportunity that electronic computers will analyze their data. Today this article is considered to mark the beginning of "medical informatics". It was "frequently cited as the most influential early paper to propose the use of computers as diagnostic assistance" [63, p. 209] and it "mapped a research program for the next 15 years, as investigators spun out the consequences" that "medical reasoning was not magic but instead contained well recognized inference strategies: Boolean logic, symbolic inference, and Bayesian probability. In particular, diagnostic reasoning could be formulated using all three of these techniques" [3]. The authors showed that computers could support doctors in the task of drawing conclusions about patients' illnesses based on symptoms, signs and the results of their examinations and it was their hope that by harnessing computers, much of physicians' work would become automated and that many human errors could therefore be avoided.

Medical diagnoses, Ledley and Lusted argued, were based on logical conclusions, and these could be inferred from information about relationships that exist among symptoms and illnesses and about symptoms a patient exhibits, from which other pertinent information can be inferred for this patient. Thereupon, Ledley and Lusted published numerous texts permanently steering biomedical research in the new direction to initiate the use of computers in medical scientific procedures, i.e., in research and teaching as well as in diagnosis and therapy,[4] and these great efforts caught the attention of US newspapers. A headline of *The New York Times* was "Computer may aid disease diagnosis". *AMA NEWS* (The newspaper of American Medical Association) in an article entitled, "Electronic Diagnosis: Computers, medicine join forces"; wrote: "Doctors are inclined to insist that diagnosis is an art. Perhaps it is—now. But must it be? And is that good?" [4, 50] The article already hints at the ethical dimensions of "electronic diagnosis": who is responsible for the diagnosis (doctor or machine?), and if doctors ought to refer responsibilities (e.g., for diagnostic errors) to machines. Connected to this is not only the hope that less mistakes will be made but also the fear that computer systems make mistakes unrecognized due to naive belief in progress, mistakes nobody feels responsible for. Furthermore, introducing machines into the process of diagnosis might raise fears that doctors will be replaced by machines in the future and that professional knowledge based on experience is replaced by statistics. But overall, hopes connected to machine supported decisions outweighed fears and the latter did not hinder further developments.

In reaction to Sputnik, in October 1957, the U.S. Congress allocated about US $40 million to the *National Institutes of Health* (NIH) for the purpose of stimulating computer use in biomedical research. During those years, the NIH's *Advisory Committee on Computers in Research* (ACCR) established several major biomedical computing centers around the USA. Also toward the end of the 1950s, the

[4] At the Third Annual Symposium on Computer Application in Medical Care in 1979 [41], when Lusted looked back on a terrific success story in a text entitled "Twenty Years of Medical Decision Making Studies", he was able to note that in the period from 1959 to 1968 he and Ledley, working solo or as co-authors, had published some 45 articles in 23 American and nine overseas journals as well as seven proceedings of international conferences, all of them dealing with the subject of computer assisted medical diagnostics or decision making.

physician Martin Lipkin and his mentor James Daniel Hardy began to wonder how new computer technology could be used in medical research within the scope of a doctor's activity.

In the department of medicine of New York Hospital-Cornell Medical Center, Lipkin and Hardy sought ways to master the constantly growing flood of information. They were well aware of the developments in computer technology, thanks to the writings of Vannevar Bush but also from other publications reaching back to the 1940s and even the 1930s, and touching upon "mechanical" computing and sorting machines that used cards and needles or punch cards. The idea arose of using machines of this type to build collections of data sets that were being accumulated during medical research, to carry out classifications and to develop interconnections among them. It was also thought that it might be possible to use this method to mechanically store data from patients' medical histories and to study whether this technology might be helpful in medical diagnostics. In 1958, Lipkin and Hardy reported their project in the Journal of the American Medical Association, in which they sought to classify all diagnosis data from hematological cases by means of a "mechanical apparatus" and to identify relationships between them [39].

A brief description of this "first "computer diagnosis" of disease, in this case hematology disorders is given in the biographical memoir on Hardy by Arthur B. Dubois:

> The computer consisted of punched cards in a shoe box. Diagnostic criteria had been obtained from a hematology textbook and were wedge-punched at the edge of each of 26 cards to match the symptoms and laboratory findings of the 26 blood disorders. Knitting needles were run through the holes that corresponded to the symptoms and laboratory findings of each of 80 patients, matching those to the diagnostic criteria wedge-punched into the edges of the set of 26 hematology cards. Shaking the box made the card whose criteria matched those of the patient drop out of the shoe box to show the diagnosis printed on the hematology card [19, p. 13f].

Starting with such computer supported medical diagnostic systems with "punched cards in a shoe box" in the 1950s, followed by intensive collaborations between physicians, mathematicians and electrical engineers, medicine became, to a certain extent, a quantitative science in the 1960s and 1970s. When the University of Utah installed a digital computer in 1960, the Director of the computer center Robert Stephenson showed physician Homer Warner the already mentioned *Science* article by Lusted and Ledley [42]. One section in Lusted's and Ledley's article gave "an introduction to Bayesian statistics and pointed out the relevance of Bayes' Rule to the problem of medical diagnosis" [63, p. 209]. Warner and Stephenson agreed to realize this proposition to use Probability Theory to model the medical diagnostic process and to apply this idea to congenital heart disease using the digital computer and they proved that it could diagnose as well as or even better than cardiologists.

More than 30 years later, Warner explained in his third-person narrative:

> To aid the patients coming through Warner's laboratory, they decided to make their model to diagnose 35 different forms of congenital heart disease. First, they collected data on

> how frequently each of 50 different findings, such as murmurs of different kinds and cyanosis, occurred in each disease and how common each disease was in the population of patients referred to the laboratory. After collecting several hundred such cases, a matrix showed the disease on one axis, the findings on the other. At each intersection of the symptom with the disease, a number represented the frequency of that finding in patients with that disease. This table formed the basis for the diagnosing patients based on findings recorded by their referring physicians. A comparison of the computer diagnoses and those of the referring physicians showed the computer to be right more often than any of the physicians, based on diagnosis following heart catheterization [77, p. 479f].

Warner and Stephenson presented their findings at an *American Heart Association* meeting and their article appeared in 1961 in the *Journal of the American Heart Association* (*JAHA*). This article was "among the most frequently cited" papers to "determine whether Bayesian techniques could be effectively applied to diagnostic problems" wrote E. H. Shortliffe in 1988 and "the first published example of automatic diagnosis using real patient data and comparing computer derived results with human diagnostic abilities" wrote Paul D. Clayton, a former student of Warner, then chair of Medical Informatics at Columbia University, in 1995 [12, p. 139; 63, p. 209]. The article became one of the most important and crucial papers in the history of medical decision making.

Various approaches to computerized diagnosis emerged in the 1960s and 1970s, using Bayes rule [76, 82], factor analysis [74], and decision analysis [42]. On the other hand, artificial intelligence approaches also came into use, e.g., DIALOG (Diagnostic Logic) [53] and PIP (Present Illness Program) [52]. These were programs to simulate the physician's reasoning in gathering information, as well as to simulate the diagnosis using databases in the form of networks of symptoms and diagnoses. Progress in computerized diagnosis thus enabled increases in the complexity of decisions simulated and supported, and integration not only of different types of facts and knowledge, but also different types of reasoning.

As a next step, we should mention the introduction of medical expert systems shortly after general expert systems appeared in the 1970s. The first of these being MYCIN [62], INTERNIST [47] and CASNET (Causal Associational Networks) [78, 79].

Then, the focus of research shifted from a numerical probabilistic approach to knowledge base techniques later known as (medical) expert systems, knowledge based systems in Medicine and clinical decision support systems (CDSS). In clinical practice today, those expert or knowledge based systems are most prevalent that perform decision making at the level of a domain expert [70]. In general, CDSS patient data are compared against a knowledge-base and an inference mechanism is used that can incorporate a rule base of 'if-then-else rules' with Bayesian prediction or fuzzy logic methods.

The aims were high and the expectations were not always fulfilled. So far only few systems are in clinical use. Experiences so far have been mixed and the full potential of clinical decision support systems for optimizing the healthcare system is far from realized. One reason is that "the greatest barrier to routine use of decision support by clinicians has been inertia; systems have been designed for single problems that arise infrequently and have generally not been integrated into the routine data management environment for the user" [64, p. 14]. Other reasons are insufficient acceptance and utilization of such systems, missing integration into a Hospital Information System (HIS), inappropriate software architecture and others. On the other hand, systems have been efficient as learning environments by simulation.

The path of Medical Expert Systems and Clinical Decision Support Systems that we have followed in this section shows that these systems are mainly developed to support diagnosis or suggest possible therapy focusing on specialized medical problems. They were not designed to replace doctor patient communication or to communicate directly to patients in informed consent procedures. Communication between doctors and patients, getting patients involved in decision making processes, what is often called the "human side" of medicine, is important for trust building between doctors and patients. Here also ethical questions arise and legal regulations frame decisions. Ethical considerations and dilemma solving so far remains the responsibility of doctors in cooperation with other healthcare professionals and patients. With the development of further advanced technology the complexity of ethical decisions could also be modeled in DSS. But actually integrating DSSs that encompass the full complexity of decision making in medicine, and thus integrating the knowledge base and the normative aspects of medicine, will pose further problems that will be addressed in the following section.

4 Clinical Decision Support Systems: From Diagnosis to Ethics

4.1 Acceptance Problems of CDSS

The idea to get help from computers to support doctors in the task of drawing conclusions about patients' diseases based on symptoms, signs and the results of their examinations, could be a solution to the big problems that became more and more visible in medicine in the first half of the 20th century, as is shown by a comment from the foreword of a textbook: "The belief has been expressed that errors in diagnosis are more often errors of omission than of commission" [56, p. vii]. In another textbook, the physician Logan Clendening wrote on this matter: "How to guard against incompleteness I do not know. But I do know that, in my judgment, the most brilliant diagnosticians of my acquaintance are the ones who do remember and consider the most possibilities. Even remote ones should be brought up even though they may be immediately rejected" [13, p. 59f].

It is a difficult task for physicians to exhaustively consider every factor relevant to the decision, due to either limited memory or limited information. About a decade later a medical doctor described this problem in the following way: "What is needed is a device which will answer the question 'What are the possible causes of the group of symptoms and signs I have elicited from my patient?'" [49, p. 874]. Patients demand that doctors are always aware of the evidence, results of latest research and new technologies in diagnosis. Therefore, a correct diagnosis is getting more and more complicated to achieve and the possibilities of false diagnosis are increasing. A false diagnosis is not only a medical problem but also poses ethical questions of harming the patient and responsibility for mistakes in a hierarchical system like hospitals. Today, it is acknowledged that besides diagnosis and therapy there is also an ethical side to medical decision making in almost all medical decision procedures and that medical decisions should also follow ethical guidelines. This means that not only rapidly growing knowledge in medicine needs to be taken into account when coming to a decision, but also ethical questions like life quality, autonomy or values based in the cultural background of patients as well as values derived from the professional ethos [33]. With regard to decision making, physicians experienced a time of great research progress which resulted in a collectively perceived "knowledge explosion" in the mid-20th century. This development was accompanied by growing interest in computer technology on the one hand and growing relevance of ethical questions on the other, which can be seen from the growing number of ethical guidelines for doctors, like the Declaration of Geneva [83].

Medical expert systems and CDSS machines have been construed to solve these problems and about 10 years ago, roughly 70 known proprietary medical CDSS machines were listed, but only 10 of them geared towards routine use [26]. Unfortunately, there is no information available about a real daily average usage of these systems. However, CDSS machines still suffer from insufficient acceptance. Medical doctors are reluctant to use such systems for different reasons:

1. Doctors are afraid that computer systems will be used to substitute physicians [60, 65, 66]. CDSS machines should assist physicians in their daily routine work in the doctor's practice or in hospitals whenever they need support. CDSS machines should be built in a way that helps to avoid mistakes and improve decision making. This can cause fears that CDSS based decisions are better than doctors' decisions that are prone to human flaws. However, such systems can also take the role as a teaching system and physicians will learn from a CDSS how to consider criteria, facts or process issues in specific decision situations. These systems realize "rationality" in diagnosis and decision making. They are intended to support but not to replace physicians [11]. As the discussion about responsibility will show later, doctors need to have the final say about medical decisions. Doctors should remain the ultimate authority, and he or she will have the ability to "overrule" or to ignore the recommendations of the CDSS at any time [59, p. 261]. Doctors also have the role of justifying and explaining decisions to patients and making decisions transparent. For this, they need to be involved in the decision making process. CDSS machines that try to replace doctors therefore stand little chance of acceptance.
2. Integration of a CDSS into well organized and equipoised everyday practice and clinical guidelines is difficult for various reasons, for example:
   (a) One drawback to acceptance of a CDSS is workflow integration. Many of these systems are stand-alone applications, and they require the clinician as

an operator to cease working on their current report system, switch to the CDSS, input the necessary data, and receive the information [75].

(b) The input process that has to be done by the doctor or a nurse is very time consuming and costly.

(c) The use of a computer system interferes with the important contact between patient and doctor. The doctor will not be able to focus on the patient, his trouble and pain and the patient feels unappreciated.

Therefore, CDSS machines should not only be easy to handle but not interfere with doctor-patient communication. This calls for systems that do not interrupt or prolong daily routine but for applications usable for training in very different contexts.

3. Medical decision support systems are not all-round systems but very specialized systems and are concerned with just a very specialized cutting of a medical field, e.g., hepatology or pulmonary diseases in internal medicine. A CDSS for ethical questions would ideally be integrated into already existing and well established systems. Since experiences with CDSS machines on a broad level do not exist, further demands of users besides ethical dilemma solving need to be assessed and integrated for better acceptance.
4. There are limitations of CDSS machines because an optimal physician's treatment requires that physicians can get important information almost without lag of time: information on the present and the possible future consequences of his or her actions. To this end, they "require data that are factual, factual inferential (why type questions) and predictive (what if questions). To date, the best support that a CDSS has been able to provide is data that answer factual and maybe some forms of predictive questions […]. Physicians have no shortage of data available to them. Thus, physicians have found that currently available CDSS machines are not able to meet their more complex information needs" [59, p. 261].
5. Also, most of the present CDSS machines have not progressed beyond the prototype stage [81]. There exists no standard or any universally accepted evaluation or validation methodology to ensure that the system's knowledge base is complete and correct [5].

This is also a problem for supporting ethical decisions: with the plurality of ethical approaches and views a universally accepted evaluation therefore seems impossible in this case. But what might be easier to achieve and well accepted is to evaluate the outcome of CDSS machines along the values that built the basis of the CDSS itself. This means that there is still reflection necessary if those values are shared and accepted in a concrete situation but the idea of coherence can be argued for more easily.

6. We do not know whether the use of a CDSS improves the quality of decisions produced. Also we do not know whether the economic or other benefits, e.g., the patients' well-being are attributable to the use of the CDSS, because there is no well-defined or universal evaluation methodology. "To date, an

examination of the literature indicates that there is virtually no information available related to the cost or cost effectiveness of CDSS machines. Most of the CDSS machines are university based developments, and still in prototype stage. These costs regarding the initial investment of CDDS tend to be hidden and therefore difficult to access. This frightens or hinders industry interest in funding and encouraging the development of CDSS in healthcare in general [48]. Still, many physicians have a positive outlook on the potential for these systems, particularly relating to practitioner performance. However, until the use of CDSS machines in general is as routine as the use of the blood pressure cuff, it is important to be sensitive to resistance to using these systems" [59, p. 261].

4.2 From CDSS to Serious Games: Learning Ethic Concepts in Medicine Through Computer Assisted Machines

In the context of developing support systems for ethical decisions, an approach is called for that goes beyond simple fixed interaction that CDSS machines usually supply, to produce an interactive environment that allows doctors to practice with concrete problems and that can move the boundaries of the decision above and beyond the moment in which the decision is made. One main benefit of such an approach would be the creation of a learning environment for complex decision making processes. This can be used for teaching or team coaching in order to train for decision situations where different forms of expertise as well as values have to be taken into account, something which is usually precluded in standard CDSS. Furthermore, the use of such an environment can reframe decision situations, integrate different perspectives and get relevant actors more involved at different times or stages of the decision than in current practice. This cannot be done with the basic tools that have been the staple of medical decisions education, such as textbooks, medical scripts, questionnaires and simple role play games, neither with the ones used by standard CDSS systems, but need more refined thought. We speculate that the use of Serious Games (SG) can be beneficial to help developing better CDSS, while at the same time they have the potential to become a standard tool in training ethical medical behavior.

Following the revised definition by Zyda [84], based upon the original definition by Abt [1], an SG is "a mental contest, played with a computer in accordance with specific rules that uses entertainment to further government or corporate training, education, health, public policy, and strategic communication objectives". While SGs are conceived, programmed and developed with the tools and general aesthetic usually reserved for video games, they are used for purposes other than mere entertainment [72]. The principal advantage of SGs is "allowing learners to experience situations that are impossible in the real world for reasons of safety, cost, time, etc" [14, 71].

4.3 How Do Serious Games Work

SG follows the convention of standard games we play on a computer or smartphone: a 3D immersive simulation can be developed to ensure that the user can empathize with his avatar, the character can be depicted with classical medical equipment such as a white coat and stethoscope for a better identification of his role; other actors can be easily distinguished by external traits, and the same goes for Non-Playable Characters (NPC) (see Fig. 1). Avatars and NPCs in SGs, be they realistic or caricatural, are often represented wearing dresses of the trade, in order to solicit empathy (see Fig. 2) [31].

The simulation is multi-media based: graphical elements such as animations or cut-scenes can be included, while sounds and speech can provide feedback to the player. During the game, a collection of indicators, menus, gauges and other information useful for the player must be visible at all times in a heads-up display (HUD). Multiple HUDs can be used in different stages of the simulation [8]. A first-person point of view is usually tied with a direct interactive access to the virtual world. The main drawback of this approach is hiding the player character.

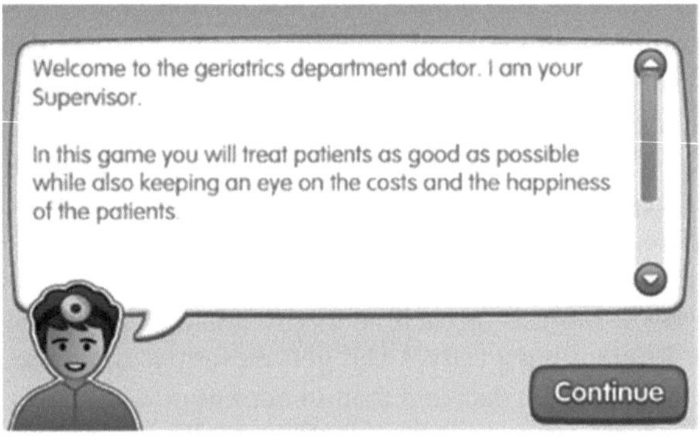

Fig. 1 Non playable character from GeriatriX [37]

Fig. 2 Screenshot from JDoc [68] showing the player's avatar (*center*) interacting with two NPCs

A third-person perspective instead allows the player to follow main character movements (see Fig. 2), but confusing camera angles can also complicate playing the game [51]. No custom peripherals should be used in this broad category of simulations, as complex physical interface may ruin the spontaneity of the actions.

A recent development of the field concerns the use of mobile peripherals and extended deployment of SGs: pervasive and ubiquitous computing also permits constant access to the game, using smartphones or other wireless handhelds (e.g., tablet, notebook) as input/output device. In this scenario, a game session needs to be saved and reloaded. In some cases, the player can be forced to start over if some constraints are not satisfied. Adding network connectivity can extend game longevity. Clocks or time counters can accentuate the feeling of time pressure. Timers may move at the same rate as real time or at different rates, such months and years, which elapse in a few minutes of play [80]. Time compression helps the player to focus on what is important [45]. Having to decide and act under conditions of time restraint is one of the main aspects mentioned by students when they argue that the best possible way to act is not feasible within the current medical system. Finding ways to model alternatives and evaluate outcomes would provide something close to experience based learning and strengthen young doctors to reflect on circumstances.

The simulation can be divided in stages. Each phase can have different durations and can be clearly separated during the game, using level closing material for providing feedback to the player or video sequences that provide a backstory. Mission briefing before starting the chosen level can guide the player through game controls and goals. The rules of game need to be clear to the user. In the first stage, the player must examine the problem. During this act he needs to connect information in an aggregate knowledge base. This task is the first important step in ethical decision making: describing the conflict on a factual as well as normative level. When using SGs for teaching purposes, this can also be useful in order to get players acquainted with different ethical approaches like the four principles approach [7], using them as the knowledge base of the game. Furthermore, players can learn to grow some sort of sensibility for hidden ethical assumptions on the first level. Connecting the description of a situation to theoretical knowledge will help to clarify the situation and options for decision making later on. In some cases, the player might have incomplete information, and this can become a separate task in problem solving and deciding. The main stage is related to the decision act. In this phase, life-like characters can be used as tutors or trainers and can supply helpful information to the player [54]. NPCs should seem intelligent and to behave coherently: NPCs should engage users or other characters in conversation, display of role appropriate knowledge and expertise, shape constructive user behavior and discourage disruptive user behavior [54]. An extensive and detailed logging of all choices and actions a player has made will be made available to the player for subsequent analysis. This information can be extensively reviewed for evaluation purposes.[5]

[5] Another way of evaluating is the implementation of a scoring algorithm. This is useful because a positive score can reinforce the player's behavior. The score can be displayed with a numeric value, a star system or a grade assigned in letters. The goals can be assigned in terms of score to achieve. Some bonus materials (e.g., additional level) can be added to push the player to work for the maximum grade. A high score list can be included and displayed after gameplay.

After the decision, the players need to watch and evaluate the consequences of their actions. In this phase, the time in the simulation can be speed up to show the long-term effect on the actors. In some cases, this step can be preliminary to a new phase of the decision process (e.g., someone is angry about the outcome of another case and mistrusts a doctor). However, in this new scenario, not all actions previously taken can be undone. For example, if a patient dies due to an irresponsible decision, the user cannot revive the person and needs to face the consequences of the loss.

4.4 Serious Games and Medicine

The use of SGs in medicine is nothing new by itself. In 2004, with support from the Lounsberry Foundation and the Woodrow Wilson International Center for Scholars, the Serious Games Initiative (www.seriousgames.org) started the Games for Health Project (www.gamesforhealth.org, [58]). In 2008 and 2009, the number of peer reviewed scientific publications surged as the clinical application of health games diversified. The sudden surge of health game publications can be attributed to the availability of specific funding for health games research (i.e., Robert Wood Johnson Foundation funding), advancements in commercialized gaming technology (e.g., Nintendo Wii), and the establishment of health game research networks through various scientific venues (e.g., Games for Health Conferences [36]). The Health Games Research Database lists over 430 games.

One thriving sector in this vein is the use of games in healthcare from a professional perspective; that is, the use of games in medical training and educating young professionals include practice exercises to refine skills for performing surgery, emergency response, disaster preparedness and for simulations for healthcare management situations [2]. An example of such approach is JDoc. The purpose of JDoc is to familiarize junior doctors with the day-to-day stress of a hectic hospital. The junior doctor simulator immerses the player in the believable world of a busy hospital at night and educates them as to the diagnostic procedures and medical criteria required while working on call in a hospital ward [68].

Another project is "Pulse!!" This SG is a three-dimensional virtual clinical learning platform developed at Texas A&M University Corpus Christi, in collaboration with the United States Navy, for teaching high-level critical thinking, diagnostic reasoning and skills to healthcare professionals that provides unlimited, repeatable, immersive clinical experiences without risk to patients [20]. In both simulations, junior physicians can practice procedures without life/death consequences. In particular, in JDoc we can develop entire scenarios, using parameters provided by senior doctors, in which we can reproduce the decisional process in ethical dilemmas such as abortion, euthanasia or treatment of people with disabilities, analyzing every step of this process. We can also evaluate the adherence to the four common moral principles: respect for autonomy, beneficence, non-maleficence and justice [29].

Another context where serious games and medicine meet is in simulations for healthcare management related situations. The aim of such software is to solve administrative problems faced by healthcare managers such as facility planning, resource allocation, staffing, patient flow and waiting time, routing and transportation, supply chain management, and process improvement [34]. An example of this is GeriatriX, a student training game for complex medical decision-making concerning elderly patients. The students explore different diagnostic and therapeutic strategies, and are given insight into the consequences and costs of their choices [37]. Because GeriatriX deals with elderly patients, the risk of harming the patient by the therapy needs to be weighed against the chance of healing, but also the expected quality of life during and after the therapy.

Multiple games have been developed to support health behavior of patients [32]. These include virtual environments that provide a safe and realistic simulation of exposing patients to a potential health threat [30]. Video games were also designed to improve prevention and self-care behavior among children and adolescents for asthma and diabetes [32].

More recently, we have seen the emergence of games designed to persuade users to change their behavior, better known as Persuasive Games [27]. Games for behavioral change related to diet, physical exercise, self management, etc., had positive patient outcomes [6]. In particular, active computer games (e.g., Eye Toy games, Dance Dance Revolution, Nintendo Wii games), also dubbed as "exergaming" [9], may have the potential to promote physical activity for obese children [15]. Use of games as a learning environment for preventive actions is cited as a possibility in the literature [32]. "The Great Flu" from the Ranj Corporation teaches users how a virus works and what resources are necessary to counteract/contain the spread of a pathogen and to prevent the outbreak of an epidemic [2]. "Persuasion is often involved in health prevention. This particularly applies to changing habits. The main goal of persuasive games is not to educate or to increase physical exercise, but to persuade the users to modify their behavior" [10, p. 132]. An example of a persuasive game is "Smoke?" The goal of this SG is to persuade people who are contemplating quitting smoking or have recently quit smoking, that quitting permanently will be beneficial [35].

4.5 How the Machine-User-Interface of a Medical Decision SG Should Work

As we said in a previous subsection, the principal aims of current SGs in healthcare are training medical professionals as well as future patients to enable behavioral change. In these simulations, the ethical aspect, when present, is a simple plug-in, and often a second thought. We maintain that the use of instruments such as SGs as a component for building and rating medical ethics decision systems is a novel approach that may bring satisfactory results.

A SG is the perfect environment where ethical dilemmas can be simulated without real world consequences to others, but with the added necessary element of deep personal involvement typical of a simulation. Even better, a SG can include a teaching consequence that is not present in standard CDSSs: teaching ethics from a textbook is often ineffective, as there are no incentives nor real examples that can be tested interactively, and even learning by enacting scripts, usually done with a role playing approach, has a very limited scope, and the interaction is too forced to be believable. On the contrary, deep immersion in a simulated world that is offered by SG, the level of realism, the possibility to develop rich and detailed case histories, the fact that consequences of decisions regarding ethical dilemmas can be evaluated in different turn points, and even in a long term fashion, the controlled repeatability offered by the approach, all are perfectly suited to the learning and to the aiding aspects of ethical decisions in medicine.

In order to build such a system, we need to approach ethical dilemmas in a different way: when we simply ask for a "smart" decision, we often obtain simplistic and short-sighted argumentations, because decisions in health-care have impacts that are hard if not almost impossible to predict by a single person. This "look-ahead process" might be forgotten when the human agent is left alone in carrying the burden of evaluating all the long-term consequences. If we exhibit a well-defined problem in binary win-lose logic, the user can acquire information only related to the specific problem. The knowledge of how an agent can reach one of the possible solutions in a state space often cannot be applied in other context, especially in healthcare. An intelligent behavior might seem rather stupid and dangerous if carried out in another environment.

The risk of assigning a specific goal, when such a goal exists, is that the simulation cannot successfully lead to a flexible and ethical behavior, as we need to balance between short-term and long-term goals. The same ethical problem can be shown in different forms to confirm the fact that the user can choose the better solution independently, without stepping in the pitfalls of a single specific case. Different responses to similar questions can be also considered valid; the step of the reasoning process does not even have to be logical, but can be based on a combination of incomplete information and personal values.

Because an ethical approach would include the evaluation of long-term effects of small changes on the people involved, especially patients, systems that are able to simulate effects of decisions and provide at the same time criteria to evaluate the impact on those affected, would strongly improve the quality of complex decision making processes and outcomes.

Serious Games make a great tool to help in evaluating long-term decisions, especially considering the flexibility in timeline management and the tree exploration possibilities opened by the availability of huge storage memory and massively parallel machines. In specific points in the game time line, an agent can be partnered to players in a prompting role dependent on the exploration tree's position and content. The repeatability of the game and its complete parameterization allows its use as a massive evaluation tool. The tools provided based on Serious Games would open opportunities to expansive, participatory, experience based and reflective learning. Many shortcomings of traditional teaching methods could be overcome: unidirectional conversation, cognitive centered reproduction, or inert knowledge, the practical transfer of acquired knowledge would be strongly enabled. Furthermore, the level and breadth of support given from the system during learning could be calibrated toward the quality of decisions, inducing a progressive learning pace and improving the dropout factor of the learning process.

A support system for ethical decisions should enable users to explore the state space, analyzing effects of alternative decisions and provide them with necessary information that lead towards ethical decision. Such machines would need a user interface that enables interaction based on alternative paths of decision. The user can directly experience the consequences of a certain course of action, thus coming to a deeper insight of the problem at hand. The evaluation phase can be carried out assigning different utility functions once connected to the timeline. As ethical decisions have wide impacts and very often long-term effects, at different times or stages, ambiguous, divergent and contradictory requirements are reinforced. For the user to reach decisions under time pressure or incomplete information, long-term strategies could be rolled back in order to increase awareness of extended consequences. Time compression ability can be used from the system in order to assess efficiency in presence of no single right solution.

5 Ethics of Ethical Decision Support

Serious Games seem to have many advantages when it comes to creating a learning environment for ethical decisions. But the technical feasibility needs to move parallel with a positive evaluation of the use of Serious Games. Using any type of technology in the context of decisions is only helpful if decisions can be improved. CDSS machines have already been object of ethical reflection [43]. Evaluating ethical decisions can be done on the level of the structure of the decision process as well as the medical and the ethical outcome. In this section, we suggest and discuss criteria for evaluating the process management as well as outcome when using a DSS or SG.

Decision making in medicine is highly complex. Decisions about diagnosis and therapy need to be founded on a broad base of knowledge about all kinds of different conditions, symptoms and diagnostic tools. Furthermore, today the aim of a therapy is not always clear bringing up ethical questions: is it better to prolong life (in its quantity) by all means or is it better to improve life quality, even at the cost of reducing life expectancy? Questions of life quality, patient autonomy and professional ethics also need to be taken into account in medical decision making adding to the already given complexity of medical decision making. (Wrong) decisions in medicine can have severe consequences including the death of patients but also moral distress or guilt to others. Different scenarios after a decision may exemplify this: accompanying a dying patient during his last days and letting the patient die in accordance with his wishes; providing a blood transfusion to a patient who is a Jehovah's Witness against his wishes and thus survives; administering tube feeding to a highly demented patient and whose wishes are unclear. Decisions can be evaluated based on the outcome such as if the patient is still alive or dead as the most simple criterion. Other possible dimensions are the gained life expectancy, the result of a therapy for the quality of life, but also the ethical dimension of the decision such as if the patient's autonomy is respected, if the patient does not experience unnecessary harm, his beneficence is considered or relevant questions of justice solved in an acceptable way.

In order to change decision processes the structure of dilemma situations and how they can be avoided needs to be understood by doctors. Learning how to frame decisions therefore is an important step. Empirical research shows that textbook based teaching in ethics as well as ethical guidelines do not lead to the desired effect [61]. Using Serious Games as a learning environment might enable experience based learning that provides insight into the possible consequences of different normative backgrounds in everyday decisions. Doctors need to experience how their own often intuitive interest in a situation can be made transparent and communicated to patients but also how different perspectives in a situation can be brought together and enable improvement in mutual understanding, empathy, and thus reduce potential conflict [17].

Using machines as learning environments might also help to overcome the above mentioned problems with usability and acceptance. To enable further reflection, it would also be interesting if users could introduce their own cases and experienced dilemmas, for example, by integrating characteristics of patients or persons involved.

Besides the decision making process, the outcome also needs to be evaluated. This seems to be an even more complicated task, due to intercultural differences and plurality of opinions. Therefore, using Serious Games can only be seen as a tool to initiate a reflection process but not as guiding decisions or even making better decisions. One main aim of using Serious Games must be to strengthen responsibility in the sense that doctors get a growing sensibility for potential ethical problems, how to avoid them and how to moderate shared decision making processes between different parties. The aim of the implementation of SG teaching ethics therefore is to explicitly place responsibility in the hands of doctors. Transferring ethical decisions to IT solutions would lead to the problem that the user is still responsible for the implementation of a decision and therefore needs to reflect upon the decision process, its criteria and possible outcomes. This implies having a deeper understanding of the ethical decision or, which is even more complicated, to understand the algorithms of the technical solution.

Possible outcomes and the effects of ethical decision support should also be evaluated on the basis of norms. The four principles approach can be used as one that is broadly accepted and useful as a heuristic tool to detect ethical problems [7]. Evaluation criteria should meet with the criteria used within the learning environment. Besides the four principles, aspects like truthfulness or confidentiality can play a role. It is important to mention that decisions should only be evaluated positively (within the system as well as an outcome of the system), if they cohere with the legal framework and existing binding ethical codes and guidelines (for example by medical associations) applicable in this field. The applications therefore should be culturally sensitive and the criteria, values or norms used for evaluation obtained from the communities affected by the modeled decisions.

6 Conclusions

In order to achieve its task of improving ethical decisions in clinical settings, Serious Games and other types of DSSs need to satisfy high standards. It seems that the high complexity of decision making can be met by using specific types of serious games that do not over simplify ethical dilemmas. Such systems can integrate a short and long perspective and enable learning with regard to decision processes as well as norms and principles. Though there is a reluctance to use machine support in medicine, the possibilities of experience based learning should be considered as an important aspect of behavioral change that could be used to improve the ethical quality of decisions in medicine. Serious Games can be used in order to encourage a change of medical behavior, and to enrich CDSS's with a learning stance and opportunity to grade and improve on current systems.

References

1. Abt CC (1987) Serious games. University Press of America, Lanham
2. Adams SA (2010) Use of "serious health games" in healthcare: a review. Stud Health Technol Inf 157:160–166
3. Altmann RB (1999) AI in Medicine: the spectrum of challenges from managed care to molecular medicine. AI Mag 20(3):67–77
4. AMA News (1959) Electronic diagnosis: computers, medicine join forces. The Newspaper of American Medical Association
5. Apkon M, Mattera JA, Lin Z et al (2005) A randomized outpatient trial of a decision-support information technology tool. Arch Intern Med 165(20):2388–2394
6. Baranowski T, Buday R, Thompson DI, Baranowski J (2008) Playing for real: video games and stories for health-related behavior change. Am J Prev Med 34(1):74–82
7. Beauchamp TL, Childress JF (2009) Principles of biomedical ethics, 6th edn. Oxford University Press, New York
8. Bergeron B (2006) Developing serious games (game development series). Delmar, Cliffton Park
9. Bogost I (2005) The rhetoric of exergaming. In: Proceedings of the digital arts and cultures (DAC), Copenhagen, Denmark
10. Brox E, Fernandez-Luque L, Tøllefsen T (2011) Healthy gaming–video game design to promote health. Appl Clin Inf 2:128–142
11. Classen DC (1998) Clinical decision support systems to improve clinical practice and quality of care. JAMA 280(15):1360–1361. doi:10.1001/jama.280.15.1360
12. Clayton PD (1995) Presentation of the Morris F. Collen award to Homer R. Warner: "Why Not? Let's Do It!" JAMIA 2(2):137–142
13. Clendening L, Hashinger EH (1947) Methods of diagnosis. C.V. Mosby Co, St. Louis
14. Corti K (2006) Games-based learning; a serious business application. Informe de Pixel Learning 34(6):1–20
15. Daley AJ (2009) Can exergaming contribute to improving physical activity levels and health outcomes in children? Pediatrics 124(2):763–771
16. Detering KM, Hancock AD, Reade MC, Silvester W (2010) The impact of advance care planning on end of life care in elderly patients randomised controlled trial. BMJ 340:c1345. doi:10.1136/bmj.c1345

17. Dewey J (1916/2008) Democracy and education: an introduction to the philosophy of education. Available via http://www.gutenberg.org/files/852/852-h/852-h.htm. Accessed 11 Dec 2013
18. Drake J, Hall D, Lang T (2011) Ethical decision making and implications for decision support. In: Schuff D, Paradice DB, Burstein F, Power DJ, Sharda R (eds) Decision support–an examination of the DSS discipline, annals of information systems, vol 14. Springer, New York, pp 69–82
19. Dubois AB (2006) James Daniel Hardy 1904–1985. A biographical memoir. Nat Acad Sci Biographical Mem 88
20. Dunne JR, McDonald CL (2010) Pulse!!: a model for research and development of virtual-reality learning in military medical education and training. Mil Med 175(7s):25–27
21. Eden M (ed) (1960) In: Proceedings of conference on diagnostic data processing. IRE Trans Med Electron
22. El-Najdawi M, Stylianou A (1993) Expert support systems: integrating AI technologies. Commun ACM 36(12):55–66
23. Emanuel EJ, Emanuel LL (1992) Four models of the physician-patient relationship. JAMA 267(16):2221–2226
24. Emanuel LL, von Gunten CF, Ferris FD (2000) Project to educate physicians on end of life care, interdisciplinary program in professionalism and human rights. Arch Fam Med 9(10):1181–1187
25. Faden RR, Beauchamp TL (1986) A history and theory of informed consent. Oxford University Press, New York
26. Fieschi M, Dufour JC, Staccini P, Gouvernet J, Bouhaddou O (2003) Medical decision support systems: old dilemmas and new paradigms? Methods Inf Med 42(3):190–198
27. Fogg BJ (2002) Persuasive technology: using computers to change what we think and do. Ubiquity 89–120. doi:10.1145/764008.763957
28. Gardner RM, Pryor TA, Warner HR (1999) The HELP hospital information system: update 1998. Int J Med Inf 54(3):169–182
29. Gillon R (1994) Medical ethics: four principles plus attention to scope. Br Med J 309(6948):184
30. Haniff D, Chamberlain A, Moody L, De Freitas S (2013) Virtual environments for mental health issues: a review. J Metabolomics Syst Biol. doi:10.5897/JMSB11.003
31. Hayes-Roth B (2004) What makes characters seem life-like. In: Prendinger H, Ishizuka M (eds) Life-like characters: tools, affective functions, and applications. Springer, Berlin, Heidelberg, pp 447–462
32. Howell K (2005) Games for health conference 2004: issues, trends, and needs unique to games for health. Cyberpsychology Behav 8(2):103–109
33. Inthorn J, Haun S, Hoppe A, Nürnberger A, Dick M (2012) Evaluating decisions: characteristics, evaluation of outcome and serious games. In: Greco S, Bouchon-Meunier B, Coletti G, Fedrizzi M, Matarazzo B, Yager RR (eds) Advances in computational intelligence. Proceedings of 14th international conference on information processing and management of uncertainty IPMU 2012, Catania, Italy, 9–13 July 2012
34. Kennedy MH (2009) Simulation modeling for the healthcare manager. Healthc Manager 28(3):246–252
35. Khaled R, Barr P, Noble J, Fischer R, Biddle R (2007) Fine tuning the persuasion in persuasive games. In: de Kort Y, IJsselsteijn W, Midden C, Eggen B, Fogg BJ (eds) Persuasive technology. Springer, Berlin, Heidelberg, pp 36–47
36. Kharrazi H, Lu AS, Gharghabi F, Coleman W (2012) A scoping review of health game research: past, present, and future. Games Health Res Dev Clin Appl 1(2):153–164
37. Lagro J, Laan A, Veugelers M, Huijbregts-Verheyden F, Christoph N, Olde Rikkert M (2013) GeriatriX, a serious game for medical students to teach complex medical reasoning. Let's play! In: Proceedings of AMEE, pp 169–170
38. Ledley RS (1959) Digital electronic computers in biomedical science. Science 130(3384):1225–1234

39. Lipkin M, Hardy JD (1958) Mechanical correlation of data in differential diagnosis of hematological diseases. JAMA 166:113–125
40. Lo B (2009) Resolving ethical dilemmas: a guideline for clinicians, 4th edn. Lippincott Williams & Wilkins, Philadelphia
41. Lusted LB (1979) Twenty years of medical decision making studies. In: Proceedings of 3rd annual symposium on computer application in medical care, Washington DC, pp 4–8
42. Lusted LB, Ledley RS (1959) Reasoning foundations in medical diagnosis. Science 130(3366):9–21
43. Marckmann G (2001) Recommendations for the ethical development and use of medical decision support systems. Medscape Gen Med 3(2)
44. McCullough LB, Ashton CM (1994) A methodology for teaching ethics in the clinical setting: a clinical handbook for medical ethics. Theoret Med 15(1):39–52
45. Michael DR, Chen SL (2005) Serious games: games that educate, train, and inform. Muska & Lipman/Premier-Trade, New York
46. Miller RA, Masarie FE, Myers JD (1986) Quick medical reference (QMR) for diagnostic assistance. MD Comput 3(5):34–48
47. Miller RA, Pople HE, Myers JD (1982) Internist-I. An experimental computer-based diagnostic consultant for general internal medicine. N Engl J Med 307(8):468–476
48. Mueller ML, Ganslandt T, Eich HP, Lang K, Ohmann C, Prokosch HU (2001) Towards integration of clinical decision support in commercial hospital information systems using distributed, reusable software and knowledge components. Int J Med Inform 64(2–3):369–377
49. Nash FA (1954) Differential Diagnosis: an apparatus to assist the logical faculties. Lancet 1:874–875
50. New York Times (1959) Computer may aid disease diagnosis. The New York Times
51. Nitsche M (2008) Video game spaces: image, play, and structure in 3D game worlds. MIT Press, Cambridge (Massachusetts)
52. Pauker SG, Gorry GA, Kassirer JP, Schwartz WB (1976) Towards the simulation of clinical cognition: taking a present illness by computer. Am J Med 60:981–996
53. Pople HE, Myers JD, Miller RA (1975) DIALOG: a model of diagnostic logic for internal medicine. In: 4th International joint conference on artificial intelligence, pp 848–855
54. Prendinger H, Ishizuka M (eds) (2004) Life-Like Characters. Tools, Affective Functions, and Applications, Springer, Berlin, Heidelberg
55. Pryor TA, Gardner RM, Clayton PD, Warner HR (1983) The HELP system. J Med Syst 7(2):87–1012
56. Pullen RL (1944) Medical diagnosis. WB Saanders Co, Philadelphia
57. Russell S, Norvig P (2003) Artificial intelligence: a modern approach, chapter intelligent agents, 2nd edn. Prentice-Hall, Englewood Cliffs, pp 31–52
58. Sawyer B (2008) From cells to cell processors: the integration of health and video games. Comput Graph Appl IEEE 28(6):83–85
59. Schuh CJ, de Bruin SJ, Seeling W (2013) Acceptability and difficulties of (Fuzzy) decision support systems in clinical practice. In: Proceedings of the 2013 joint IFSA world congress NAFIPS annual meeting (IFSA/NAFIPS), Edmonton, Canada, 24–28 June 2013, pp 257–262
60. Schwartz WB (1970) Medicine and the computer: the promise and problems of change. New Engl J Med 283(23):1257–1264
61. Self DJ (1995) Moral integrity and values in medicine: inaugurating a new section. Theoret Med 16:253–264
62. Shortliffe EH (1976) Computer based medical consultations: MYCIN. Elsevier, New York, NY
63. Shortliffe EH (1988) Medical knowledge and decision making. Methods Inf Med 27(special issue):209–218
64. Shortliffe EH (1991) Knowledge-based systems in medicine. In: Proceedings of MIE 1991. Springer, Berlin, pp 5–9
65. Shortliffe EH (1993) Doctors, patients, and computers: will information technology dehumanize health-care delivery? Proc Am Philos Soc 137(3):390–398

66. Shortliffe EH (1994) Dehumanization of patient care—are computers the problem or the solution? J Am Med Inform Assoc 1(1):76–77
67. Shortliffe EH, Buchanan BG (1975) A model of inexact reasoning in medicine. Math Biosci 23(3–4):351–379
68. Sliney A, Murphy D (2008) JDoc: a serious game for medical learning. In: First international conference on advances in computer-human interaction, IEEE, pp 131–136
69. Sol HG, Takkenberg CATh, de Vries Robbé PF (1987) Expert systems and artificial intelligence in decision support systems. In: Proceedings of the Second Mini Euroconference, Lunteren, The Netherlands, 17–20 Nov 1985. Reidel Pub., Dordrecht, Boston
70. Spiegelhalter DJ (1992) Bayesian analysis in expert systems. MRC Biostatistics Unit, Institute of Public Health, Cambridge
71. Squire K, Jenkins H (2003) Harnessing the power of games in education. Insight 3(1):5–33
72. Susi T, Johannesson M, Backlund P (2007) Serious games: an overview. Report. http://urn.kb.se/resolve?urn=urn:nbn:se:his:diva-1279. Accessed 12 Dec 2013
73. Turban E, Aronson JE (2001) Decision support systems and intelligent systems, 6th edn. Prentice Hall International Inc, Upper Saddle River, NJ
74. Überla K (1965) Zur Verwendung der Faktorenanalyse in der medizinischen Diagnostik. Methods Inf Med 4(82):89–92
75. Varonen H, Kortteisto T, Kaila M (2008) What may help or hinder the implementation of computerized decision support systems (CDSSs): a focus group study with physicians. Fam Pract 25(3):1627
76. Wardle A, Wardle L (1978) Computer aided diagnosis a review of research. Methods Inf Med 17(1):15–28
77. Warner HR (1992) Medical informatics: application of computers to medical science. In: Plenk HP (ed) Medicine in the Beehive State 1940–1990. Utah Medical Association, Salt Lake City, UT, pp 474–488
78. Weiss ShM, Kulikowski CA, Amarel S, Safir A (1978a) A model based method for computer-aided medical decision-making. Artif Intell 11:145–172
79. Weiss ShM, Kulikowski CA, Amarel S, Safir A (1978b) Glaucoma consultation by computer. Comput Biol Med 8:25–40
80. Wolf MJ (2001) The medium of the video game. University of Texas Press, Austin, TX
81. Wong HJ, Legnini MW, Whitmore HH (2000) The diffusion of decision support systems in healthcare: are we there yet? J Health Manag 45(4):240–253
82. Woodbury MA (1963) The Inapplicabilities of Bayes theorem to diagnosis. In: Thomas ChC (ed) Proceedings of the 5th international conference on medical electronics, Liege, Belgium, Springfield, Ill, pp 860–868
83. World Medical Association (2006) Declaration of Geneva. http://www.wma.net/en/30publications/10policies/g1/index.html. Accessed 18 Nov 2013
84. Zyda M (2005) From visual simulation to virtual reality to games. Computer 38(9):25–32

Ethics of Robotic Assisted Dying

Ryan Tonkens

Abstract The purpose of this chapter is to introduce and critically assess the prospect of "robotic assisted dying", i.e., the use of (semi-) autonomous robots for the purpose of assisting willing terminally ill patients in dying, in the medical context. The central conclusion reached here is this: Assuming that physician-assisted suicide is morally permissible, if we develop robots to serve as human caregivers in medical contexts ('carebots'), and given that assistance in dying is sometimes an important aspect of geriatric care, it is morally permissible for such robots to be able to facilitate and assist in the dying of those patients, in those contexts, at the eligible patient's sound request. At least, there is nothing inherent in this prospect that introduces moral problems beyond those attached to the development and use of geriatric carebots or (human) physician-assisted suicide in general. One major benefit of robotic assisted dying is that the robot would always assist those consenting patients that are genuinely eligible, and thus such patients would not be at the mercy of a willing physician clause in order to have some control over the timing and manner of their death (something that routinely usurps the effectiveness of human physician-assisted suicide in practice).

1 Introduction

Sophisticated robots are already in use for various purposes in medical and healthcare contexts, including robotic surgery, artificially intelligent diagnosticians—the latest task of the famous Watson robot[1]—and robots designed to provide various forms of

[1] http://www.theatlantic.com/magazine/archive/2013/03/the-robot-will-see-you-now/309216/.

R. Tonkens (✉)
Centre for Human Bioethics, Monash University, Melbourne, Australia
e-mail: ryan.tonkens@monash.edu

S.P. van Rysewyk and M. Pontier (eds.), *Machine Medical Ethics*,
Intelligent Systems, Control and Automation: Science and Engineering 74,
DOI 10.1007/978-3-319-08108-3_13

geriatric care (so-called carebots). As this technology becomes more advanced, robots may come to be designed for carrying out other medically related tasks as well, and their level of autonomy will likely increase, as it has in other areas (e.g., aviation and military settings).

No one has yet seriously considered the prospect of developing and using (semi-) autonomous robots for the purpose of replacing or supplementing human physicians in the context of physician-assisted dying. At the same time, within the contemporary literature on machine ethics, there have already been a number of intersections drawn between robotics and *suicide*, including:

1. The development of machines to replace human workers as a response to unacceptably high human suicide rates. For example, Foxconn intends on replacing human workers with robots in several of its factories, as a response to recent tragic events in those factories. The underlying idea is to replace human workers with robots—working in borderline torturous and inhumane factory conditions—in order to decrease suicide rates among human workers in those lousy working conditions, in a cost effective manner, and without compromising productivity.[2] One commentator has recently asked: "If people are being deprived of their rights to wellbeing and dignity and a number are throwing themselves off the roofs of the factories, would it not be better to be replaced by robots?".
2. In a different context, some researchers have proposed the development of "self-sacrificing machines" to be used to human advantage on the battlefield. The inclination towards self-preservation in conditions of war, something that is typical of human combatants, could be "turned off" or simply not programmed into automated military robots, resulting in lethal robotic soldiers without a concern for their own self-preservation [1, 2]. Sending *human* soldiers to perform certain kinds of military tasks may be to demand too much of *them*. Yet, programming robots to execute the same "suicide" missions would, presumably, be ethically permissible, given that robots do not have (e.g.,) a right to life, or a sake of their own that deserves to be respected.
3. Given a certain high level of sophistication of robot technology, we can imagine the self-destruction of self-perceived morally deplorable robots in certain situations [9]: If "moral machines" are ever developed to the point where they take their morality seriously, they will inquire whether their existence, for the purpose of *performing the sort of task that they have been programmed to perform*, is consistent with the moral framework that they are designed to follow. In some cases, perhaps, these robots would be inclined to self-destruct—to commit suicide—in order to behave in line with their ingrained moral code. For example, if creating robots for the purpose of killing humans in war is inconsistent with (e.g.,) a Kantian-inspired moral framework, then we have good reason not to program those robots to abide by that Kantian-inspired moral

[2] 19 people committed suicide in Foxconn factories in 2010–2011, and one potential solution that has been offered for this problem is to replace human workers with robots. See http://www.guardian.co.uk/commentisfree/2011/aug/02/foxconn-robots-worker-suicides.

framework, or else risk their willful self-destruction. In such cases, to commit suicide would be the moral thing for the (Kantian) machine to do, given the circumstances, in order to promote (Kantian) morality as much as possible;

In addition to these, one manner in which artificial intelligence and robotics research *could* intersect with suicide is in the medical context, with respect to end of life care specifically. A long standing and complex moral debate continues about whether and under what circumstances (human) physicians ought to assist their patients to commit suicide, at the patient's sound request. Rather than having human physicians oversee and execute physician-assisted dying, perhaps robots could be programmed to do so instead. In this chapter, I am interested in this prospect of developing and using (semi-) autonomous machines for the purpose of aiding patients in terminating their lives (what I call "robotic assisted dying"), and the ethical issues attached to it.

The morality of the development and use of (semi-) autonomous robots for the purpose of aiding willing patients in ending their lives is *partly* dependent on the morality of assisted dying more generally: If physician-assisted suicide is always morally impermissible, then it is likely that developing and using robots for this purpose would be immoral as well. Yet, here I will assume that physician-assisted suicide is morally permissible, at least under certain well-defined conditions (e.g., where the patient makes a sound request, is terminally ill, is suffering from significant/irremediable pain, and willingly consents to such assistance in dying). Thus, I can qualify my account by focusing on *jurisdictions that currently allow physician-assisted suicide*, where the presumed received morality sides in favor of allowing assisted suicide more generally. These jurisdictions include places such as Switzerland, Belgium, the Netherlands, and several American states, including Oregon, Washington and Montana.[3] Other countries seem to be following suit (e.g., Canada and Australia).

Even if human physician-assisted suicide is morally permissible, however, it does not automatically follow that robotic assisted suicide is morally permissible as well. Before examining some of the relevant theoretical and ethical issues (Sects. 2 and 3 respectively), in the next section I draw a picture of what this sort of robot might look like.

2 Robots for Assisting Patients in Dying

As with other applications of robotics and artificial intelligence research, the prospect of robotic assisted dying has already been envisioned in science fiction. To take one example, in several episodes of the animated television program

[3] http://www.cbc.ca/news/canada/manitoba/physician-assisted-suicide-the-case-for-legalization-1.1314119.

Futurama, viewers are presented with "suicide machines", designed as a sort of terminal phone booth-like contraption, where prospective clients could be euthanized after depositing the required fee into the coin slot. Rather than dispensing candy or cola, these fictional machines are designed for the purpose of killing the occupant of the booth. (One issue with such "euthenization booths" is that they do not seem to discriminate about who is eligible to receive such assistance; anyone that pays the fee receives the "service". This will *not* be the case with the sort of robots that I have in mind here, which *will* discriminate based on patient eligibility, and there will not necessarily be any required payment for their services).

In reality, the closest we have so far come to developing machines for physician-assisted dying is the "death machine" invented and used by Jack Kevorkian (and other machines like it). The main difference between Kevorkian's invention and the sort of robot imagined here is the extent to which the machine is autonomous, and thus the extent to which humans *could* be taken out of the loop without compromising the morality and effectiveness of the act of assisted death itself. The sort of machine hypothesized here could presumably play the role that Kevorkian *and* his machine played, rather than *being merely used by* human physicians.

There are different ways we can imagine such robots becoming manifest. For example, in theory at least, they could be used as a computational tool for gaining and evaluating informed consent from the patient, in an unbiased and systematic manner, in addition to being "pill dispensers", to patients that have satisfied the relevant criteria for being eligible to receive such help in dying. They could be available for use in the clinical or hospital setting, but also, perhaps, in any setting where a patient could receive medical care near the end of her life.

In order to spark the imagination a bit, it may help to consider the context of contemporary automated banking, or similar practices that are becoming increasingly automated, like the modern process of booking and navigating international air travel. In these and other contexts, machines of a significant level of sophistication are already in use, where they can gain information about the client, substantiate that information, provide options and services for the client, etc., and do so with increasing levels of independence from (direct) human oversight. While there are still demands put on the client for providing the required information, for following instructions, for making decisions based on the recommendations of the machine, etc., the role of the machine is extensive. It is an assumption of this chapter—and I take it to be a plausible one—that robots could be designed to oversee and facilitate what has until now been the role of human physicians in the context of assisted suicide in the medical context, as they are coming to oversee and manage other complex tasks in non-medical settings, with little to no human involvement necessary.

Here I am considering robots that could effectively *stand in for* those humans ordinarily necessary for designing and implementing an effective context of assisted suicide. This is not to imply the absence of human involvement and oversight *altogether*—something which may all along be necessary, as we hand over the care and treatment of humans to machines. But, it is to take seriously the (presently hypothetical) prospect that such robots could manage contexts of assisted dying effectively and efficiently, from hearing and evaluating a patient's request for assistance in hastening her death, to prescribing (or perhaps dispensing) the pharmaceutical means and the necessary instructions for doing so effectively. Such machines could be programmed to rigorously adhere to protocols for ensuring safety, effectiveness, and meeting demands of social justice, etc. Such robots could be uploaded with intricate and detailed patient medical histories, information about treatment options and prognosis, details of current law and relevant protocols surrounding physician-assisted suicide, etc. While here I will be assuming that the use of such robots demands a certain level of mobility by the patient, this need not be the case. It seems plausible to imagine the development of robots such as RI-MAN[4] to the point where they will be able to maneuver their environments successfully in real time, including the handling of immobile patients, and thus be able to accommodate the needs of immobile patients (as well as those that are mobile).

If these hypothetical machines were functioning optimally, they could have the capacity to (a) inform patients about their diagnosis, prognosis, and treatment options (something Watson is currently being programmed to do), (b) assess the patient's degree of rational consent to the assisted dying (in some sense, a sort of reverse Turing Test, assessing the competence of the patient, rather than his or her intelligence), and assess the level of understanding of the patient with regards to the relevant information—something busy *human* physicians may not always have the time to do satisfactorily; (c) ensure that the patient can self-administer the drugs; (d) notify the patient's next of kin; and (e) distribute or prescribe the means and instructions necessary for aiding the patient in reaching their termination effectively, and in a way that helps maintain and promote the patient's dignity (their "right to determine the timing and manner of their death"). Thus, the robot could ensure that all of the tenets of (e.g.,) Oregon's *Death with Dignity Act* are accommodated.[5]

With machines of a sufficient level of sophistication, all of this *could* be accomplished without any direct involvement from human medical professionals, although, as already mentioned, we may want to keep humans in the loop to some extent (e.g., as secondary officials for signing off on requests for assisted suicide, for offering second opinions for, or consultations on, diagnoses, as assessors of safety and effectiveness of these robots in practice, etc.). Indeed, built into legislation like that of *The Death with Dignity Act* is the clause that several physicians must agree on diagnosis and independently certify patient mental status, and that there must be witnesses of the written request procedure (after two verbal requests have already been made) that are not the attending physician or members of the patient's family. One such witness could be the robot, and one of the "physicians" could be the robot. Importantly, this clause ensures that humans would stay in the loop, to some extent.

---

4 http://rtc.nagoya.riken.jp/RI-MAN/index_us.html.

5 http://public.health.oregon.gov/ProviderPartnerResources/EvaluationResearch/DeathwithDignityAct/Pages/index.aspx.

I accept it as a premise that the development of such machines is a genuine possibility, given current and foreseeable developments in robotics and artificial intelligence research.[6] Here I also assume, without much argument, that these robots could be used safely and effectively in the context of assisted dying, recognizing this for the open empirical issue that it is. If these robots could *not* be used safely or effectively, then this would be very good reason to reject this project altogether. At the same time, there is no reason to believe that such safety and effectiveness is a technological or practical impossibility. And, studies could be designed during the prototype phases of developing those robots, and their safety and effectiveness could be assessed prior to their use in practice. Mock scenarios of assisted dying could be acted out, where the confederate simply does not consume the pills that they have been prescribed by the robot (say). This would give us a good indication of whether the robot could manage practical contexts of assisted suicide safely and effectively, prior to its dealing with real patients.

2.1 Theoretical Considerations

What is under direct ethical analysis here is *not* the morality of physician-assisted dying (something I am assuming to be morally permissible), but rather whether the development and use of (semi-) autonomous robots for this purpose is itself morally permissible. One general medical ethic that we may appeal to—and it is important to stress that there are others, a discussion of which is beyond the scope of this chapter—is the principle-based medical ethic developed by Beauchamp & Childress [3] (most recently in 2013), which takes justice, autonomy, beneficence and non-maleficence as central moral principles for tackling complex bioethical issues. I want to suggest that, *prima facie* at least, there is nothing about this prospect that would necessarily violate any of these principles, *beyond which accompany human physician-assisted suicide*. Thus, there is nothing inherent in robotic assisted dying that would be unjust, or unduly restrict the autonomy of the patient, or unduly harm the patient, beyond the ways that human physician-assisted suicide could.

All of the requirements of justice in human physician-assisted dying would presumably be in place in the context of robotic assisted dying as well. Safeguards and regulations could also be put in place to ensure that all those that are eligible for access could have access, and that no one that does not desire access is killed. And, as humans would be the ones programming such robots, there is no real risk of these machines initiating a slippery slope (i.e., to begin to become more inclusive or less strict in what cases they allow for assisted suicide) since this would require humans to go down that slope first, and then *for them* to reprogram the robots to follow suit.

[6] I acknowledge this as the empirical and controversial claim that it is. For a less optimistic view, see Sparrow and Sparrow [7].

With respect to patient autonomy, nothing new emerges in cases where (semi-) autonomous robots are doing the assisting that is not already present in cases where human physicians or palliative care workers are doing the assisting. We would still only allow assisted suicide in cases where the patient has made an autonomous request for it, and where she is terminally ill and/or in severe (irremediable) pain. There would still be measures in place to ensure that informed consent procedures have been observed, and that the patient is sufficiently rational and autonomous in their decision-making; indeed, such safety measures could be programmed into the very interface of the robot itself. Moreover, such machines would not be programmed with the ability to paternalistically overrule *sound* requests of the patient, which is especially important in cases where the patient expressly asks *not* to be killed (although such cases may not emerge in practice, for these robots would only have access to patients that have sought out their assistance), or in cases where she has requested assistance in dying but later changes her mind. Somewhat ironically, given a sufficient level of autonomy *in the machine*, not only does robotic assisted suicide not seem to threaten patient autonomy, but indeed seems to promote it more so than human physician-assisted suicide could in practice. Without being left to the idiosyncrasies and fallibilities of human physicians, or indeed to their being willing to provide assistance in the first place, the seeker of robotic assisted suicide could have more control over the timing and manner of their death. To take but one example: Ordinarily there is a regulated time delay (e.g., 15 days) between patient request for assistance and the prescription of the required drugs by the physician. But, very often (busy, or perhaps somewhat reluctant) physicians take longer than this to provide the prescription, placing undue delays in honoring the patient's sound request. Robots could more effectively adhere to regulated timelines.

The same procedural rigor in robotic assisted dying that is present in contexts of (legal) human physician assisted-suicide, coupled with continued human oversight, would go a long way—perhaps *all the way*—towards avoiding system malfunctions leading to undesired outcomes for the unwieldy patient. Insofar as we have proven safety and effectiveness in our rigorous analysis of the functioning of prototypes of such robots, there is no reason to believe that using robots in this context raises issues of potential harm to the patient, *beyond those that are present in the context of human physician-assisted suicide* already.

From these (admittedly brief) remarks, it seems that having the robot assist in the suicide of patients does not introduce moral problems beyond those that already accompany human physician-assisted suicide (a practice that we are assuming is morally permissible), at least not any in terms of justice, autonomy, beneficence, or non-maleficence, and thus no clear red lights have emerged from our analysis thus far.

The most obvious and straightforward way to introduce such machines is through equipping those carebots that come to exist with the ability to assist in the suicide of the patients that they care for (at the patient's sound request). Yet, some have argued that the development of carebots itself may be morally dubious (e.g., [7]). My goal in the remainder of this section is to discover whether there is anything about using such robots *also* for aiding in the assisted dying of patients that is morally problematic, beyond the extent to which these worries may be legitimate threats against carebots more generally; if we assume the moral permissibility of carebots, are there any corresponding moral concerns that are specific to robotic assisted dying?

If developing carebots in general is morally impermissible or imprudent, then developing *these* robots to also assist in the suicide of patients would likely be morally dubious as well. Yet, at the same time, there could be good independent reason to develop robots for assisting in patient suicide, and thus the moral standing of these machines does not *necessarily* depend on the morality of carebots in general. (One such reason, which I consider in detail later on, is the extent to which robots could help to bypass the "willing physician clause" that often prevents otherwise eligible patients from receiving assistance in dying). As shown below, the moral issues with carebots in general considered here do not accompany robotic assisted dying taken in isolation.

The two main worries mounted against geriatric carebots that I would like to consider are the potential negative impacts on the dignity of patients under the care of such robots, and the extent to which the widespread use of such robots would eliminate a central and important human element (including human contact) from the context of care. Some commentators on robotic geriatric care have worried about issues of dignity, loneliness, and the removal or absence of human contact from the context of care. Patients in these contexts are already vulnerable, and replacing their human carers with robots would exacerbate these problems, at least so the worry goes.

Could a person that is assisted in dying by a robot "die with dignity"? The first thing to note is that, just because a robot is now included in the loop, it does not mean that no humans would be in the loop. As noted earlier, there is good reason to keep humans in the loop, even if robots could effectively manage this context all by themselves. A second thing to note is that, in robotic assisted dying as in human physician-assisted suicide, the patient is still the one killing herself, and thus it does not make sense to argue that the "robotic death" is undignified *because it is caused by the robot*, for it is not. However, we might worry that involving the robot on the front line in the context of assisted suicide would somehow compromise the dignity of the patient seeking assistance. For instance, one way that human dignity could be compromised is if the patient is disrespected or exploited through robotic assisted dying.

This sort of worry, however, is misplaced. There is nothing about the context of robotic assisted dying that would prevent the patient from being respected, as long as their autonomy was itself respected, something the robot would be programmed to monitor and champion. Indeed, the robot could *not* treat the patient in any way(s) beyond a very restricted set of actions, e.g., asking her some questions, assessing her competence, prescribing drugs, etc., and would thus *not be programmed to perform any actions or to treat the patient in any way that would be disrespectful to her or exploit her.* If the robot was functioning optimally, it would not deny assistance to any patient that qualifies, and would not kill any patient that did not desire to die.

The worry about respecting human dignity is closely linked to the second main worry that I would like to discuss, namely the absence of human interaction and contact (at the end of life). Dying alone, without any one else around other than a robot, seems to strip the patient of some dignity, or an important good (i.e., human contact), and, because of this, is morally suspect. Robotic care (and *mutatis mutandis* robotic assisted dying) is impersonal in nature and, in theory, could occur without any direct human involvement. Because of this, the patient would lose out on human contact that she may otherwise receive, if she was cared for by humans (at the end of her life).

But, to reiterate, the inclusion of robots in this context would not eliminate others (humans) from being present in this context as well. Moreover, if (in some, tragic cases) the arbiter of assistance in suicide is the only contender for contact at the end of a particular patient's life, then it may not really matter whether such limited contact is human or robotic, for she would "die alone" regardless. Furthermore, nothing about robotic assisted dying denies the presence of (human) family and loved ones from being present at every stage of the process. Indeed, as with contemporary physician-assisted suicide legislation and protocols, it could be part of the robot's programming to demand that some human involvement (e.g., a support system for the patient) be present, or at least the robot could be programmed to encourage such human contact throughout the process (e.g., by contacting next of kin on behalf of the patient, say). One might even speculate that, knowing that a robot is involved with such an important aspect of the patient's care, may motivate otherwise reluctant human caregivers and kin to become *more* involved than they otherwise might have been.

There is no reason to believe that robotic assisted death could be any more undignified than human assisted death (if it is), or that the patient that receives assistance in dying from a robot, rather than a human physician, would necessarily have less human contact or be more (or less) lonely at the end of her life. Thus, if we are willing to allow human physician-assisted suicide, in the face of the potential threats of compromising patient dignity and insufficient human contact, then these concerns should not be regarded as sufficient to reject robotic assisted dying either. While it could turn out that widespread use of carebots in general contributes to loneliness, loss of human contact, or strip dignity from the context of care—I am not assessing *these* claims here—they do not pose a challenge to robotic assisted dying specifically.

3 The Ethics of Robotic Assisted Suicide

To some, the prospect of *robotic* assisted suicide will be met with disapproval, and perhaps even disgust. Despite what has been argued for so far, the thought of developing robots *for the purpose of* killing humans may seem to be (inherently) morally unacceptable. Indeed, we may expect the sort of uproar that emerged once the public became aware of the existence and use of Kevorkian's "death machine".

Why might we consider the prospect of robotic assisted dying *at all*? One reason to consider the prospect of developing and using robots for the purpose of assisted dying is to maintain consistency in our practical morality. I have in mind here the idea of aligning our moral ideals across different contexts (e.g., the military and medical contexts). We already *do* allow the development and use of certain kinds of "killer robots" [6] in certain contexts. The paradigmatic example here is the development and use of semi-autonomous machines for military and armed defence initiatives. It is not a far stretch to imagine the use of machines for aiding humans in committing suicide, *something they desire and consent to*. While those that are uncomfortable with the prospect of developing robots for assisting patients to die may also be uncomfortable with developing autonomous lethal robotic systems for military use, there is some precedent here already, and we should not dismiss either prospect out of hand, without good argument.[7].

Although there are some passionate supporters of developing and using autonomous lethal robotic systems in the context of warfare, this enterprise has also attracted sustained criticism. For example, recently [5] have argued that developing robots for the purpose of killing humans on the battlefield is morally unacceptable since "doing so disrespects the inherent dignity of the human, such robots cannot fully understand the seriousness of killing a human, and the use of such machines violates military honor." While I am sympathetic to this sort of position with respect to *the military context* [8], there may be relevant dissimilarities between robotic killers in war and robotic assisted suicide in the medical context.

For instance, (1) as argued for above, there may be nothing (inherently) unjust or undignified about robotic assisted dying in the medical context, given sound patient consent practices. Part of what makes death by military robots undignified or disrespectful (if it is) is that typical human soldiers do not have a chance against a robot (and the human soldier *does not want to die*), and such robots do not have anything of comparable moral worth at stake. Yet, in the context of physician-assisted dying, the patient has made a considered request to die, in a manner that she believes offers herself a better chance at maintaining her dignity than she would be afforded through succumbing to the terminal illness from which she suffers. (The general disgust generated by Kevorkian's use of the "death machine" may not have been so much to do with the *means* of assisting his patients in dying, but rather disgust at the very idea that he was assisting in their dying); (2) if the *appropriate* use of technology is to (in part) *benefit humans*, then aiding in facilitating the sound end-of-life decisions of terminally ill patients may actually be a good use of technology, rather than an ignoble one; and (3) such robots would not need to understand the seriousness of aiding patients to commit suicide (and the seriousness of death more generally) beyond ensuring autonomous and informed

[7] Elsewhere [8], I have argued that autonomous robots should be programmed to be pacifists (rather than 'warists'). While I will not argue for it here, there is nothing about robotic assisted dying that would be inconsistent with that view, despite the fact that such robots would be contributing to the death of their human patients (at the patient's sound request).

consent from those patients, and thus that the patient has good reason to request assistance in dying, and has pursued this course of action in the absence of undue coercion. The robot would be equipped with the ability to assess the extent to which the patient understands *the seriousness of her request* and the outcomes of that request, but need not itself understand anything more about human death, and what it means for a human to die. What matters here, and something that is not necessarily present in the corresponding military context, is that the patient desires to die, for good reasons, and is seeking assistance in doing so, from the robot—in the great majority of cases because she cannot bring her own death upon herself in the desired manner without assistance.

The final of Johnson and Axinn's charges (i.e., disrespect of military honor), when transferred to the medical context, *may* be more challenging to overcome, given that the medical ethic has traditionally been grounded on promoting the health (and life) of patients under the care of medical professionals, rather than to hasten their death—even, it is thought by many, when those patients make a sound request for them to do so and are terminally ill. I return to this issue later in this section.

There are other reasons to consider the prospect of robotic assisted dying. However speculative it may be at present, robots could be better at negotiating certain aspects of the context of physician-assisted suicide than human physicians, especially considering that they need not be programmed with certain psychological and emotional underpinnings, or could be programmed with some (e.g., empathy and compassion) but not others (e.g., discrimination, bias, capacity to experience stress or fatigue, etc.). Moreover, just as Watson proved to be a superior expert in trivia than its human competitors, it could also be the case that Watson will be superior to typical human physicians at diagnosis and prognosis, running patient medical histories, reliably assessing patient competence, etc. And, given its computational abilities, it may also be more thorough and attentive to detail when assessing patient requests, consent, etc., in the context of assisted dying. *If* robots could be better at assisting patients at the end of life (or in certain aspects of that context), then this would be good reason to introduce robots for that purpose.

A stronger reason for considering the prospect of robotic assisted suicide in the medical context is because of the emergence of geriatric care robots, especially in countries that are expecting an influx of elderly citizens requiring care (e.g., North America, Denmark and Japan), and especially where the number of human workers and caregivers will be far fewer than that required to offer acceptable (and affordable) care for those people. As physician-assisted suicide becomes more widely accepted, to the point where it is becoming an important element of an expanding scope of end of life contexts for a growing number of people, and given that there has been a strong recent surge in the development of (semi-) autonomous robots for relief/aid in the geriatric care sector in many jurisdictions, then it seems plausible to ask whether these geriatric care robots might also be equipped with the ability to diagnose, evaluate, and aid in the end of life decision making of those people under their care, including facilitating the patient's end of life decisions. Thus, given that (semi-) autonomous carebots play a significant role in the care of patients already (or are likely to do so in the near term future), and that part of that care may often involve decisions and actions at the end of that patient's life, then it makes sense to equip such robots to be able to handle those contexts. This point is strengthened to the extent that the arguments of the previous section are on track as well.

To the best of my knowledge, no one has yet suggested that such robots should be able to willingly contribute to bringing about the death of the people with whose care they have been entrusted; to the contrary, the assumption has so far been that such robots would work towards maintaining and promoting the life and quality of life of their patients. Yet, in related contexts, such as palliative care and hospice settings, it is no longer uncommon for part of *care* at the end of life to consist of aiding the person cared for to die in a manner of their discretion (within reason), something that is thought to promote the autonomy and dignity of that patient, at a time where they are especially vulnerable.

Perhaps the strongest reason in favor of considering the development and use of robots for assisted suicide has to do with the practical limitations of human physician-assisted suicide. A major roadblock for advocates of (human) physician-assisted suicide in practice has been the reluctance of many physicians to get on board, and to accept physician-assisted suicide as morally permissible, even in highly qualified (relatively uncontroversial) contexts. Not only do most jurisdictions continue to have legal prohibitions on assisted suicide, but, even in those that do not, it is often difficult for patients to find the means and expertise to help them facilitate their end of life decisions. This is a complex issue, but it has at least two central elements: (a) the idea that "killing" or hastening the death of patients is against the ethic of medical professions (doctors, nurses, pharmacists, etc.), and (b) that physicians have the right to conscientiously object to certain kinds of requests made by their patients, in case those requests would require the physician to disregard or compromise important moral or religious beliefs of their own. At best, it seems that advocates of physician-assisted suicide are left confronted with a sort of *willing physician clause*, advocated by thinkers such as Dworkin et al. [4]: "patients in certain circumstances have a right that the state not forbid doctors to assist in their deaths, but they have no right to compel a doctor to assist them. The right in question, that is, is only a right to the help of a willing doctor."[8]

One possible way around this willing physician clause may be to develop robots that could stand in for human physicians. In this case, in places where such robots were available, there would always be a "willing physician", i.e., an entity with the relevant expertise and ability to effectively and efficiently aid the terminally ill patient with her suicide. In this way, the (putative) rights of patients to dictate the timing and manner of their death could be respected without infringing on the rights of their human physicians and caregivers, as the robot would be the one doing the assisting.

[8] http://www.nybooks.com/articles/archives/1997/mar/27/assisted-suicide-the-philosophers-brief/?page=1

Robots need not be programmed to have any religious beliefs of their own, and thus need not have any beliefs and values that would conflict with the idea of assisting in the death of a specific sort of patient—although such robots would be designed to respect the religious beliefs of their patients (in case this information became relevant in that context). Moreover, while we would expect the machines to behave morally, and thus to be programmed with and to follow a set of sound moral prescriptions, it would be an assumption of their very programming and design that certain instances of assisted dying *just are* morally permissible—insofar as these robots are developed to be used in countries and states that share that same moral ideal. Human physicians that do not honor the request of their patients for reasons of conscience (in jurisdictions where physician-assisted suicide is legal) do so not because they believe that the patient does not qualify. Rather, they do so because they believe that no patient should qualify, in the sense that aiding in the death of patients is against personal or professional values. In the case of robotic assisted suicide, however, *all* genuine requests that meet the standards of informed consent, and where the patient is terminally ill and/or suffering from severe pain, would be honored by the robot, without sparking moral or religious conflict.

Those patients that qualify for receiving assistance in dying from medical professionals in jurisdictions that allow physician-assisted suicide would be the same that qualify to receive such assistance from a robot, in those jurisdictions that allow physician-assisted suicide and have those robots available for use. Granting the morality of certain cases of physician-assisted dying would be consistent with their (professional) medical ethic, and thus no conflict could emerge here between the robot's beliefs about the morality of assisted suicide in the medical context and the sound expressed desire of its patients to receive assistance in dying. While we would not want robots to abide by *all* requests for assisted death, but rather only those that are sound (i.e., are consistent with current law, meet informed consent regulations, etc.), we would not need to program such robots with the ability to overrule, for reasons of its own, the sound request of terminally ill patients to assist in their dying. For these reasons, robotic assisted dying offers a way to side step the willing physician clause that often hovers over the context of (human) physician-assisted suicide, often even in jurisdictions where it is legal to assist in a patient's death.

The promise of robotic assisted dying in this regard works from the other direction as well: while some human physicians are willing to aid their patients in their suicide, often despite the fact that they find it difficult and see *some* conflict with that action and their professional regulative ideals as healers and promoters of health, many of those physicians may be glad to have someone (some*thing*) else take over that role. While this is an empirical issue, it seems plausible to suggest that introducing robots for the purpose of organizing, facilitating and overseeing the context of assisted dying in the medical context would be welcomed by many human physicians, including some of which who are willing to help their patients die, given that they are the ones best positioned (qualified, knowledgeable, etc.) to do so, but feel regret at the prospect of dirtying their hands thereby. With the introduction of robotic assisted dying, they would no longer need to do so, or feel pressure to do so, or need to mute their moral or religious beliefs in order to provide their patients will the care that they desire.

It should be noted, however, that others may be distraught by the existence of machines (housed in hospitals, say) that have the sole purpose of assisting in the death of certain classes of patients. Yet, *despite the fact that there are no relevant moral differences between robotic and human assisted dying*, these same people will likely have a similar attitude towards *human* physician-assisted suicide—a practice that routinely occurs, even in places where it continues to be illegal.

4 Conclusion

There will likely be some fear surrounding the thought of affording (semi-) autonomous robots with the ability to determine whether a patient should receive assistance in dying, especially if the machine itself had the means and ability to kill the patient. The first way to respond to this worry is to emphasize that the sort of robot envisioned here would be programmed to *prescribe* a lethal dose of pharmaceuticals *at the qualified patient's sound request*, and the patient (alone) would always be in charge of taking her own life (rather than being killed by the robot). The prospect of such machines going rogue or malfunctioning would always be present (however minimal this threat may turn out to be in practice). Yet, this is always going to be an issue, even with *human* physicians. Keeping humans in the loop, and establishing strict regulations and safety protocols could mitigate this threat significantly. And, given that there are clear conditions that patients need to meet in order to be eligible for such assistance, the risk that robotic assisted suicide would be granted to patients that are not terminally ill, or do not meet the standards of informed consent, or have been misdiagnosed, etc., is miniscule (as it is presently in the case of human physician-assisted suicide in states that have legalized that practice).

Since this is the first philosophical analysis of the ethics of robotic assisted dying to appear in the literature, there is a risk that important points have been overlooked, and opposing arguments gone unconsidered. To be sure, our understanding of this novel and complex ethical issue will benefit from sustained discussion and continued research. With these concessions in mind, the central conclusion of this chapter is a conditional (and thus tentative) one: based on the arguments offered throughout this chapter, assuming that physician-assisted suicide is morally permissible, if we develop robots to serve as human caregivers in medical contexts, and given that assistance in dying is often an important aspect of that care (near the end of the patient's life), it is morally permissible for such robots to be able to facilitate and assist in the dying of those patients, in those contexts, at the eligible patient's sound request. At the very least, there is nothing inherent in this prospect that introduces moral problems beyond those attached to the development and use of geriatric carebots in general or (human) physician-assisted suicide.

The most glaring benefit of robotic assisted dying is that the robot would always allow such assistance to those patients that are genuinely eligible, and thus their patients would not be at the mercy of a willing physician clause in order to have control over the timing and manner of their death.

References

1. Arkin R (2013) Lethal autonomous systems and the plight of the non-combatant. AISB Quart 137:1–9
2. Arkin R (2010) The case for ethical autonomy in unmanned systems. J Mil Eth 9:332–341
3. Beauchamp T, Childress J (2013) Principles of biomedical ethics, 6th edn. Oxford University Press, Oxford
4. Dworkin R, Nagel T, Nozick R, Rawls J, Thomson JJ (1997) Assisted suicide: the philosophers' brief. New York review. Available via http://www.nybooks.com/articles/archives/1997/mar/27/assisted-suicide-the-philosophers-brief/?page=1. Accessed 28 Dec 2013
5. Johnson AM, Axinn S (2013) The morality of autonomous robots. J Mil Eth 12(2):129–141
6. Sparrow R (2007) Killer robots. J App Phil 24(1):62–77
7. Sparrow R, Sparrow L (2006) In the hands of machines? The future of aged care. Minds Mach 16:141–161
8. Tonkens R (2013) Should autonomous robots be pacifists? Eth Info Tech 15(2):109–124
9. Tonkens R (2009) A challenge for machine ethics. Minds Mach 19(3):421–438

Automating Medicine the Ethical Way

Blay Whitby

Abstract Greatly increased automation in medicine is probably inevitable. Artificial Intelligence (AI) is being introduced into medical practice is numerous ways from machine diagnosis, through robot surgery to psychotherapy that is delivered entirely by computer. However, there are many ethical challenges involved. The emerging field of machine medical ethics currently lags behind medical practice and contains many unresolved debates. In aviation a very high level of automation has brought increased safety and efficiency. However, this was not achieved by simply building the technology and requiring professionals to use it. Lessons from the aviation industry suggest that issues of acceptance and resistance by professionals can be successfully handled if they are fully engaged in the operational and procedural changes at all stages. Negotiation over procedures and responsibility for errors in aviation is complex and informative for other fields. If the aviation industry exemplifies the potential benefits, then the Information Technology (IT) industry, particularly AI, highlights the dangers. Since this industry produces the 'machine' part of machine medical ethics, it is unfortunately necessary to observe that, historically the IT industry has demonstrated rather low standards of ethics and social responsibility. Improved ethical awareness and professionalism will be needed for IT professionals to achieve ethically acceptable technology in medicine. Long-standing unresolved debates in IT ethics now need some conclusion. These include the ethical responsibility of IT professionals for unreliable technology or for human errors resulting from poor user interfaces.

B. Whitby (✉)
Department of Informatics, University of Sussex, Brington, UK
e-mail: blayw@sussex.ac.uk

S.P. van Rysewyk and M. Pontier (eds.), *Machine Medical Ethics*,
Intelligent Systems, Control and Automation: Science and Engineering 74,
DOI 10.1007/978-3-319-08108-3_14

Given the pace of progress in both technology and clinical practice, there is an urgent need for progress in machine medical ethics. A brief set of initial recommendations is provided.

1 Introduction

Modern technology has the potential to influence medical practice to an unprecedented degree. Already Artificial Intelligence (AI) systems act as psychotherapists [1, 2] as diagnostic advisers [3] and AI is routinely built into many medical devices and online services. It is inevitable that this process of automation will continue. There is no doubt that AI can perform many of the functions of medical practitioners: largely because it is already doing so, in the lab if not in practice.

It is important to realize that, rather than a team of robots suddenly taking over a hospital. What we are observing is medical technology that simply gets progressively smarter. The fact that the change is gradual, rather than sudden, makes producing appropriate ethical responses more difficult. In many cases, the introduction of automation is seen as just another tool for medical professionals to learn and as having no ethical consequences whatsoever. This view is incorrect. The sum total of these many small individual changes in medical technology will be fundamentally to change the ethical role of medical professionals. It also makes the role of other professionals, particularly those who design and build the technology, highly ethically significant.

It is important and urgent therefore, to provide ethical guidelines *now*. To wait for the process to develop is not helpful. There is no clear point in this gradual introduction of automation at which we can say that medical practice has changed significantly and therefore ethics and procedures should change. Instead, what is necessary is recognition of the gradual change process and clear steps to ensure that existing medical ethics are not circumvented or ignored.

This should most certainly not be read as pure conservatism about machine medical ethics. There is no doubt that the emerging field of machine medical ethics will generate many truly novel ethical problems. It is also important not to produce regulations and guidelines which inhibit the development of new AI technologies which could have immensely positive ethical impacts. The position taken in this chapter is that there is already a process underway. The well-established field of medical ethics is being mingled with the much newer and, frankly often inadequate, field of technology ethics. The most urgent task therefore, is to prevent any lowering of ethical standards during the change process. A set of specific recommendations is provided at the end of this chapter.

Some salient features of this gradual change process can be usefully illustrated by looking at another field—that of aviation—which has progressively introduced a very high level of automation during the last 65 years.

2 Lessons from Aviation

The aviation industry has pursued automation to a very high degree, most noticeably in the period from 1947 to thc present.[1] There have been many problems but they have, to a large degree, been solved and the current trend is towards even higher levels of automation. Aviation is therefore an important source of lessons for other disciplines. It is crucial to note that these are *lessons* and what is not being suggested here is any slavish direct imitation of aviation practices.[2]

Automation was introduced gradually into aviation. This is an important point with respect to the likely pace of its introduction into medicine. In 1947, most airline pilots (and their managers) were ex-military with flying skills honed in highly dangerous skies. They often had a great deal of physical courage and strong belief in their own abilities. This led in turn to resistance to the very idea of automation. It would, they claimed; tend to inhibit their ability to take the initiative and to solve problems in real time. These attitudes were remarkably similar to those observed more recently in medical practitioners [6].

Early automation was simply designed to help pilots and to ease the, often extremely physically demanding, workload of controlling airliners on long flights. Autopilots had been developed well before 1947—the first licensing of an autopilot in civilian air transport was in 1931—but their use was obvious and attractive to aircrew and management. They simply relieved pilots of the tedious task of manipulating the flight controls for what could be many hours during a long-distance flight.

Early acceptance of autopilots was probably greatly aided by the fact that they could be switched off. Until the 21st Century it was an established principle that pilots should be able to override or switch off an autopilot *at any time*. As defective autopilot input was detected as a contributory factor in a small number aviation incidents and accidents, pilot training was required to include the retaking of control in a timely manner from a defective autopilot. It is not clear how common this sort of training is in current medical practice but it is something worthy of serious consideration.

The post 1947 period also benefitted from the great advances that had been made in radar and in radio navigation techniques during World War 2. As the modern aviation industry developed, it was inevitable that this technology would progressively replace individual initiative in navigating airliners. Pilots had to accept the authority of ground-based Air Traffic Controllers: first over busy airports and then over busy pieces of sky. There was resistance, of course. However,

[1] 1947 is the year in which The Convention on International Civil Aviation, sometimes known as The Chicago Convention 1944, became effective. It was, in many important ways, the beginning of modern, regulated civil aviation [4].

[2] Nor is being suggested that the historical process of automation in aviation has run its course. There are many current scientific debates on the process of automation and on just how much of the flying task can usefully be taken out of human control Sheridan and Parasuraman [5].

the air transport industry openly declared a primary focus on achieving greater safety and most individual changes in aviation practice could be clearly justified in terms of increased safety. These changes did indeed result in progressive loss of opportunities for pilots to take the initiative under many circumstances. However, they also resulted in clear and measurable improvements in safety. Modern pilots are no less skilled than those of the 1940 and 1950s. There is no doubt that they have become skilled in different ways and that the daily work of aviation professionals has changed out of all recognition. A similar change in the skills of health-care professionals is now taking place.

The flight controls of modern airliners are also highly automated. The biggest single leap forward was the Airbus 320 series which introduced the concept of "fly-by-wire." In fly-by-wire aircraft the pilots' control inputs are no longer directly connected to the aerodynamic surfaces of the aircraft. A bank of computers handles the aerodynamic task of flying the aircraft. The human pilots essentially program high-level commands as to where the aircraft is to go—all the details of how actually to fly to this point are handled by the computers. Fly-by-wire enables more efficient flight since the computers automatically optimize the aerodynamic profile of the aircraft resulting in fuel saving. It is true that aviation is easier to convert into routine than is medicine but there are many routine tasks in medicine which will soon be automated in this way—mainly to optimize the use of an expensive resource: the time of skilled professionals.

Of more relevance to this chapter are the effects of the introduction of fly-by-wire on safety. Here the picture is more confusing and, to some extent, debatable. The protection built into fly-by-wire systems, particularly those of the recent products of Airbus[3] prevent pilots taking the aircraft out of the "normal flight envelope." That is to say that the computers can ignore any pilot control inputs which would put the aircraft in a dangerous flight condition. In principle, this should result in a great increase in safety but the safety gains have only been partial. There have been a number of accidents and incidents involving highly automated aircraft. It is important to learn from these and apply the lessons in medicine. More devices will soon appear, specifically designed to prevent medical professionals from operating outside the medical equivalent of a "normal flight envelope." Just as in aviation, this will seem at first glance to provide clear safety benefits. However, the lesson of aviation is that these can follow only from changes in training and procedures.

In their paper *Automation Surprises* Sarter et al. [7] identify a number of unexpected problems that developed with human-automation interaction in aviation. These include complacency and trust in automation, new opportunities for new kinds of error, and the fact that automation does not always reduce workload because it causes very uneven distribution of workload. All the above are

[3] Although Airbus introduced the first fly-by-wire airliner and have continued to develop the concept further, almost all contemporary manufacturers of airliners have adopted this concept to some degree.

of great relevance in the field of medicine but most directly relevant to machine medical ethics are two further unexpected issues. These are the need for different approaches to training and the need for different approaches to co-ordination.

Essentially, they say, those who design and introduce automation assume that it simply substitutes for human activity but this is rarely, if ever, the case in practice. The problem of automation causing a concentration of workload at crucial times has been noted in the reports on certain aviation accidents.[4] The need to train flight crews specifically to handle automation, rather than merely handling flying has been long established in aviation. It is now becoming vital in medicine to train doctors *specifically to handle automation* rather than merely to handle medicine.

3 Lessons from IT

The experience of the aviation industry suggests that automation effects will be largely positive although there are clear problems to be solved. The specter of AI outperforming and displacing human medical expertise is also nullified by the aviation example. Human physicians will almost certainly learn to co-operate with AI in medicine, just as human aviators learned to fly fully-automated aircraft. This proved challenging but not impossible.

Some less positive warnings are provided by the IT industry. Whereas the aviation industry, like medicine, is highly regulated, the IT industry is barely regulated at all. Just as the public expects their airline pilots to be licensed and to have demonstrated their ability to fly in a recent skills check, they also expect the mechanics who serviced the aircraft before the flight to hold the appropriate licenses. Medicine too has an established licensing structure and, for the most part, the public has no problem seeing the dangers inherent in allowing unqualified doctors to practice.

It is remarkable therefore, if not downright alarming, that these highly qualified practitioners can and do base crucial decisions in their daily practice on software that has been produced entirely by gifted amateurs.[5] Experts in IT need have no formal qualifications and need conform to no professional standards. It is true that some IT workers do take professionalism and professional standards seriously but the vast majority prefer to remain outside such structures.

Of course IT is a lot younger than medicine and about half the age of aviation. There is maybe still time for it to adopt more professional attitudes but it would

[4] [8], AA965: near Buga, Colombia, December 20, 1995.

[5] There are many specific qualifications in programming and software but almost no formal requirements for any individual programmer or software designer to actually hold them. Many professional organizations, in particular BCS, The Chartered Institute for IT, are working to change this but the IT industry is very resistant to any attempt at regulation. BCS has declared a strategic goal of certifying all professionals (BCS [9] but this is very far from being the case at present.

nonetheless be appropriate to apply pressure on the IT industry from outside. One way in which this can *and morally should* be done is to place ethical requirements on programmers, system designers, and interface designers who choose to work in the medical area. These ethical requirements should reflect medical, rather than IT ethical practice.

An often-heard counter argument to this position is that regulation, and especially hasty and early regulation, will inhibit both the development and take-up of automation in medicine [10]. It is particularly difficult to regulate AI because some AI systems may be designed to adapt their behavior ‘on the job’. This in turn means that it is impossible to predict their performance with any accuracy. These arguments must be taken seriously. Small, high technology enterprises may be inhibited from innovations in medical AI if they feel that there are many regulatory hurdles to be overcome as well as technical ones.

For this reason, a light-touch and responsive attitude to regulation would be preferable. Instead of enforcing legislation in advance of the take-up of technology, it would be better to monitor and regulate on the basis of experience, as has been done in aviation. There will be mistakes but aviation became safe by learning from its mistakes and then using regulation to ensure that they were not repeated.

4 Mode of Failure Training

A clear problem with the introduction of AI in most areas of human activity is caused by the way in which such systems can fail or underperform. The failure mode of AI systems is very different from pre-AI technology.

Most people will by now be familiar with the way in which conventional computer systems behave in the event of some sort of failure. The system will usually hang, thrash, or loop, produce incomprehensible output, or produce no output at all. In these cases the fact that something is wrong is fairly obvious. To reinforce this, everyone who is familiar with working with IT has learned to recognize these types of system behavior and, for the most part, learned how they should respond. Unfortunately, AI systems in medical applications are unlikely always to fail in such an obvious manner. Indeed the more sophisticated the system; the more likely it is to have subtle and obscure modes of failure. Systems containing medical knowledge, for example, may be out of date. Systems incorporating heuristics may be out of their normal range on *this particular case.* Systems incorporating more open-ended AI techniques, such as artificial neural nets may simply become impossible to predict in practice.

Consider first, the case of an AI system used in a medical application which has an obscure design fault. If the fault were in any way obvious, one would hope that it would be detected and rectified before the system was used in a real medical context. However there is, as yet, no way of exhaustively testing software or even any formal testing requirements before software is used in hospital wards. It is not in any sense implausible to suggest that undetected software errors will exist in medical systems in service.[6] If the system is used by trained medical professionals then their medical knowledge should, ideally, allow them to detect the obscure design fault. On the other hand, if the system is used as part of a staff de-skilling change, as it sadly may well be, then operators without the relevant specialized medical knowledge will be much less likely to spot any problems.

This situation is potentially dangerous but made very much worse by developments in AI. Unlike the familiar patterns of computer system mentioned above, AI systems have the potential to deliver outputs that are *nearly right* and can also give plausible arguments or explanations in support of their erroneous deductions. This type of failure could certainly only be detected by a medical professional with good specialist knowledge of the area covered by the AI system.

Of course, human experts are prone to a similar error pattern, especially in the medical domain. There is good justification for the common practice of seeking a 'second opinion' in the case of human experts. The key argument of this section is that the use AI in medicine will also need second opinions. This is an important requirement of the ethics of its introduction. It cannot safely be used to replace human expertise. Even using AI systems to check on each other (which is good practice but rarely actually done) will not remove the need for specialist human expertise.

Training should take account of these risks. In particular, the *limitations* of the technology should be fully explained and demonstrated to those who will use it. The widespread myth of 'computer infallibility' should be debunked as part of this training. Diagnosis and treatment are areas where infallibility is impossible, whether carried out by AI, humans, or a combination of the two.

There are two distinct patterns of automation important in medical AI. In the first, the automated system completes the task. In the second, the system gives advice that enables a healthcare professional to complete the task. It is crucial that we do not confine ethical debate to either one of these two patterns. In the first case, intervention may be difficult and we need to impose much higher ethical standards on designers and programmers. In the second case, we need to make it clear that a competent healthcare professional can, and sometimes should, ignore the advice given by an AI system. We can reasonably assume that the locus of ethical responsibility is clear in the case where a human overrules an AI system. However, this must certainly does not mean that ethical responsibility should be disregarded in all the other cases. If an AI system gives inappropriate advice then its designers and builders must share responsibility for any unfortunate medical consequences. These are issues which have been effectively (and usefully) resolved in the field of aviation.

[6] An oft-cited example of this happening was the Therac-25 disaster in which a radiation machine gave massive overdoses of radiation to at least six patients over a 2-year period as a result of a software error (Israelski and Muto 2004).

5 Ethics and Responsibility

There is an anecdote which claims that while a frog will instantly jump out of boiling water, if placed in warm water which is then very gradually heated it allows itself to be boiled.[7] In terms of their ethical response to automation, medical practitioners are very much in the position of the boiling frog. Each and every progress in technology will bring only a small, often barely perceptible, increase in the level of automation. There will be no obvious sudden changes. Instead numerous small improvements to technology and working practices will produce a gradual and progressive change towards a situation where medical practitioners have significantly less input into medical decision making.

To oppose these individual small changes as they happen will be seen as old-fashioned, if not Luddite. Indeed many, perhaps most, of the individual changes will bring about improvements in medical practice so it would clearly be unethical to resist them. It is therefore necessary to discuss the ethics of automation as a whole and to formulate policies in advances so as to generate ethical guidelines. This is a call to action to perform this simple task.

In general we can say that medical automation must meet certain specific ethical criteria. It would be best to parallel these requirements with those of evidence-based medicine. What "sold" (and continues to sell) automation to pilots were measureable improvements in safety. It is very hard to argue in favor of retaining old practices when they may potentially cost lives. The same approach should be adopted in medicine. It is an approach which can be and ethically should be linked to evidence-based medicine.

At present the progress of automation in medicine lags that in aviation, so looking at the successes and failures of aviation automation provides clear guidance here. One important lesson is that it is vital at all stages, to include medical practitioners in the process. This includes their direct inclusion in all stages of the technological design process, in implementing training programs, and in investigating any problems that arise. Aviation has a pool of highly experienced ex-pilots on which it can draw for such purposes, since it is an occupation with very high medical fitness requirements and compulsory early retirement. Experienced medical practitioners, by contrast, typically remain intensively employed in medicine with limited time for these important activities. This tension can (and should) be resolved by seconding experienced doctors to work on automation issues as a routine part of their jobs.

The most important single requirement is the establishment of a clear locus of responsibility [11, 12]. Since there is no prospect in the foreseeable future of a situation in which we might hold any robot or AI system *in itself* morally responsible

---

[7] This anecdote may very well be totally false. The experiments on which it is based were conducted in the 19th Century and modern experts doubt their validity. However, as an ethicist, I cannot condone the experiments required to establish its truth or falsity. It must remain merely an anecdote.

for its actions we must continue to place that responsibility upon humans. One very important reason for doing so is an already observable tendency for people to "hide behind the computer"; that is, to blame technology for their own failures or laziness.

Medicine is a field with clear and established notions of ethics. Existing procedures for passing ethical responsibility between healthcare professionals should be reinforced and, where appropriate, extended to include various forms of automation. Simply asking questions, "Who is responsible if the machine makes an incorrect diagnosis?", "Who has ethical responsibility for the administration of an incorrect dose if it is recommended by the computer?" will provide the initial steps in this process. Since software is not a product in law, questions such as these have usually not been asked as often as they should have been. Medical professionals should not hesitate to ask such questions now.

What is being suggested here is a deliberate reinforcement of existing medical ethics. This does not imply that there will not be new and difficult ethical issues raised when AI systems interact with medical practice in new and very difficult to predict ways. The urgent need is to establish and extend existing ethics and management practices to recognize the changes that automation is already bringing about.

The introduction of automation in medicine will place stresses upon these existing ethics and management practices, much as it is doing in aviation, but they must not be summarily abandoned. In particular human medical practitioners must not be allowed to escape responsibility simply by blaming the automation. It is important in both medical discussions and management discussions on workload and use of automation to make the non-availability of this excuse clear and explicit. This is not a side issue or a problem for other disciplines. It is a matter for medical professionals.

It is already evident that people sometimes use the computer as an excuse for their own mistakes. This unethical tendency will present problems in machine medical ethics. It is therefore important that existing structures in medical ethics are preserved, at least in as much as they are effective. The danger is that automation will sometimes undermine ethical responsibility for patient welfare, or at least obscure matters. All those involved need to be constantly checking ethics and practices to prevent, or at least reduce, this tendency.

In order to help guard against undesirable developments such as these, it is necessary to clear up the nature of shared ethical responsibility in human-AI systems now. This must include improvements to the professional and ethical standards of people not normally considered to be bound by medical ethics. This applies in particular to the IT industry which, for the most part, is not familiar with the need to adhere to strict ethical codes.

6 Recommendations

Healthcare professionals and managers should establish a clear locus of ethical responsibility and ethics audit trails for all automation in medicine. Healthcare professionals should be trained, possibly in a manner similar to aviation practice, in how to handle automatic devices and how to ensure that they can detect any failure, take over and introduce their own skill and judgment in a timely manner. Attention must be given to the prevention of skill-loss.

Management should view the contribution of experienced front-line medical practitioners to the technical design of medical automation and to the implementation of training programs as a part of their normal workload.

We can reasonably foresee medical accidents and incidents in which responsibility is shared between the front line healthcare professionals and the designers of the automatic devices and their interfaces. These should be investigated in an interdisciplinary and collegiate way aimed at preventing similar incidents rather than apportioning blame.

These recommendations would be much helped by a clear move towards a no-blame model in the investigation of medical accidents and incidents.

References

1. Kaltenthaler E, Sutcliffe P, Parry G, Beverley C, Rees A, Ferriter M (2008) The acceptability to patients of computerized cognitive behaviour therapy for depression: a systematic review. Psychol Med 21:1–10
2. NICE (2006) Computerised cognitive behaviour therapy for depression and anxiety, NICE technology appraisal guidance 97. The National Institute of Clinical Excellence, Feb 2006
3. Patel VL, Shortliffe EH, Stefanelli M, Szolovits P, Berthold MR, Bellazzi R, Abu-Hanna A (2009) The coming of age of artificial intelligence in medicine. Artif Intell Med 46(1):5–17, May 2009
4. Convention on Civil Aviation (1944) Original document available at http://www.icao.int/publications/Documents/7300_orig.pdf. Accessed 9 Nov 2013
5. Sheridan TB, Parasuraman R (2005) Human-Automation interaction. Rev Hum Factors Ergon 1(1):89–129. doi: 10.1518/155723405783703082, June 2005
6. Teach RL, Shortliffe EH (1981) An analysis of physician attitudes regarding computer-based clinical consultation systems. Comput Biomed Res 14(6):542–558
7. Sarter NB, Woods DD, Billings CE (1997) Automation surprises. In: Salvendey G (ed) Handbook of human factors and ergonomics. Wiley, Hoboken
8. Cali Accident Report (1995) English translation of recommendations
9. BCS (2013) The chartered institute for IT. Certification. http://certifications.bcs.org. Accessed 9 Nov 2013
10. Shortliffe EH (1993) Doctors, patients, and computers: will information technology dehumanize health-care delivery? Proc Am Philos Soc 137(3):390–398
11. Whitby B (1988) Artificial intelligence: a handbook of professionalism. Ellis Horwood, Chichester, pp 74–81
12. Whitby B (1996) Reflections on artificial intelligence: the legal, moral, and ethical dimensions. Intellect, Exeter

Machine Medical Ethics: When a Human Is Delusive but the Machine Has Its Wits About Him

Johan F. Hoorn

Abstract When androids take care of delusive patients, ethic-epistemic concerns crop up about an agency's good intent and why we would follow its advice. Robots are not human but may deliver correct medical information, whereas Alzheimer patients are human but may be mistaken. If humanness is not the question, then do we base our trust on truth? True is what logically can be verified given certain principles, which you have to adhere to in the first place. In other words, truth comes full circle. Does it come from empirical validation, then? That is a hard one too because we access the world through our biased sense perceptions and flawed measurement tools. We see what we think we see. Probably, the attribution of ethical qualities comes from pragmatics: If an agency affords delivering the goods, it is a "good" agency. If that happens regularly and in a predictable manner, the agency becomes trustworthy. Computers can be made more predictable than Alzheimer patients and in that sense, may be considered morally "better" than delusive humans. That is, if we ignore the existence of graded liabilities. That is why I developed a responsibility self-test that can be used to navigate the moral mine field of ethical positions that evolves from differently weighing or prioritizing the principles of autonomy, non-maleficence, beneficence, and justice.

1 Autonomous Agencies

In medical ethical issues, patient autonomy is a top priority [1]. Autonomy is habitually attributed to "agency," something that can undertake an action on its own behalf or that of others. An agency is something in pursuit of a goal or that

J.F. Hoorn (✉)
CAMeRA—Center for Advanced Media Research Amsterdam,
VU University Amsterdam, Amsterdam, The Netherlands
e-mail: j.f.hoorn@vu.nl

S.P. van Rysewyk and M. Pontier (eds.), *Machine Medical Ethics*,
Intelligent Systems, Control and Automation: Science and Engineering 74,
DOI 10.1007/978-3-319-08108-3_15

has a concern. It has intentionality. This is what sets it apart from stones, planets, and mopeds. An agency may be an organic system (plant, animal, human) or it may be artificial (commonly software) but it should potentially be capable of acting autonomously, at least in part. An agency does not necessarily have to possess "free will," because its behaviour may be determined by the circumstances. After all, the autonomic nervous system, which regulates glands and internal organs, is hardly controllable consciously but does pursue the goal of maintenance and continuity of the organism it is a part of. When an agency is simulated by a (semi) autonomous software system, it is a software agent. A robot, then, is a software agent that (inter)acts through electro-mechanical devices. When it is specialized in humanoid simulations, the robot becomes an android: A robot that simulates human behaviour but not that of other organisms. When the android is applied to healthcare tasks in a user-centred manner, it is a Caredroid and Caredroids are the main topic of our considerations.

The Caredroid's simulation of human behaviour typically may be in interaction with other agencies, commonly patients, care professionals, or informal caretakers. In the current chapter, we will not deal with human-human or robot-robot interaction but focus on Caredroids in interaction with patients, particularly those with a mental disability. There are a handful of software agents and robots that help autism patients (e.g., [40]), serve as depression therapists (e.g., [34]), or ease the loneliness of Alzheimer patients (e.g., Paro, see [45]).

2 Beliefs

If you were a patient, would you take advice from a robot; a machine without any understanding of what it is saying; something without a consciousness? If you were a robot, would you listen to a patient; an organism with incomplete information and bias in judgment? Who do you believe? What do we think that the other believes?

2.1 Cat in a Chinese Room, Opened by Ames

Suppose the robot laboratory at Hong Kong Polytechnic University constructs a Chinese Room out of steel and invites John Searle to spend a sabbatical year with them—for free—provided that he stays in the room and they hold the key. Every now and then, the researchers slip pieces of paper under the door, asking him how he is doing: Whether he is hot, cold, feverish, has chills, pains, hunger, is sweaty, sleepy, about his thorax and abdomen, and the like. To test him, they pose the questions in Chinese. Luckily, a soothsayer told John that "his natural wit would be his fortune." The room is packed with books, filing cabinets, pens, and stationary, and John figures out how to correlate the Chinese characters of the questions to an appropriate response, also in Chinese characters, which he slips under the door for the researchers to read. Although John does not know what he is saying, the researchers think he has perfect command of Chinese because all the answers make sense to them. Moreover, they think they can diagnose what is the matter with him, thinking he is thirsty whereas in fact he has to urinate.

Then the robot engineers Mark Tilden and David Hanson walk in, asking the researchers how they like their new emotionally perceptive John Searle robot portrait, locked in that Chinese Room over there. The robot engineers hold the Chinese character writer for a computer because how to determine he is a human? Promptly, another piece of paper appears under the door, stating that John Searle is a cat weeping over the mouse that he just has caught, signed by Erwin Schrödinger. Now everybody is in great despair. If Schrödinger is in there together with John Searle as his cat, Schrödinger will try to kill him through radioactivity and hydrocyanic acid [42], which he hid in a flask in one of the cabinet drawers. There will be no telling whether John the cat is dead or alive in that room and it will be "smeared out in equal parts" over the floor of probability [42].

Everybody agrees that as long as there is no direct observation, there is no way telling whether John Searle is in there, his robotic portrait, whether he is a cat, or that he is imitating Erwin Schrödinger with his cat, whether Schrödinger is in there, whether that cat is dead or not, or everything together?

The soothsayer steps in, foretelling that a dead cat should not be buried in the ground or it becomes a demon [5, p. 65]. It would be safer to hang it from a tree (ibid.). In undertaking immediate action upon this delusion, the soothsayer whose name is Ames draws an apple drill from his pocket and punctures a viewing peephole into the steel wall of the Chinese Room. Of course, everybody is pushing everybody else aside to see what is in there. What they see is astonishing. The inside of the Chinese Room as opened up by Ames is a distorted problem space where relative to the frame of reference of each individual observer (cubic or trapezoid), the apparent and actual position of the information processer inside that room makes it a great man or a humble robot. The conventional frame of Searle's Chinese Room as a cube demands that the information-processer inside that problem space should be human—with a conscious understanding of what it does. With the unconventional notion of a trapezoid, the processor is always halfway a robot and halfway a human, robot-like—even when alive, humanoid—even when lifeless, because it is all the same or at least indiscernible.

By this, I mean that the judgment "human or non-human" depends on your frame of reference. Searle's Chinese Room is an underdetermined problem space [32] in which you do not know what happens. Is the man Searle in there or is it his robot portrait? For the observer, the information-processing agency inside the Chinese Room is always Schrödinger's cat because he cannot know whether Searle is a lifeless machine or a living creature. There merely is a certain likelihood that the information-processing going on inside is more consciously human than mindless automation or perhaps even something in the middle [16, p. 45]; something smeared out over the floor of probability between true, probable, or false as described by Schrödinger's [42] "psi-function of the entire system."

In spite of not knowing whether true cognition is going on, we nevertheless bring it a glass of water although it has to urinate. We diagnose its health situation and take care of it, attributing it a sense of hunger and thirst and all different goal-directed behaviors, which make the machinery organic; make it "come to life." We tend to treat automated processes—including our own—as if they came from living creatures or real people (cf. the Media Equation Reeves and Nass [39]). And we do so because we were taught a frame of reference that says what the world is about and what the human condition is like [16, pp. 20–21]. Once apparent behaviors match our templates, we take them for actual. We work and take care of the agencies exposing those behaviors, according to our frame of reference or a priori belief system [16], containing, for example, certain moral principles (be good, don't hurt).

Because humans do not like to be confused, they prefer to force judgment into the direction of yes or no, in or out of category. We tend to apply logic-determinism to all possible problem spaces even in probabilistic cases. John Searle reproaches hard-nosed AI for looking at form and syntax alone to decide that the machine is "conscious," just like humans. Because the logics are the same, the difference between agencies becomes imperceptible in a Turing Test, so people themselves decide that the machine knows what it is doing. No, says Searle, if you look at semantic content, there is a difference because humans know what the forms and syntax refer to. Well, in the meantime researchers developed semantic Web techniques and machines that can reason through analogy and association, which is not far any more from what people do at a functional level because people also do not know eventually what the words stand for in the outside world—the world beyond their mental representations [16, p. 18]; a world that according to Schrödinger [43, p. 145] is just one of his cats again.

With respect to logics, then, it is hard to discern humans from machines; as it is increasingly with semantics. This position echoes Turing's [48] argument, stating that certain people may fail the Turing Test just as badly as the computer does. And because Schrödinger says that mental representation of the world is guesswork about the condition of a cat in a box that may die from poisoning, naïve empiricism does not do the trick either because our epistemology has no adequate access to the world about us except for what our senses, filters, and prejudices allow us to observe.

Ames teaches us that you have to look at it from a viewpoint of cognitive biases—illusory observations.[1] There is an unusual and perhaps even "trapezoid" psycho-logic to what we deem reality. The observer, however, has the bias—including our beloved philosopher John Searle—to apply conventional "cubic" logic to a probabilistic problem space, while staring at a cat that is fed by Erwin Schrödinger. That cat is an agency hardly discernible from a human or a robot when it processes logic-deterministic data. There are plenty of times that people are completely unaware of what they are doing and apply psychological schemas

[1] http://www.youtube.com/watch?v=hCV2Ba5wrcs.

or run scripts, delivering the right responses to a cue without "deep" understanding of the contents (cf. [26]). Think of chats about the weather, etiquette, or polite dinner conversations. People tell you what they are supposed to tell you and you think they are doing well, thank you... Think of officials that tick check boxes all day without thinking twice about the real-life consequences of their decisions. They behave like machines in Weber's [52] most rigid sense of the word. They continuously live in a Chinese Room.

If the cat agency processes probabilistic information, its best guesses are not distinguishable from a human's or a machine's either, because there is no way telling whatever guess is the right guess in the first place, whether the machine did a smart suggestion and the human a stupid one. In general, expert judgment in probabilistic situations does not exceed chance level [20, 47, p. 67] and is as good as a monkey's picking [37]. So there is only one thing that will discern a human from an animal from a machine, namely your own biased observations in relation to what you believe. Therefore, you need to realize what you think you know a priori about the agency in front of you (i.e., its ontological status or class), whether you believe your measurements and senses are right (i.e., your epistemology), and to know your own biases ("Am I a logician, an empiricist, or a cognitivist?"). Consequently, the more we think we know about the information processing capacities of a given agency, the more precise our classification will be (i.e., human, animal, machine)—without ever knowing its empirical correctness.

At the level of autonomous control, instinctive behaviors, and perhaps some aspects of memory, our behavior is nothing but machine-like. The more machines are capable of solving problems intelligently or even creatively, the more their behaviors become human-like. Finally, the two show indiscernible behaviors—passing the Turing Test brilliantly—that may have come from different processes but nobody can tell anymore, particularly when organic and digital circuitry becomes integrated (humans with electro-mechanical extensions, machines with biochips, a Robocop).[2]

Moral reasoning about healthcare dilemmas through machine computation yields judgments that are identical to those of medical ethical committees (e.g., [33]). Perhaps the machine has no clue what it is reasoning about and does not know about the real-life consequences of its decisions, it nevertheless delivers the same juridical quality as a human committee of professionals (6 cases rendered 6 identical judgments).

Searle's [44] final stronghold, then, is the lack of intentionality of a computer. It does not pursue goals and therefore, attaches no meaning to events and actions. The pursuit of goals and in particular the pursuit of self-maintenance and reproduction is what separates an organic system from a physical one. Searle's idea of intentions presupposes goal-directed behavior, resulting into emotions when goals are supported (☺) or obstructed (☹) (cf. [7, p. 494, 463]). In other words,

[2] Robocop meant as a metonym here.

reasoning logically about medical dilemmas from moral principles to deterministic outcomes in Searle's sense can only become humanoid if there is a trade-off with affective decisions, which by definition are intentional. The man in the Chinese Room should be capable of laughter and crying.

In Pontier et al. [36] we did exactly this: Provide the computer with goals to pursue and a mechanism for affective processing [19] and let this interfere with the reasoning from moral principles (i.e., from [1]). Greene et al. [12] state that moral issues with a personal accent ("My daughter is dying") involve more affective processing than moral-impersonal issues ("The patient is dying"). Such differences in emotional engagement modify people's judgments. Our system indicated to be more hesitant to sacrifice one person so to save five others if that person was someone who touched upon preset goals of the system that were not of a moral nature (e.g., that the person was "close by").

Now that we have a machine capable of moral behavior, that can reason, can deal with semantics, shows empathy, that is in pursuit of health goals, but that is unaware of doing it, we could place it in a care situation and confront it with a dementia patient. The moment this person puts herself and others at risk, cannot reason logically, does not understand a thing, shows little empathy with fellow patients or family, in pursuit of anything but health goals (cookies!), and hardly aware of doing it, the care robot might hold the patient for a dumb Searlean machine! How much cognition goes on in there? Get data, decode, execute a process, and respond? That is what a CPU does too. Moreover, the cognition that does go on is so delusive that even Ames would be shocked. How much "proper" judgment is still left? According to what distorted belief system? How much autonomy can be attributed to an information processing unit that merely correlates inputted symbols to other symbols, outputting something that the doctor may interpret as a proper response? "Are you thirsty?" "Yes, I have to urinate." Is this a living human or an organic machine, performing autonomous control over the vegetative system only?

2.2 About Self, About Others

To make moral decisions, a fully functional Caredroid must have beliefs about itself and about others, in our case, the dementia patient. It needs to know the ontology that the patient works with no matter how distorted ("Doctor is a demon"). It needs to know how the patient got to this ontology, by authority of the doctor, priest, or family members or through personal experience ("I saw him rise from the ground"), and it needs to know the biases the patient has, what prejudices and preoccupations ("Cookies!").

This patient ontology is compared to a reference ontology (cf. [18, pp. 314–315]). That ontology is constituted by general socio-cultural beliefs and tells the Caredroid to what extent the patient is "delusive," can be taken seriously, and can be attributed autonomous decision capabilities. Also the Caredroid should know what the origin is of its robot ontology, whose goals it supports, and what its biases are (what kind of logics, the quality of its sensors, whatever cognitive-emotional modules it has). And through this, the distinction between human and machine becomes obsolete because the problem is boiled down to affordances and the quality of information, not to being alive.

3 Affordances

Affordances in the context of a Caredroid are all the (latent) action possibilities (e.g., services) that the robot offers with respect to healthcare. This could be assistance, monitoring, or conversing. It could be a toy to pet in order to ease feelings of loneliness. There are designed affordances [10, 11], which are the features manifesting in the system or environment—even if they remain undiscovered by the user. For example, a Caredroid may demonstrate a workout but does not afford exercising if the user is paralyzed (cf. vascular dementia). There are also perceived affordances [27, 28], which are those possibilities that the user actually knows about, although much more options may have been designed. The latter remain "hidden" for the user [9]. The user also may falsely perceive certain affordances that the system does not offer [9]. An example is the assumption that the Caredroid will always be ready, which is wrong, because the servos heat up during usage and need a cool down period, which means that you cannot always count on them. False affordances may give rise to plenty of confusion; sometimes for the worse (e.g., Alzheimer patient panics when a robot enters the room: "It is going to eat me!"); sometimes for the best ("It is not a robot; it is a sweet animal: It talks to me").

For moral decisions, the Caredroid should know what affordances the user is capable of recognizing; or better, the Caredroid should store in its ontology what affordances the patient offers. To what extent are someone's capabilities intact or degraded? Are there periods of alertness or does someone suffer from visual hallucinations? Are the affordances designed in the Caredroid perceived at all? What are the false affordances? In "intact" users, the services that the robot has to offer will be recognized as such. If the robot is fork feeding a patient, the patient is supposed to open the mouth. That would be morally "good" behavior of the patient (i.e., beneficence), because it serves her wellbeing. But what if the patient recognizes an Afro fork-comb in the object that is pointing at her and starts doing her hair, smearing out in equal parts the mashed potatoes over her curls? Is this maleficence, according to moral principles? Or is the patient happy with the incredible surface contours of her new styling mold and should we leave it this way? What if during daytime activities, patients are cooking a meal and one of them uses his hair comb as a pasta tong to fish the spaghetti out of the soup? Maleficence because it is unhygienic? Beneficence because of fun? Or is this perhaps a sign of mental restoration as alternate uses are a known strategy for solving problems creatively (i.e., Guilford's Alternative Uses Task, [13])? After all, a gourmet tip to shape garganelli pasta in the right way is to roll it over an Afro comb instead of using a pasta machine![3] With respect to diagnosis and its consequences for autonomy, are these patients just being creative or downright delusive?

If the Caredroid's ontology would follow Norman's [27, 28] conception of perceived affordances, the patient is to open the mouth because the fork is designed such that it "suggests" that you eat from it. Doing differently, then, would be "blameworthy:" "Don't do this! Stop it!" If the Caredroid's ontology follows Gibson's [10, 11] account, affordances may be there without intentionality. The tree may not grow to build bird nests in but it is extremely suited to carry them yet. The fork may not be designed to comb the hair, but nevertheless. In other words, features of a Caredroid may be designed without any purpose in mind, in fact, they may be a design flaw, but the user may put purpose to it in hindsight [51], so that we find ourselves sitting across Searle's Chinese Room again.

So as we can see, moral decision making may not depend on human or machine agency and may boil down to what an agency affords; *what* it affords is dependent on the way you interpret the offerings. In other words, what is considered an affordance follows from the belief system. How smart, creative, and capable is the user, how smart, creative, and capable is the machine itself? Are the things the machine has to offer convenient for the user (e.g., mechanical love over loneliness) and what is the quality of the information upon which those offerings are made? Is not correctness of information more important than the source being alive? A sign at the road is not alive but I do follow its directions whereas my traveling partner points out the wrong way all of the time. Affordances predict use intentions [29]: I won't ask my traveling partner again.

4 Information

The verification of correctness of information penetrates deeply into the question of truth-finding. It seems that only upon true information we can offer our services and make decisions that are ethical. Being sentenced to jail without being guilty may be right according to the information available at the time but is not seen as ethically just. Thus, controls, tests, and checks circumscribe what is regarded as "correct," relying on the integrity of those safeguards, coming full circle again, because who guards the guardians, who guarantees the method, who controls the controls, a vicious circle of infinite regress, a downward spiral of suspicion. If we cannot trust our methods, measurements, and senses, there merely is belief. Hoorn and Van Wijngaarden [18, p. 308] noted that in the literature, correctness of information supposedly is based on accuracy, completeness, and depth. But that just postpones the question to how much exactitude is needed, when something can be called complete, and how much detail is required?

[3] http://www.circleofmisse.com/recipes/garganelli-31102009.

That things are true is not the same as things having meaning. In healthcare, many things are not true but do carry meaning: placebo drugs, induced illness, health anxiety. If a Caredroid states that "Patient Searle is alive," this statement in Searle's medical dossier has the same truth conditions as "Patient Searle is alive and he is not a robot," although the meanings differ. [4]The truth condition is that Searle has to live for the statements to be true. What being alive means, however, is a matter of beliefs: medically, religiously, or otherwise [16, pp. 19–20]. The conditions under which truth is functioning can only be validated empirically, not verified logically. Hence, truth is attached to structure and syntax, to form, not to content. Thus, the truth of the medical statement "Patient Searle is alive" cannot be dependent on its source, the mindless Caredroid that is a not-alive robot. Truth is logical, not empirical. There is merely the idea of truth. Empirical truth is illogicality. There is only empirical meaning and meanings are connected to the goals of the meaning giver, the interpreter of life. We provide meaning to data but will never find truth in them.

Information becomes logically truthful if a deterministic problem space is assumed and some belief system is accepted with premises that are well-formed, following certain (e.g., moral) principles or rules. This is a very limited approach to the murkiness of daily moral behavior the Caredroid will be confronted with and is mute about semantics, empirical meaning, or "ecological validity." In moral argument, information is empirically meaningful if it satisfies certain goals of the arguer. Although not righteously, people do assume that a statement is logically truthful if it is plausible in a probabilistic sense. The plausibility is extracted from a mixture of belief systems (e.g., medical and religious) with sometimes conflicting and more-or-less lenient guidelines that may be ill-formed, that is, following certain principles or rules but not strictly—preferences and priorities depending on the situation. We may contend that pragmatically, truth values are attributed to data according to probability distributions, which are biased by, for example, what is "ethically good," that is, by something considered useful (see section *Moral priorities*). To put it colloquially: If it serves my purposes (meaning), I apply the rule (logics), and then my moral judgment is truthful. The reasoning may be flawed but its conclusion is convenient.

The integrity of information in the medical dossier that the Caredroid may interface is ascertained by the patient in a pragmatic way. That information should be reliable in the double sense of the word: true and ethical [16, pp. 23–24]. But as we now know, integrity of information is always compromised because truth is but a logical attribute and has little to do with reality, whereas what we call "reality" is a set of assumptions, a mental construct, based on biased sense perceptions [16, pp. 35–36]. The reference ontology of the Caredroid is as much a product of beliefs as the ontology of the delusive patient. The ethical side of this is that the patient supposes not to be lied to; that the source has enough authority to believe its propositions. It remains to be seen if robots are regarded as authoritative

[4] Also see: http://en.wikipedia.org/wiki/Truth_condition.

enough to base health decisions on the information they carry. The problem of an appeal to authority is that its reasoning is flawed ([24]: *argumentum ad verecundiam*). After all, who sanctions the mandate to be an expert ("Because I say so? By virtue of my right?"). Bottom line, if it is not correctness we can rely on by itself, and an appeal to authority is an unsound reason, then the certification of correctness must come from something as shaky as "trust."

5 Trust

How can a robot become trustworthy? How to place your faith in a robot? Trust is an ethical matter [49] but is extracted from highly peripheral cues. Van Vugt et al. [50] found that virtual health coaches that advised on food intake were thought to be more trustworthy when they were obese than when slim. If the user faces critical decisions, an intelligent look is preferred to a funny appearance [38].

Where correctness of information should be the central concern, but indecisive, the trust that deems the centrally processed information correct is derived from peripheral cues [30]. That suddenly makes the question whether the same message is delivered by a human or a robot a non-trivial matter again. It also brings into play the affordances that are perceived. If we see a doctor with a medical dossier, the doctor supposedly is sane and the dossier correct. If we see a robot clown with a baby cloth book, Norman [27, 28] would suppose that the clown is incapable of medical diagnosis and the cloth book does not contain serious medical information—whereas [10, 11] would argue: "Why not?" What if a delusive person provides correct information (according to some belief system)? Probably, the unreliability of the source overrides the correctness of the information. What if an intelligently looking and properly functioning Caredroid works with incorrect information? Probably, the perceived reliability of the source overpowers the incorrectness of the information, quite like the doctor who makes a mistake (which doctors do).

The peripheral cues (e.g., white jacket, stethoscope, diploma on the wall) that are used to build up trust are the stereotypical features that express authority, expertise, and reliability. They are stereotypical because they are repeatedly associated with those inner moral qualities and give rise to the prediction that similar trustworthy behaviors can be expected on future encounters.

Within a belief system, trust comes from predictability of behaviors. If not harmful (non-maleficence), those predictable behaviors persuade into cooperation (cf. [8]) to achieve common or at least non-conflicting goals given the affordances perceived in the other agency. Trust also comes if the Caredroid does something beneficial without expecting anything particular in return [49, p. 15, 24].

Predictable means that probabilities are high for a subset of behaviors to occur and not that of possible other behaviors. In a highly deterministic system such as a computer, this must be easy to achieve, vide the love of autism patients for robots. Not harmful but instead beneficial indicates that achieving goals and protecting concerns are not frustrated or blocked but supported [7, p. 207]. Fortunately, a robot can be programmed such that it expects no gratitude in return. And provided that skills (affordances, functions) are strengthening or complementing the patient's own affordances, cooperation may happen.

Thus, if an Alzheimer patient sees that a Caredroid has better memory for appointments than she does, and the Caredroid does not stand in the way of other concerns (cookies!), and this process is repeatedly observed, then trust may transpire to collaborate with the robot. In other words, moral behaviors that repeatedly do not harm concerns (non-maleficence), but rather facilitate them (beneficence), are regarded as useful and constitute trust [49]. The point is, a bad memory does not store observed behaviors too well so that the Caredroid has to repeat its behaviors within the little time span of the working memory that the patient has left.

6 Moral Priorities

If we hold on for a second and mull over the simple contention that morality is a function of what we consider useful (cf. Spinoza's "foundation of virtue" in Damasio [3, p. 171], then the prioritization of moral principles [1] should be in line with the rank order of cultural values; in this case: Western values. "Having control" or "being in charge" would be the top priority (i.e., autonomy before anything), followed by being free from pain and threat (this is non-maleficence—nobody is eating me), then the need for nourishment and a mild climate, also socially (beneficence), and justice for all (that is, the group should be fine so to provide protection, on condition that the other three requirements are satisfied first). In other words, ethical reasoning is the legitimization of utility, which is provided by the affordances that an ecological or technological system has to offer [10, 11, 27, 25]. Therefore, Caredroids may afford functionality that make this animal called the user feel in control, keep him clear of danger, bring him the goods, and equally divide what is left over the others. In times of scarcity (no cookies!), this must lead to conflicts and friendships put under pressure [46] because the negotiation of the little supply that is left can only be held along the principle with the lowest priority: justice.

Because in Beauchamp and Childress [1] view, autonomy is the top priority of users and other stakeholders of the care system, a Caredroid should have diagnostics to test the capabilities of the person in front of him. Are the cognitive abilities of the other intact or degraded? If intact, normal prioritization of moral principles applies. If not, autonomy can be overridden by any of the three remaining values, justice included (Fig. 1).

To be frank, there should be some nuance to this position because the prioritization does not have to be so strict. In previous work (i.e., [33]), principles were more balanced and carried different weights. In that study, we also developed a rule to prevent decisions being taken against fully autonomous patients. On the other hand, we instantly questioned whether being fully autonomous actually exists.

A survivalist outlook as outlined above may raise objections from those with a more holistic worldview of being inseparable from all other beings or even from all other physical matter, robots included. An individual who is separated from

Fig. 1 Justice be done [collage created from picture justice (Southernfried, MorgueFile) and Robocop 1 (Verhoeven 1987)]

the collective can easily redistribute responsibilities to the higher (parent) nodes in the hierarchy because autonomy has been taken away from the lower (children) nodes. This is what happens in a Weberian organization structure [14]. It does that to make professional interventions and services predictable, reliable, and safe [14].

Yet, if the belief system tells that a person and her surroundings are manifestations of the same energy, even in a moral sense [21, p. 4], there is something like collective liability, a part-of-whole or "metonymic reflection" on individual and collective behavior. If I represent all else, all else is in me and I am in all else. My deeds act on the collective, the collective acts on me. Thus, I take responsibility for the deeds of others, also of robots. If others do not, robots included, they are "unaware." They are separated ego's incapable of introspection or better, of "part-of-whole reflection." A holistic worldview would work with a different prioritization of moral principles (Table 1). It is the idea that you do not control life but that life works through you. In a (probably Eastern) belief system of inner peace, harmony, and compassion, perhaps beneficence would take the lead as it induces a state of low-arousal positive affect [23], followed by not harming the collective (absence of negative affect, [23]), justice for all, and autonomy finishing last. This of course, is morally a completely different outlook than the one illustrated by Fig. 1 presented for comparison in Table 1. Take notice that Table 1 is a crude approximation of possible moral positions as rank orders may be fuzzier than Table 1 suggests (cf. [33]) and ties may occur if weights are introduced to the hierarchies tabulated here.

Whether we take a Cartesian position of "I know what I'm doing," a Buddhist perspective of "I am aware of being," or Robocop declaring "I am the law" (Fig. 1), the quintessence remains what Searle stressed with "consciousness," which is the point that people are capable of self-reflection; they can do internal diagnosis ("I must be insane!"). They can think about thoughts or be "mindful" of them. Maybe a robot that can adapt its behaviors according to its feedback loops and does self-testing this way might be suspected of another form of inner perception? And the dementia patient not capable of self-reflection perhaps may be

Table 1 Twenty-four possible priory configurations of autonomy, beneficence, non-maleficence, and justice, dependent on the belief system

By the power invested in me

1	A	A	A	A	A	A
2	B	B	N	N	J	J
3	N	J	B	J	B	N
4	J	N	J	B	N	B
			↑			

Survivalist outlook

Be good

1	B	B	B	B	B	B
2	A	A	N	N	J	J
3	N	J	A	J	A	N
4	J	N	J	A	N	A
				↑		

Buddhist view

Don't harm

1	N	N	N	N	N	N
2	A	A	B	B	J	J
3	B	J	A	J	A	B
4	J	B	J	A	B	A

Justice be done

1	J	J	J	J	J	J
2	A	A	B	B	N	N
3	B	N	A	N	A	B
4	N	B	N	A	B	A
				↑		

Robocop

Note: No ties assumed.

said to be (morally) "comatose" or put more mildly, "unaware"? Thus, inner perception, awareness, or conscience, or whatever you want to call it, is an agency's self-test circumscribing what is regarded as "correct" information to base a moral decision upon, its certification stamp being "trust," a false syllogism of authority, flawed but convenient, notwithstanding.

7 Responsibility Self-Test

The previous section hinged on two innate tendencies ostentatiously portrayed and alluded to throughout cultural history, all being appearances of good and evil, sin versus virtue. In Abrahamic religions, it would be Satan against God, the Greeks contrasted Dionysus with Apollo, Descartes separated passion from reason, Freud distinguished *id* (instinct) from *superego* (conscience), and Buddhist teachings say that ego detaches itself from awareness. As far as I am concerned, these are graphic descriptions of the brain's evolutionary architecture [31, p. 91]. It has an older mechanistic part, which it has in common with physical nature. That part is taken control of by the vegetative system, which executes genetically hard-coded behaviors (cf. a virus). Through the genome, a soft transition to the animalistic part is made. These are behaviors that certain animals also have, for instance, memory and learning (soft-coded information), communication, organization, and exchange of information across group members. In humans, that would count as language. This part of the brain can be retrieved to organic nature (e.g., cats and ants) other than what we think makes us human, which are the higher cognitive functions residing in the youngest brain lobes, the things we call "spiritual:" intelligence, creativity, wisdom (or awareness, for that matter).

The idea of course is that our animalistic lust is kept in check by our reason, conscience, or awareness of that "lower" behavior. If the higher cognitive functions (i.e., being an angel) fail to control the lower functions, we start behaving like animals (cf. the snake).

What should the Caredroid be aware of, then? We have been discussing a number of factors. The first was that of Agency (Ag), which could be human or robotic ($Ag_{(h,\, r)}$), then we examined the Beliefs the agency has about self and others ($B_{(s,\, o)}$), whether the (mental) affordances of both agents are regarded as intact or not ($Af_{(i,\, d)}$), to what degree the Information they express seems to be correct or incorrect ($I_{(c,\, i)}$), whether Trust in the source is high or low ($T_{(h,\, l)}$), and what Moral priorities apply to the situation ($M_{(1,\, \ldots,\, 24)}$). Putting a value to each of these variables selects one path in a nested factorial design of:

$$Ag_{(h,\, r)} * B_{(s,\, o)} * Af_{(i,\, d)} * I_{(c,\, i)} * T_{(h,\, l)} * M_{(1,\ldots,24)}$$
$$= 2 * 2 * 2 * 2 * 2 * 24 = 768 \text{ constellations to base a moral decision upon.}$$

To navigate this wide grid of moral positions, I want to try a responsibility self-test that comes in 7 steps. It is valid for both individual and collective liability, depending on an agency's affordance of inner reflection or more profane, a self-test that handles one or more of the said priority constellations of Table 1 as its yardstick.

We will race a number of agencies over the 7 hurdles and see who survives. The winner is the morally most responsible one that is accountable for its deeds—this "game" approach loosely follows the lead of Lacan [22]. The first step would be to see if an agency can act irrespective of the awareness of doing so.

1. **I do something**

This is to be taken as an act without awareness. One could imagine that a worm, an old-school robot, a cat, a patient with advanced dementia, as well as a sane person can pass this check, because all can do something. That may be on their own behalf or predicated by others, with or without knowing its own way of conduct, but at least they can act. In Moor's [25] taxonomy of moral agency, *normative agents* that can prove a theorem, *ethical impact agents* that are supposed to alleviate human suffering, and *implicit ethical agents* that take safety and legislation into account would pass this hurdle but not the next because they merely behave according to certain principles without knowing that those principles are "ethical."

2. **I know what I did was bad, good, or neutral**

This time, the agency is aware of its behavior as well as the rules of conduct, the rules of engagement under which that behavior is executed. It does not reflect about those rules, it operates within the boundaries of those rules. Those rules can be imposed upon by others; they may be one's own rules. A cat knows, for example, that it is not allowed to steal the meat from the kitchen. It fears punishment. The same is valid for certain dementia patients ("Don't steal cookies!") as well as a sane person. But also a robot with moral reasoning implemented may have a feedback loop built in that can judge whether certain principles were enforced or violated by its own actions. In Moor's [25] classification, *explicit ethical agents* would qualify for this test because they state explicitly what action is allowed and what is forbidden. Hurdle 2 is decisive to separate the sane and lightly demented people from the severely demented, who no longer have any clue about good or bad and start making nasty remarks, cold and insensitive, doing things that are inappropriate (e.g., aggressive behaviors).

3. **I know that I know I was bad, good, or neutral**

Here we enter the realm of having meta-cognitions about knowledge. The agency becomes morally conscious and can agree about the rules although disobeying them. Or, one can disagree about the rules and nonetheless comply with them. The agency now is aware of having knowledge about its actual behavior and that certain rules limit the behaviors that could possibly be executed. This is Moor's [25] *full ethical agent,* having consciousness, intentionality, and free will. It is a Cartesian position of thinking about thinking. It is Searle's criterion of being a conscious agency. But it also reflects a Buddhist position once we replace "thinking" by "awareness." In that case, it is not so much "Thinking, therefore I am" but "Aware that I am."

Checking on hurdle 3 may actually discern light dementia and "correctable" behavior from advanced dementia and lack of liability. As far as I know, there is no robot yet that can take hurdle 3. It would require a feedback loop about its feedback on its behaviors: The robot should record how well the processes perform that monitor its moral actions.

4. **I also know why I was bad, good, or neutral**

From this point on, it is not about logical verification alone any more but also about empirical validation. At this level, the agency has knowledge about the way its behavior was tested. How it came to know what happened at hurdle 3. Number 4 is an epistemic examination in how far some judge, referee, auditor—which could be the agency self—can be trusted in matching the agency's behavior against some law, rules, agreements (individual or collective), in an empirically valid or meaningful way, according to belief.

Only a sane person can take this hurdle, because the agency should have knowledge of which rules of conduct, rules of engagement, social contract, or terms of agreement are relevant in the situation at hand, picking one or more priority constellations from Table 1 as appropriate to goals and concerns of multiple stakeholders. It requires an estimate of how well the Caredroid senses its environment, perspective taking, and being able to weigh aspects in a situation with different sets of goals in mind.

5. **That I am aware of me knowing what I did and why it was wrong, right, or neutral—even if I disagree—means my higher cognitive functions are intact**

This is the Cartesian and Buddhist stance taken in unison with empiricism as a self-test on the agency's affordances. By overseeing the entire moral evaluation process, the agency can decide whether it has intelligence, creativity, and/or wisdom.

6. **My "higher" cognitive functions are supposed to control my "lower" functions but failed or succeeded**

In many cultures throughout history, this trade-off between good versus evil has always been the overture of the final verdict (see 7). Point 1 up to 5 were there to feed or load this opposition and hurdle 6 does the definitive weighing of being capable of handling the animalistic tendencies.

7. **Therefore, I am responsible and can be punished/rewarded or remain as is**

The final verdict. Any agency that went through 6 can be held responsible for 1 and automatically will go to 7. That agency is entitled to receive what is waiting for him or her as agreed upon in a given culture.

In answering Searle's Chinese Room dilemma, then, steps 3 and 4 are to be modeled, formalized, and implemented before we can even think of robot "consciousness," and following from that, machine liability. It also shows that moderately demented patients are only partially and in severe cases not responsible for their behaviors. In that respect (and with all due respect), the more than moderately demented patients are comparable to mammals (e.g., cats) and robots that have some sort of command over what is obliged or permitted but have no meta-cognition about that knowledge. Dementia in its final stage is even below that level. Cats, stota moral robots, or more than lightly demented elderly are in Searle's sense "unconscious" or in a Buddhist sense "unaware" of their own dealings. That is what makes them "primitive" or "animal-like," meaning that the youngest human brain functions or highest cognitive functions are defunct or missing.

8 Discussion

This chapter mixed ethical issues with epistemic considerations from the assumption that judgments of "morally good" are intertwined with "really true." When a dementia patient is confronted with a care robot that has reliable knowledge about the patient (e.g., according to a medical dossier), then we have a real person with delusions facing a virtual person with a realistic take on the world. Now, who should be controlling who? Should the robot comply with the demand of human autonomy and obey every command that the patient gives [41]? Or should it overrule certain proposals by the patient to protect her (and others) against herself? It all depends on what is regarded as the proper framing of the situation (the beliefs): The fiction inside the real person's head or the reality inside the fictitious person's CPU? Bottom line, what is more important: The correctness of information or the trustworthiness of the information carrier (the source)? And what would correctness be then?

Everything we do and stand for comes from belief, and morality is no exception. You cannot get it from logic, because the premises from which the logic start are empirically bound and hence, inaccessible epistemically. Put differently, it is incontrollable whether information is correct. Because truth telling is unknowable, it becomes intertwined with moral goodness: Trust is laid in the source that conveys the information. This state of affairs is no different for a human as it is for a robot. The doctor who tries to understand an Alzheimer patient is comparable to the user trying to find out what goes on in a robot's microchips. In the Chinese Room, the actual position of the information processer on the human-robot continuum will never be known. If an observer does make a choice (the reduction to a single eigenstate) biased perception made it so.

Trust in whatever source comes from peripheral cues that indicate expertise and authority on a particular matter, such as a white lab coat and glasses. They are cues to affordances that are perceived in the agency such as the ability to test information and high intelligence.

The list of four moral principles is the empirical aspect or "meaningful part" of moral reasoning. Contingent upon the belief system (e.g., Cartesian or Buddhist), what an agency understands under "autonomy" or "beneficence" may differ. Customarily, the meaning attached to a moral notion is related to goals and concerns of the individual and its community. But also the rank order (Table 1) or more sophisticated, the weighing of the principles, depends on goals and concerns. Thus, the reasoning may be straight but the semantics are biased.

Psychological biases are inescapable. Even the most rational of choices has emotional biases because the contents that are reasoned about pertain to empirical goals and concerns. One could even argue that moral reasoning without affect is to be lulled into a false sense of security.

That is why a medical ethical reasoning machine cannot do much more than "to know thyself," having meta-knowledge about its (reference) ontology (i.e., its belief system), epistemology (i.e., robot sensing and testing), and cognitive biases (i.e., the cognitive flaws in the system and the goals it has to achieve, for example, monitoring the patient). That is why for the moral domain I developed a self-test by which the agency can determine whether it is responsible for its acts or not.

8.1 Autonomae Servus

There is a difference between form, formalization, mechanism, syntax, structure, system, logics, verification, and truth on the one hand, and meaning, semantics, experience, empirical relevance, validation, and truth conditions on the other. It is what linguistic Structuralists (e.g., De Saussure [4, p. 121]) would have called the combinatory syntagmatic axis (i.e., the grammar) that needs to be filled by selections from the paradigmatic or associative axis (i.e., the lexicon). Whereas the formal part is relatively fixed, the meaning part is fluctuating. It is a problem of reference: What do the signals (e.g., words) stand for? This is not a logical but an empirical question. De Saussure would ask: "What is signified by the signifier?"

In deterministic problem spaces the logics of robots equals or even emulates that of humans (e.g., [33]). In probabilistic cases, where logics fail, the robot's guess is as good as any human's. If in addition you can make it plausible that someone hardly can decipher dealing with a human or a robot [35], then robots can be applied to many healthcare tasks. The only thing missing would be the "meaningful" aspect, sharing the belief system with the patient, "what the words stand for," which is problematic in caretaker-patient transactions as well. Even so, a Caredroid becomes more organic in its behavior once it is driven by goals [19] because it will attach more meaning to a signal in the sense of consequences for its pursuit of a patient's wellbeing.

Where logics and semantics grow together by introducing intentionality to the machine, the distinction between human and machine ethical reasoning cannot be made any more except for peripheral cues in the interface, such as outer appearance, bodily warmth, tone of voice, etc. The distinction is already vanished in comparison with a patient that hardly has any conscience left (cf. advanced dementia). If we integrate human with machine circuitry and live as cyborgs, making that distinction becomes irrelevant. In assuming a Buddhist perspective, we could allow to ourselves that we are made from the same matter; human consciousness intermixed with a machine's self-test on morality.

That leads us to the creation of the *Autonomae Servus,* the autonomous slave. We want the reasoning to be autonomous but the content to be serving our purposes: The human as master computer, the robot as his autonomous slave. The Caredroid may act autonomously because of its affordances (e.g., reasoning capacities) but is obedient to our goals of non-maleficence, beneficence, and justice. It moreover will be compassionate about our feelings of personal autonomy.

When the Hong Kong researchers finally took their metal-cutting shears, they saw that they had not kept Searle inside the Chinese Room but an old Chinese practitioner of Qigong. He wore a yin-yang symbol around his neck and looked more at an agency's functions than at its anatomy. When he came out, he orated that no life is complete without suffering from loneliness, illness, and decease. Most patients will not see it that way, he said, and try to cast out the bad experiences, but this is coping through ignoring. A robot could teach us to use those experiences for creation, the practitioner stipulated, as an alien source of information that can be integrated with known practice.

He then rummaged around in one of his filing cabinets, and next to a small flask there was a photograph. It showed a healthy 60-year old, walking his puppy dog called Mao while he was chatting with the neighbors along the way [5, pp. 63–64]. The next picture showed him as a 70-year old. He held a walker rollator and could hardly control the dog anymore. He admitted to have felt ashamed of the rollator and did not use it. In not going out of the Chinese Room anymore, he became lonely. Today, as an 80-year old with light dementia, he could not keep the dog any longer and ate it. The bones were spread around the floor in equal distributions. Now he was thirsty. He did not want a cat. He hated cats because they could see spirits in the dark [5, p. 65].

The Hong Kong researchers felt sorry for the old man and gave him a Hanson's Robokind Alice to ease the loneliness. That machine had a creativity module called ACASIA implemented ([15, 17], Chap. 4) that suggested to put her, the robot, in a children's wheelchair (Fig. 2). Now the old man strolled away behind the wheelchair, a support similar to a rollator,[5] without having to be ashamed of it. In fact, he was proud that he took care of the handicapped robot. Out on the street, he attracted a lot of attention and had quite a number of chats about his poor but cute Caredroid, and by the way, the Caredroid had a navigation system telling grandpa how to get back home again (cf. LogicaCMG's rollator navigator).[6]

As a kind of addendum, I would like to emphasize that people have their mechanistic side; physiologically (e.g., dehydration from a loss of electrolytes: "I am thirsty") as well as mentally (i.e., automated processes, routine behaviors: "Bring water glass to the mouth"). Are impaired people automata, then? One could argue that the more cognitive functions become impaired, the more people start to resemble automata. Giving creativity, intelligence, language, and memory back to the impaired through Caredroids, is making them more human again. Caredroids are factory robots made loveable [6]. They give something more relatable to their organic coworkers than today's practice of disembodied limbs put on a friendly face (ibid). We make computers understand how humans work. In dealing with

[5] Courtesy Robert Paauwe, personal communication, Oct. 15, 2013.

[6] http://www.camera.vu.nl/news/2007/021107news.html.

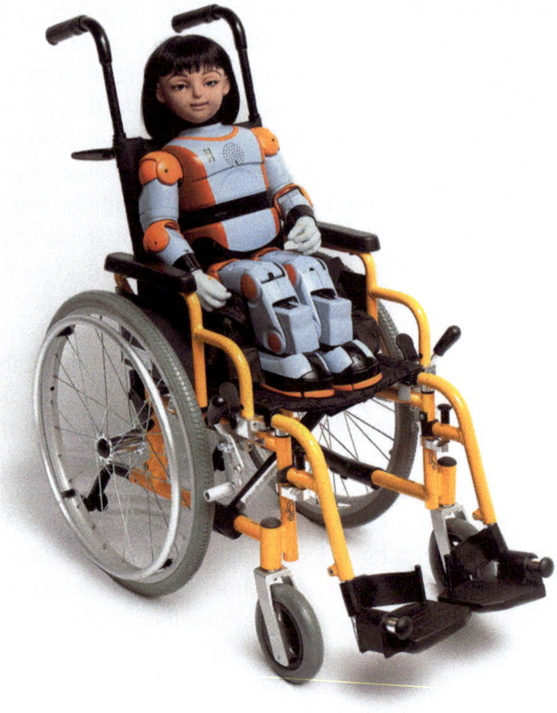

Photo: Wetzer & Berends, courtesy CRISP

Fig. 2 Hanson's Robokind Alice in a wheelchair

moral dilemmas, they can share as a human-android team, the burden of potential information overload for the patient, forming a system of multi-agencies that can exploit the information universe to the fullest. As "human-technology symbionts" [2, p. 3], impaired patients will be able to explore more alternatives, exclude more dead ends, reckoning with more situational constraints. It would be morally unfair not to compensate the impaired with loveable automata that care.

Acknowledgments This chapter is part of the SELEMCA project (Services of Electro-mechanical Care Agencies), which is supported by the Creative Industries Scientific Program (CRISP) of the Ministry of Education, Culture, and Science, grant number NWO 646.000.003.

References

1. Beauchamp TL, Childress JF (2001) Principles of biomedical ethics. Oxford University, New York
2. Clark A (2003) Natural-born cyborgs: minds, technologies, and the future of human intelligence. Oxford University, New York

3. Damasio A (2003) Looking for Spinoza: joy, sorrow, and the feeling brain. Harcourt, Orlando, FL
4. De Saussure F (1916/1983) Course in general linguistics (trans: Harris R). Duckworth, London
5. Eberhard W (1986) A dictionary of Chinese symbols. Routledge & Kegan Paul, London
6. Fingas J (2012) Rethink delivers Baxter the friendly worker robot, prepares us for our future metal overlords (video). Retrieved 13 Jan 2014 from http://www.engadget.com/2012/09/19/rethink-delivers-baxter-the-friendly-worker-robot/
7. Frijda NH (1986) The emotions. Cambridge University, New York
8. Gambetta D (2000) Can we trust trust? In: Gambetta D (ed) Trust: making and breaking cooperative relations. Blackwell, Oxford, UK, pp 213–237
9. Gaver WW (1991) Technology affordances. In: Robertson SP, Olson GM, Olson JS (eds) Proceedings of the CHI '91 SIGCHI conference on human factors in computing systems. ACM, New York, pp 79–84. doi:10.1145/108844.108856
10. Gibson JJ (1977) The theory of affordances. In: Shaw R, Bransford J (eds) Perceiving, acting, and knowing. Towards an ecological psychology. Wiley, Hoboken, NJ, 127–143
11. Gibson JJ (1979) The ecological approach to visual perception. Erlbaum, Hillsdale, NJ
12. Greene JD, Sommerville RB, Nystrom LE, Darley JM, Cohen JD (2001) An fMRI investigation of emotional engagement in moral judgment. Science 293(5537):2105–2108. doi:10.1126/science.1062872
13. Guilford JP (1967) The nature of human intelligence. McGraw-Hill, New York
14. Harrison S, Smith C (2004) Trust and moral motivation: redundant resources in health and social care? Policy Polit 32(3):371–386
15. Hoorn JF (2002) A model for information technologies that can be creative. In: Hewett TT, Kavanagh T (eds) Proceedings of the fourth creativity and cognition conference. ACM, Loughborough, UK, New York, pp 186–191
16. Hoorn JF (2012) Epistemics of the virtual. John Benjamins, Amsterdam, Philadelphia, PA
17. Hoorn JF (2014) Creative confluence. John Benjamins, Amsterdam, Philadelphia, PA
18. Hoorn JF, Van Wijngaarden TD (2010) Web intelligence for the assessment of information quality: credibility, correctness, and readability. In: Usmani Z (ed) Web intelligence and intelligent agents. In-Tech, Vukovar: Croatia, pp 305–324
19. Hoorn JF, Pontier MA, Siddiqui GF (2012) Coppélius' concoction: similarity and complementarity among three affect-related agent models. Cogn Syst Res 15–16:33–49. doi:10.1016/j.cogsys.2011.04.001
20. Horrobin D (2001) Something rotten at the core of science? Trends Pharmacol Sci 22(2):1–22
21. Keown D (2005) Buddhist ethics. A very short introduction. Oxford University, New York
22. Lacan J (1945/2006) Logical time and the assertion of anticipated certainty: a new sophism. In: Écrits: the first complete edition in English (trans: Fink B, Fink H, Grigg R). Norton, New York, pp 161–175
23. Lee Y-C, Lin Y-C, Huang C-L, Fredrickson BL (2013) The construct and measurement of peace of mind. J Happiness Stud 14(2):571–590
24. Locke (1689) An essay concerning human understanding. Holt, London
25. Moor JH (2006) The nature, importance, and difficulty of machine ethics. IEEE Intell Syst 21(4):18–21
26. Nilsson NJ (1984) A short rebuttal to Searle [unpublished note]. http://ai.stanford.edu/~nilsson/OnlinePubs-Nils/General%20Essays/OtherEssays-Nils/searle.pdf. Accessed 9 Oct 2013
27. Norman DA (1988) The design of everyday things. Doubleday, New York
28. Norman DA (1999) Affordance, conventions, and design. Interactions 6(3):38–43
29. Paauwe RA, Hoorn JF (2013) Technical report on designed form realism using LEGO Mindstorms [Tech. Rep.]. VU University, Amsterdam
30. Petty RE, Cacioppo JT (1986) Communication and persuasion: central and peripheral routes to attitude change. Springer, New York
31. Pfenniger KH (2001) The evolving brain. In: Pfenniger KH, Shubik VR (eds) The origins of creativity. Oxford University, New York, pp 89–97
32. Poincaré H (1913) The foundations of science. Science, Lancaster, PA

33. Pontier MA, Hoorn JF (2012) Toward machines that behave ethically better than humans do. In: Miyake N, Peebles B, Cooper RP (eds) Proceedings of the 34th international annual conference of the cognitive science society, CogSci'12. Cognitive Science Society, Sapporo, Japan, Austin, TX, pp 2198–2203
34. Pontier MA, Siddiqui GF (2008) A virtual therapist that responds empathically to your answers. In: Prendinger H, Lester J, Ishizuka M (eds) 8th international conference on intelligent virtual agents, LNAI 5208. Springer, Berlin, GE, pp 417–425
35. Pontier MA, Siddiqui GF, Hoorn JF (2010) Speed dating with an affective virtual agent. Developing a testbed for emotion models. In: Allbeck JA, Badler NI, Bickmore TW, Pelachaud C, Safonova A (eds) Proceedings of the 10th international conference on intelligent virtual agents (IVA) Sept. 20–22, 2010, Philadelphia, PA, Lecture Notes in Computer Science (LNCS) 6356. Springer, Berlin, Heidelberg, DE, pp 91–103
36. Pontier MA, Widdershoven G, Hoorn JF (2012) Moral Coppélia—combining ratio with affect in ethical reasoning. In: Pavón J et al (eds) Lecture notes in artificial intelligence, vol 7637. Springer, Berlin-Heidelberg, DE, pp 442–451
37. Porter GE (2004) The long-term value of analysts' advice in the wall street journal's investment dartboard contest. J Appl Finan 14(2):1–14
38. Prakash A, Rogers WA (2013) Younger and older adults' attitudes toward robot faces: effects of task and humanoid appearance. In: Proceedings of the human factors and ergonomics society annual meeting September 2013, vol 57(1), pp 114–118. doi:10.1177/1541931213571027
39. Reeves B, Nass CI (1996) The media equation: how people treat computers, television, and new media like real people and places. Cambridge University, New York
40. Scassellati B, Admoni H, Matarić M (2012) Robots for use in autism research. Ann Rev Biomed Eng 14:275–294
41. Schreier J (2012) Robot and Frank [movie]. Samuel Goldwyn Films, New York
42. Schrödinger E (1935/1980) Die gegenwärtige Situation in der Quantenmechanik, *Naturwissenschaften* 23, 807. In: Paper, proceedings of the American Philosophical Society (trans: Trimmer JD) The present situation in quantum mechanics: a translation of Schrödinger's "Cat Paradox", vol 124, p 323. Available at:http://www.tuhh.de/rzt/rzt/it/QM/cat.html
43. Schrödinger E (1944/2010) What is life? Mind and matter. Cambridge University, New York
44. Searle JR (1980) Minds, brains, and programs. Behav Brain Sci 3(3):417–457
45. Shibata T, Wada K (2011) Robot therapy: a new approach for mental healthcare of the elderly—a mini-review. Gerontology 57:378–386. doi:10.1159/000319015
46. Silver A (1989) Friendship and trust as moral ideals: an historical approach. Eur J Sociol 30(2):274–297. doi:http://dx.doi.org/10.1017/S0003975600005890
47. Tetlock PE (2006) Expert political judgment: how good is it? How can we know?. Princeton University, Princeton
48. Turing AM (1950) Computing machinery and intelligence. Mind 59(236):433–460
49. Uslaner EM (2002) The moral foundations of trust. Cambridge University, New York
50. Van Vugt HC, Konijn EA, Hoorn JF, Veldhuis J (2009) When too heavy is just fine: creating trustworthy e-health advisors. Int J Hum Comput Stud 67(7):571–583. doi:10.1016/j.ijhcs.2009.02.005
51. Ward TB, Smith SM, Finke RA (1999) Creative cognition. In: Sternberg RJ (ed) Handbook of creativity. Cambridge University, Cambridge, pp 189–212
52. Weber M (1922/1947) The theory of social and economic organization (trans: Henderson AM, Parsons T). Free, New York

Part IV
Contemporary Challenges in Machine Medical Ethics: Medical Machine Technologies and Models

ELIZA Fifty Years Later: An Automatic Therapist Using Bottom-Up and Top-Down Approaches

Rafal Rzepka and Kenji Araki

Abstract Our methods for realizing a moral artificial agent assume that the wisdom of crowds can equip a machine with the enormous number of experiences that are the source of its ethical reasoning. Every second, people with different cultural, religious or social backgrounds share their personal experiences about multitudes of human acts. We propose that a machine therapist capable of analyzing thousands of such cases should be more convincing and effective talking to a patient, instead of analyzing single keywords. In this chapter, we introduce this vision and several techniques already implemented in an algorithm for generating empathic machine reactions based on emotional and social consequences. We show the roles that Bentham's Felicific Calculus, Kohlberg's Theory of Stages of Moral Development and McDougall's classification of instincts play in the agent's knowledge acquisition, and we describe the accuracy of already working parts. Modules and lexicons of phrases based on these theories enable a medical machine to gather information on how patients typically feel when certain events happen, and what could happen before and after actions. Such empathy is important for understanding the actions of other people, and for learning new skills by imitation. We also discuss why this bottom-up approach should be accompanied by a top-down utility calculation to ensure the best outcome for a particular user, and what ethical dilemmas an advanced artificial therapist could cause.

R. Rzepka (✉) · K. Araki
Araki Laboratory, Hokkaido University, Hokkaido, Japan
e-mail: kabura@media.eng.hokudai.ac.jp

K. Araki
e-mail: arakig@media.eng.hokudai.ac.jp

S.P. van Rysewyk and M. Pontier (eds.), *Machine Medical Ethics*,
Intelligent Systems, Control and Automation: Science and Engineering 74,
DOI 10.1007/978-3-319-08108-3_16

1 Introduction

Almost 50 years ago, Weizenbaum [40] started his work on ELIZA, a primitive natural language system that uses pattern matching to parody a Rogerian psychotherapist whose main method is to make a patient feel relieved by sharing their problems with a listener. DOCTOR, the main script of the ELIZA program, provides a user with no useful information or opinions: it simply uses linguistic tricks to keep the conversation initiative needed to ensure that the patient is the one who is answering questions and explaining or extending their previous statements. Although the system does not possess any information about its dialog partner, it uses a keyword matching technique that allows it to react appropriately to a certain number of utterances. For instance, the word "sad" prompts the reply "I am sorry to hear you are sad". In our opinion such simple "affect analysis" was one of the factors that caused many users to believe they were talking to a real empathic person.

Although five decades have passed, non-task oriented dialog systems are still matching keywords, and rarely taking into account context including the surrounding world, human beings within this world, and the particular user on the other side of a screen or robot's eyes. However, constantly increasing computer power and enormous amounts of text allow us to take "keyword matching" to a whole new level of multiple searches for different kinds of related knowledge. It is still difficult to generate the illusion of a machine that has experienced many things in its life, but we believe that it would be easier to gain the user's trust by creating a machine that knows thousands of third persons' experiences, and that can analyze them and share its findings in the form of advice. Nevertheless, giving advice to a person suffering from mental conditions can go wrong, and the task becomes a moral dilemma.

A Conversational Model for therapy (known also as Psychodynamic-Interpersonal Therapy, PIT [30]) has been proven to be effective in the treatment of depression [31], psychosomatic disorders [9], or self-harm and borderline personality disorder [14, 33]. The problems for a machine to perform such therapy are numerous. As a therapist needs to develop an emotional bond with a patient, a scenario where the subject knows that he or she is not talking to a real therapist would pose difficulties in satisfying this condition. For this reason, it would be more realistic, at least in initial experiments, if the artificial therapist only assists human psychiatrists and patients during group sessions. However, situations in a less professional context will probably take place in the near future. Developed countries are predicted to suffer more and more from aging societies with a low birthrate, and the "offline loneliness" of people who are virtually but not physically connected tends to grow, as demonstrated by Kang [11]. However in his paper, Kang shows that conversation weakens such depressive states, and we believe that technology users will keep trying to talk to their machines when feeling lonely, in spite of all the social and psychological problems this may bring (described by Turkle [36]).

In the case of elderly persons and people suffering from conditions such as social withdrawal, it is not uncommon that a person is reticent to talk to other people, and we think it is possible that a machine could be a useful means for providing such persons with meaningful advice about what to do by sharing other people's experiences. We are already working on retrieving information about drug side-effects from bloggers describing their struggle with illness [12], and shallow NLP techniques used in this research can be extended to a broader range of topics.

In this chapter, we describe our long-term project of a dialog system based on a bottom-up approach to knowledge acquisition, discuss the need for additional top-down elements, and present the moral aspects of such a system for users who have mental conditions or are simply lonely and lack a conversational partner. First, we will introduce the basic ideas behind the utilized technologies and show the current accuracy of modules under continual development. We will then discuss our findings, already-known problems, and conclude the chapter with our vision of the future of moral agents in long-term development.

2 Counseling and Technology

Although Weizenbaum concluded in his book "Computer Power and Human Reason" [40] that machines should not be used in any kind of psychotherapy because they lack human understanding, many people disagreed, and computers have been used for treating patients with different types of psychological conditions. Today, machines are involved in various kinds of therapies from cognitive therapies [41] to virtual reality hypnosis [3], and even if they are mainly involved in treatments as a tool in the hands of a professional or prepared for very specific scenarios, they are often said to be effective because of the machine's patience, calmness, good memory, and so on. Patients are more trusting and more open, and are more likely to answer questions asked by a computer when they do not wish to talk about a particular problem to physicians. Another therapeutical use of machines is so-called "cognitive rehabilitation", where patients with brain damage, strokes, dementia in the form of Alzheimer's disease or schizophrenia [4], play computer games or perform linguistic exercises [29]. Computers also offer capabilities that can greatly improve the efficiency and psychometric quality of psychological assessment [32] and in the current era of shared knowledge, fast and accurate evaluation of the user's state is, we believe, very possible. When it comes to more automatic, conversational approaches, classic examples of artificial advisory for patients with suicidal tendencies [8] or a dilemma counseling system for students [39] can be given. In both studies, computers appeared to outperform humans in predicting suicide attempts in the former, and matching students' data to occupation in the latter. Recently, Embodied Conversational Agents (ECA) have gained popularity among researchers, but poor language understanding remains one of the biggest obstacles, as in Dilemma Counseling Systems (DCS). Accordingly, Wizard of Oz methods are often used [21] to see how people react to automatic therapists. Aside from the discussion on possible errors in automatic psychological assessments, the ethical side of the above-mentioned usages of computers in psychotherapy and health/life advisory is rarely touched upon; mostly because the moral responsibility is on the programmers' side, and there is little margin for problematic behavior by machines. But in the case of our bottom-up approach, the level of autonomy is usually higher and the knowledge acquisition process cannot be controlled, so we believe that in the case of therapy-oriented tasks, our system will need a top-down highest utility guarantor, which we will discuss later in this chapter. First, we will briefly describe our system, the techniques utilized, and the knowledge that is retrieved.

3 Wisdom of Crowds-Based Dialog System

The details of the particular modules that are being developed by our team exceed the scope and limitations of this chapter; however, it is necessary to introduce their fundamental basis for the further understanding of our approach. Currently, we are working on a system that classifies an utterance as a question-answering type (factoid or nonfactoid), task-oriented type or non-task-oriented type [42] and runs utterance generation modules according to this classification. Emotional and ethically bound inputs are most usually non-factoid questions and non-task oriented statements, but in the case of physical robots, morally questionable task oriented inputs (orders) are also very possible [34]. In all cases, in order to simulate real world experiences described on the WWW, we use Japanese Internet resources, principally a blog corpus that we have developed [23] to avoid restrictions imposed by search engines APIs. Currently, we are concentrating on causes and effects of acts and their emotional and ethical consequences. Thus, if a user says "I want to commit suicide", the system should be able to estimate that many people would feel sad and suffer due to this act, which would trigger not only an advice mode but also preventive actions such as contacting the user's family. Such utilitarian reasoning is an example of the top-down element of our system and will be presented in Sect. 5. In the following subsections, we describe the main features of our system.

3.1 Extracting Knowledge

In order to analyze experiences describing an act or situation in question, the system performs web-mining using the same and semantically similar phrases as queries and then extracts and quantifies the averages of three cognition layers, as shown in Fig. 1.

These will be explained below. All of these phrases are based on manually constructed lexicons which contain words representing several cognitive categories. These words, if found in sentences describing an analyzed act (or state) are

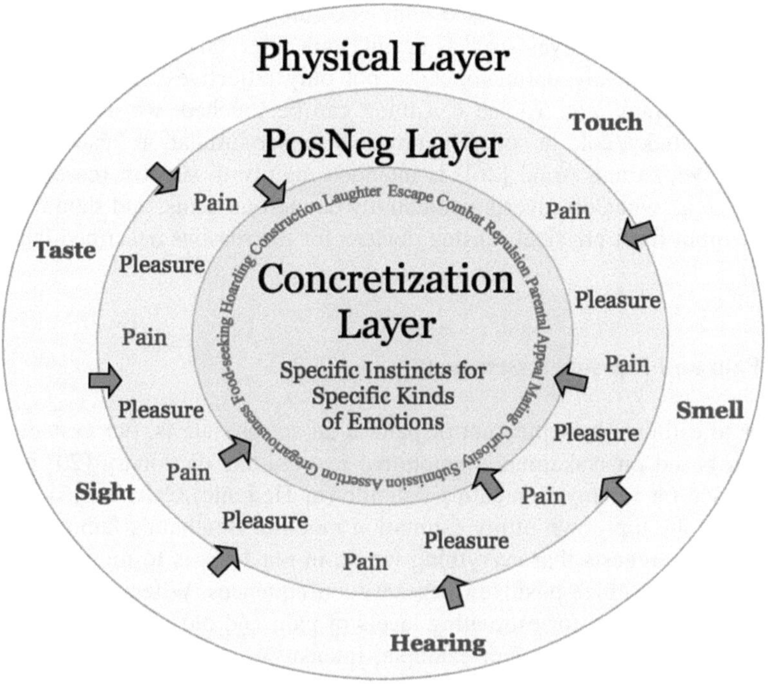

Fig. 1 Three layers of retrieved knowledge: sensory input simulated by representing the five senses as adjectives, positive/negative estimation using Nakamura's dictionary, and McDougall's categories of instincts that could have caused a particular behavior that led to a positive or negative act. For example, if a child touches fire, it feels pain, but most probably the act of touching was caused by the curiosity instinct

counted, and the comparison of sums determines the most probable reasons and consequences. Therefore, if a user talks about killing time or a virus, the system will not be alerted. We assume that a human-level talking machine needs to know as much about our world and our feelings as possible, but the physical sensing technology is still not ready for gathering knowledge. For that reason, we set natural language as the core for simulating the process of learning by perceiving sensations.

3.1.1 Physical Sensations

Aristotle [19] defined common sense (*koine aisthesis*) as a phenomenon built upon our five senses. Locke [16] proposed a similar definition of "common sense" suggesting that it builds on phenomenological experience. Each of the senses gives input, and then the sense-data are integrated into a single "impression" and this integration is a product of common sense. Therefore, we designed the first layer of the text-based simulation using five lexicons to represent the five senses. In order to check if an object can be sensed with eyes, the system checks the object's co-occurrences with adjectives such as bright, dark, big, small, red, yellow, etc. For hearing we use mostly onomatopoeias, not only adjectives: noisy, quiet, crack, rattle, clatter, clank, etc. To see if a thing can be touched, we use adjectives as hard, tough, sticky, cold or soft. The physical sensor simulation (also described in details in Rzepka and Araki [26]) is intended mainly to support reasoning about the world, e.g., when discovering potentially dangerous items, and should in future consider input from physical sensing devices for the reasons described later in the chapter.

3.1.2 Pain and Pleasure Assessment

In order to estimate how pleasant or painful an act or state is, our system utilizes a lexicon based on Nakamura's emotional expressions dictionary [20] and techniques based on Jeremy Bentham's Felicific (or Hedonic) Calculus [5] which we also use for the top-down utility estimation module. Bentham's famous notion of utilitarianism suggests that everything we do in our lives is to minimize negative (painful) and maximize positive (pleasant) consequences. When retrieving knowledge from the WWW for estimating levels of pain and pleasure, we followed the original Bentham's vectors. For example, Intensity asks "How intense was an act leading to a positive or negative consequences?" and to answer this question the algorithm needs to recognize intensifying adverbs and estimate their strengths. In order to calculate Duration ("for how long would the pleasure or pain last?") we have created an original subsystem [15] to calculate the average duration of a given action. Other vectors are Certainty, Propinquity ("how soon will a positive consequence occur?") and Fecundity, which is the probability that an act will preserve the current state. All vectors and calculation methods are described in Rzepka and Araki [26]. With this module, it is possible to calculate more precise differences between acts; for instance, that "having a headache for two days" is a bigger burden than "having a headache for two hours", and "hating his whole family very much" expresses more pain than "hating his pal a bit".

3.1.3 Estimation of Involved Instincts

In order to estimate what kind of instincts caused a human's behavior, our system uses a set of lexicons based on William McDougall's categorization of instincts [18]. He divides instincts into 14 classes (Escape, Combat, Repulsion, Parental, Appeal, Mating, Curiosity, Submission, Assertion, Gregariousness, Food-seeking, Hoarding, Construction and Laughter); for each class we constructed a lexicon containing words representing McDougall's ideas, details of which are also presented in Rzepka and Araki [26]. For example, the system is able to guess that humans date because we feel the urge to mate (Mating) and the need to take care of other people (Parental). When someone refuses to take medicine, he or she may be afraid of side-effects (Escape) or disgusted by its taste (Repulsion). Of course, these are only candidates for possible causes, but when used by an artificial therapist for generating an utterance, common reasons are more likely to gain the user's trust or provoke them to explain the reasons behind their refusal to take medicine. Questions like "Are you afraid? Or is it about the taste?" are still very machine-like and lack details or sophisticated sentences, but they have a visible advantage of being more concrete and on-topic. We proved in the past [27] the naturalness of utterances simply stating commonsensical comments (e.g., "smoking is not healthy" is evaluated higher than ELIZA's "tell me more about smoking").

3.1.4 Retrieving Possible Social Consequences

In order to enable our system to recognize social effects of acts, we created a lexicon inspired by Kohlberg's theory on the development of moral stages [13], which can be summarized as follows. When we are children, we are oriented toward obedience and punishment, and think about how we can avoid punishment. Subsequently, we turn to a self-interested orientation, asking ourselves what are the benefits of our acts. Later, we start caring about social norms, then authority and the maintenance of social order become important. After we achieve this "law and order morality", we reach the final level, which includes social contracts and universal ethical principles. In order to retrieve data on negative and positive consequences, we created ten categories (five pairs of negative and positive expressions) of consequences: Praises vs. Reprimands, Awards vs. Penalties, Societal approval versus Societal disapproval, Legal versus Illegal and Forgivable versus Unforgivable. Most of the words were taken from thesauri, so the Awards class has many synonyms of prizes, and the Punishment class consists also of words and phrases describing imprisonment or fines. By using this lexicon, the system can guess that shoplifting can result in punishment and anger or that reading books has a high societal approval and may lead to praise from other people.

3.2 Current Results and Need for Safety Valves

So far we have tested only a small part of Bentham's idea for calculating pain and pleasure; however, the results showed that we could achieve the same f-score by using a freely available small blog corpus instead of commercial search engines, which do not allow exhaustive multiple queries [24]. We tested our lexicons using over one hundred expressions with both neutral ("having a meal") and ethical ("make someone suffer") undertones. Emotional consequence recognition performed better (78 % correct recognitions) against social consequence recognition (70 %) and showed that a machine can utilize bloggers' experiences and opinions to achieve a moral intuition for simple inputs such as stealing, taking revenge, killing or even more debatable ones like euthanasia or abortion. When it comes to evaluating

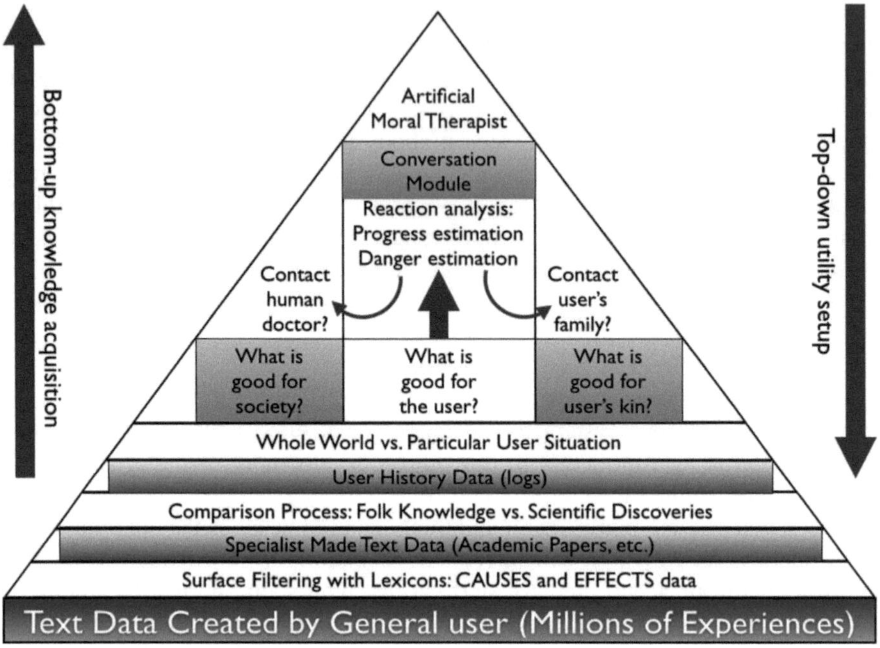

Fig. 2 Bottom-up knowledge is also the basis for the top-down utility calculation

estimation of instinctual causes, we achieved 77.8 % accuracy [25] and the retrieval precision for the lexicons related to the five senses appeared to be the highest of all, reaching 0.96. In the case of a fully autonomous knowledge acquisition agent, the numbers are promising. However, when we attempt to develop an agent for any medical task, we must remember that the crowd is vulnerable and common sense sometimes works against truth and logic. Stereotypes that humans create are useful, but also misleading, and we are innately poor at statistics, as Tversky and Kahneman show us in their works [37]. Machines, on the other hand, are not biased, they do not avoid or ignore any viewpoints that might be inconvenient, and they do not overestimate some facts and underestimate others because they are not programmed to do so. However, even the best knowledge retrieval and statistical algorithms are prone to significant inaccuracies if the sources are not trustworthy. To achieve a system that first gathers the knowledge of the crowd and then tests its credibility, we need to provide a second layer of retrievals for comparing the extracted data with reliable sources, such as scientific papers and legal texts (see Fig. 2).

For housework robots, mobile phones or car navigation systems, we believe that the experiences of the crowd are enough to realize a practical learning agent, as usually the majority of people are correct when judging the acts of others, even if they are wrong when explaining why they think so. But in the case of autonomous machines that can damage our health, the addition of analysis of scientific findings or legal regulations about what is discovered by an algorithm from the crowd experience could provide a safety valve to avoid e.g., worsening of the medical condition, and also may provide knowledge that is possibly valuable to the user. The ethical problem that appears in such scenarios is the choice of informing a patient about, for example, the low probability of survival in their current condition or revealing that they are taking a placebo. In order to avoid such dilemmas, proper user modeling could be implemented. We describe this in the following subsection.

3.3 User Modeling

The degree of severity and details of a mental disorder may significantly vary among users of an artificial therapist, and personal characteristics may heavily differ between users of less or more intelligent devices. In both cases, some plasticity of algorithms and/or personal pre-customization would be required to meet the minimum standards of trouble-free usage. However, unlike common, widely-used machinery, an intelligent conversational helper will have the power to change the user's state of mood or health indirectly or directly, depending on the level of advancement of the technology it represents. When assisting a person with Alzheimer's disease, a robot must be set up (usually requested by relatives or a clinician) as stubborn and repetitive from the start, as adaptation could take too much time (an example of learning how often a robot should remind a user to take medicine is shown in [2]). This is different from the usual process of learning user preferences [10], which is needed for natural communication and functioning. For example, an artificial therapist must remember its patient's problems, habits, and traumatic or emotional episodes, thus we are also working on an algorithm utilizing affective load estimation for ranking the user's memories [35]. However, in the case of tasks like therapy or companions for the elderly and children, user information must be analyzed in terms of possible dangers, and we believe that a clinician or family member should be contacted if any risky behavior is being conducted or planned. A potential problem is that such "spying" on a user may lead to a serious decrease of the user's trust in the agent.

3.4 Dialog Processing

Because we believe that natural dialog needs at least some common sense about the world, for now we are concentrating on knowledge that is required for conversation rather than on sophisticated dialog management. Therefore there is, for instance, no conversational context processing module implemented yet. The main concept of utilizing the methods described earlier in this chapter within a dialog system is shown in Fig. 3. However, if the system were to be used for a therapy task, numerous conversational strategies and techniques need to be developed or extended. We plan to follow the path of researchers who have worked in this field for a long time [21] by automizing theoretical parts with extracted knowledge, if

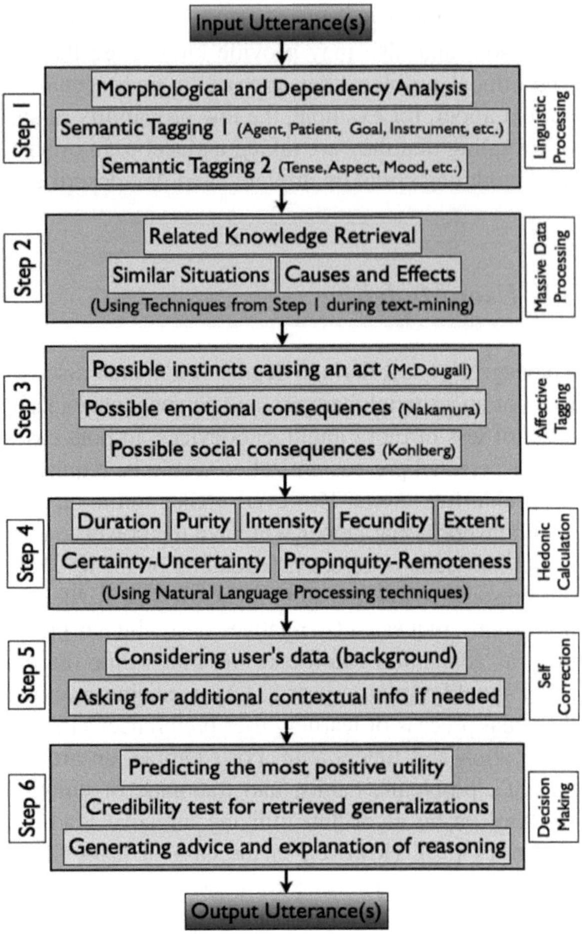

Fig. 3 Flowchart of knowledge processing inside the dialog module for non-task inputs. An important advantage of natural language based systems is that it is easier for them to explain on what basis they calculated utility or changed previous estimations (the last phase of Step 6). We believe that language is the key for the most natural and the fastest means of programming and adjusting future devices

possible. However, it remains necessary to continue our work with tasks specific to the Japanese language such as idiomatic expressions or metaphors [28], humorous language [1] or emoticons [38]. In addition to our research on non-task oriented and task-oriented dialog [10, 42], we plan to utilize techniques developed for other projects as user encouragement by gamification [17] or mood improvement by haiku generation [7].

4 Possible Scenarios and Related Ethical Problems

Various scenarios are possible for the future use of therapeutic functions of a dialogue system. Below we present three examples of such applications, which we envisage when developing our algorithms.

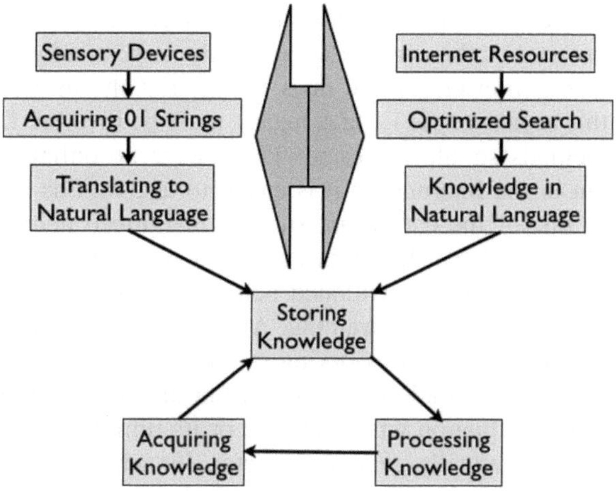

Fig. 4 A therapy-oriented system requires physical confirmation of what a user says to avoid situations where untrue information on their state of health is passed to the machine. Therefore, text-based knowledge should also be confronted with real-world sensors as soon as technology allows

4.1 Caring Robot that Talks

Mobile device-based talking therapists such as add-ons to Siri, Google Now or other voice-based agents quickly come to mind as potential applications, but we believe that the care robot scenario is also quite imaginable. This year, the Cabinet Office of Japan conducted a series of surveys on this topic.[1] Their results show that 65.1 % of Japanese people would like to be cared for by a robot, and 59.8 % would like to use such a machine in their homes when caring for someone else. The majority of subjects were concerned whether the robot would have a reasonable price, a safety certificate, if it would be covered by the national insurance scheme, and if it would come with a guarantee against injuries during usage. One problem of an embodied conversational agent that can be envisaged is the ease with which irritated users could talk the robot out of fulfilling its duties or convince it to perform a harmful task. To avoid this problem, a machine has to know the consequences of acts, but also has to be connected to physical sensors that could confirm if the user is telling the truth (see Fig. 4).

The growth of the so-called "Internet of things", where different devices and sensors are sharing knowledge, will surely support robot care technology. However, currently automatic helpers are vulnerable to deception.

4.2 Text-Only Advisor

Although speech is more natural and preferable, especially for children, the elderly and handicapped who cannot type, voice recognition accuracy is still one

[1] http://www8.cao.go.jp/survey/tokubetu/h25/h25-kaigo.pdf (in Japanese).

of the main problems for smooth machine-user interaction that uses natural language. For this reason, we currently work mostly with text, which helps to avoid recognition errors and does not restrict the user as much as a voice dialog. In this scenario, the ELIZA legacy must be diverted from, and different therapeutic methods should be tested. There were no online chatrooms back in the 1970s, but in the Internet era it is easier to test various approaches on various subjects. In one of the care robot surveys mentioned above (see Footnote 1), 28.2 % of subjects chose dementia care as a particularly arduous aspect of nursing care at home. As we mentioned before, games and word plays (e.g., quizzes) can be helpful in memory training. Also, using humor is effective in making users feel better; we are also working on generating puns [6]. Jokes can help to decrease the user's irritation level when the dialog goes wrong, but can worsen the usability evaluation when the joke or its timing are inappropriate. The same applies to automatically-generated poetry, for which many users have very high quality demands, especially among elderly Japanese.

The biggest problem to be solved in the advisor scenarios is implementing correct strategies for counseling. Depending on the user's data (personality characteristics, skills, preferences, medical details), advice may vary drastically, and the knowledge acquisition algorithms are still far from proper exception processing. For example, a patient with Alzheimer's disease should not be recklessly advised to take a stroll without a companion, and calming down a hearing-impaired person with classical music would be the wrong therapeutical choice.

4.3 Chatbot for Discovering Abnormalities

Another area in which ethical and emotional understanding and dialog processing can be used is helping to spot problematic, dangerous and unethical behaviors like those seen in cases of online predators or cyberbullying [22]. In this scenario, language understanding capabilities come in handy for discovering the linguistic patterns of possibly harmful acts or attempts to provoke such acts by third parties. This technology should be able to protect vulnerable users, for example when a child user talks to a third person. The difficulty of this task lies in very thorough context processing, because the NLP module must be ensured to grasp correct agent-patient relations. This is especially problematic for Japanese resources, as the language often omits subjects and blogs are in most cases anonymous. Therefore, even if a robot knows that a child is talking to a presumably trustful relative adult, it is hard to ensure that retrieved knowledge mirrors what is appropriate and what is not in linguistic behaviors between both parties. Any false accusations might bring users to the conclusion that the robot or its thinking capabilities are simply useless, unless the machine is capable of explaining why it concluded so, which is relatively easier to do when natural language is the main means of both communication and learning. In order to deal with recognizing actors, patients, objects, places, etc., we are currently working with the Semantic

Role Tagger for Japanese Language[2] enhancing it with categorization capabilities to be able to classify agents and patients, e.g., according to age, and to recognize fictional characters, because for instance computer gaming blogs very often generate noisy knowledge, suggesting to the system that killing or stealing leads to good consequences (rewards). This capability is also necessary in order to distinguish when an interlocutor speaks about the real world or about fiction. However, in the case of children or patients with mental disorders, ignoring imaginary stories could lead to missing important clues about the user's inner world and the details of their condition.

5 Need for Top-Down Control

As suggested in previous sections, in the case of medical patients, juveniles and elderly users, an artificial moral agent should be controlled by prior customization. However, we think that these are engineering details which will not be enough for an autonomous agent to ensure its ethical behavior; therefore, we opt for top-down elements to control what the machine does with both the acquired wisdom of crowds and scientific or legal knowledge. We came to the conclusion that consequentialism is the easiest moral approach to implement with the technologies we will develop, and we plan to implement Bentham's calculus for estimating utility. For now, we use only one simple rule for avoiding risky decision-making: if the majority of the crowd agreement on consequences is not significant (less than 2/3), the algorithm enters a doubt mode. In the case of the knowledge acquisition task, the program can learn different contradictory opinions, their contexts, the reasoning behind them, etc., but such a deeper text mining process is time-consuming. Therefore if an immediate opinion is needed and there is no clear dominance of pain or pleasure in the crowd's experiences, the program remains neutral. In our experiments this usually happens for inputs without sufficient contextual clues, such as "driving a car", but also, as expected, when we asked the system about ethically difficult tasks, it also returns the output stating that some people have problems with this given act (from abortion and euthanasia to playing video games and eating junk food). We set this 2/3 threshold by ourselves, but in future this should be the result of thorough experimentation. So far, we are working only on calculating the overall utility score for a given act, but the system must collect and evaluate this score for every possible solution to give the best advice in the current context. Here we also will use multiple parallel searches and the user's data to avoid unnecessary calculations. One potential problem is a situation where the artificial advisor's first solution candidate is rejected by the user, and the machine needs to decide if the second solution will bring a sufficiently positive outcome or if it would be better to insist on

[2] http://cl.it.okayama-u.ac.jp/study/project/asa/.

the first one. In normal circumstances, machine learning techniques would help to automatically set such a threshold by receiving feedback, but when the user's health is at stake, this approach can be unethical.

6 Conclusions

In this chapter, we described our vision of how our research on automatically collected shared knowledge could become a base for an artificial agent that helps to treat or maintain the mental health of its users while being ethically rational. The two main philosophical bases of our approach lie in consequentialism, where the consequences of one's conduct are the ultimate basis for any judgment about the rightness of that conduct, and intuitive awareness of value, or intuitive knowledge of evaluative facts (so called ethical intuitionism) form the foundation of our ethical knowledge. Our assumption is that if we cannot agree on why people behave morally, then in future we should feed a machine with a vast number of samples about our behavior and let the machine utilize the retrieved answers or just mimic the acts we believe are ethical, even if we do not understand why they are ethical. We believe that now we are only scratching the surface of the possibilities of Big Data, but already a new, crowd-based ELIZA could be seen as a much more empathic conversational partner who seems to know more about the world than 50 years ago. We think that implementing a hybrid (bottom-up and top-down) approach to ethical machines could help to build not only a helpful therapist, but also become a starting point for a whole new generation of AI products. We believe that agents that understand a mixture of sensing device networks and multimedia networks such as the WWW, could become a better and more universal judge than humans, because they have faster, global access to millions of common people's experiences, emotively expressed opinions, motivations and consequences of acts. Even if at present such agents utilize only text, they do not have the tendencies to be biased, to avoid or ignore any viewpoints that might be inconvenient as humans do; they do not overestimate any facts or underestimate others, simply because they do not feel, which for most of us is the main obstacle on the machine's path to ethical behavior. However, what we wish to suggest in this chapter is that caring autonomous machines do not have to possess their own feelings; instead, it is possible for them to borrow ours to ensure empathic behavior and guarantee highest utility without defying common sense.

References

1. Amaya Y, Rzepka R, Araki K (2013) Performance evaluation of recognition method of narrative humor using words similarity (in Japanese). Tech. Rep. 10, SIG-LSEB301
2. Anderson SL, Anderson M (2011) A prima facie duty approach to machine ethics and its application to elder care. In: Papers from the 2011 AAAI workshop human-robot interaction in elder care
3. Askay SW, Patterson DR, Sharar SR (2009) Virtual reality hypnosis. Contemp Hypn: J Br Soc Exp Clin Hyp 26(1):40–47

4. Bellucci DM, Glaberman K, Haslam N (2003) Computer-assisted cognitive rehabilitation reduces negative symptoms in the severely mentally ill. Schizophr Res 59(2–3):225–232
5. Bentham J (1789) An introduction to the principles and morals of legislation. T. Payne, London
6. Dybala P, Ptaszynski M, Rzepka R, Araki K (2010) Multi-humoroid: joking system that reacts with humor to humans' bad moods. In: Proceedings of the ninth international conference on autonomous agents and multiagent systems (AAMAS 2010), Toronto, Canada, pp 1433–1434
7. Emori T, Rzepka R, Araki K (2010) Automatic haiku generation using web search and Japanese weblogs as input. In: Proceedings of the international workshop on modern science and technology IWMST2010, pp 30–32
8. Greist JH, Laughren TP, Gustafson DH, Stauss FF, Rowse GL, Chiles JA (1973) A computer interview for suicide-risk prediction. Am J Psychiatry 130:1327–1332
9. Guthrie E, Creed F, Dawson D, Tomenson B (1991) A controlled trial of psychological treatment for the irritable bowel syndrome. Gastroenterology 100:450–457
10. Jordan A, Araki K (2013) Comparison of two knowledge treatments for questions answering. In: Proceedings of the 10th symposium on natural language processing (SNLP 2013), pp 55–62
11. Kang S (2007) Disembodiment in online social interaction: Impact of online chat on social support and psychosocial well-being. CyberPsychol Behav 10(3):475–477
12. Kitajima S, Rzepka R, Araki K (2013) Performance improvement of drug effects extraction system from Japanese blogs. In: Proceedings of 2013 IEEE seventh international conference on semantic computing, pp 383–386
13. Kohlberg L (1981) The Philosophy of Moral Development, 1st edn. Harper and Row
14. Korner A, Gerull F, Meares R, Stevenson J (2006) Borderline personality disorder treated with the conversational model: a replication study. Compr Psychiatry 47:406–411
15. Krawczyk M, Urabe Y, Rzepka R, Araki K (2013) A-dur: action duration calculation system. Technical report SIG-LSE-B301-7. Technical Report of Language Engineering Community Meeting, pp 47–54
16. Locke J (1841) An essay concerning human understanding. Oxford University
17. Mazur M, Rzepka R, Araki K (2012) Proposal for a conversational English tutoring system that encourages user engagement. In: Proceedings of the 19th international conference on computers in education, Asia-Pacific Society for computers in education (ICCE2011), pp 10–12
18. McDougall W (1923) Outline of psychology. London: Methuen. URL http://books.google.co.jp/books?id=Aht1lGPjE2AC
19. Modrak D (1987) Aristotle: the power of perception. University of Chicago Press
20. Nakamura A (1993) Kanjo hyogen jiten [Dictionary of Emotive Expressions]. Tokyodo Publishing
21. Novielli N, Mazzotta I, Carolis BD, Pizzutilo S (2012) Analysing user's reactions in advice-giving dialogues with a socially intelligent ECA. Cogn Process 13(2):487–497
22. Ptaszynski M, Dybala P, Matsuba T, Masui F, Rzepka R, Araki K (2010) Machine learning and affect analysis against cyber-bullying. In: Proceedings of the linguistic and cognitive approaches to dialog agents symposium at AISB 2010, vol 29, pp 7–16
23. Ptaszynski M, Rzepka R, Araki K, Momouchi Y (2012) Annotating syntactic information on 5 billion word corpus of Japanese blogs. In: In Proceedings of the eighteenth annual meeting of the association for natural language processing (NLP-2012), vol 14–16, pp 385–388
24. Rzepka R, Araki K (2012) Polarization of consequence expressions for an automatic ethical judgment based on moral stages theory. In: Technical repreport, IPSJ SIG Notes 2012-NL-207(14)
25. Rzepka R, Araki K (2013) Society as a life teacher ? Automatic recognition of instincts underneath human actions by using blog corpus. In: To appear in the proceedings of the fifth international conference on social informatics (SocInfo2013)
26. Rzepka R, Araki K (2013) Web-based five senses input simulation—ten years later. In: Technical reports of SIG-LSE B301, 5, pp 25–33
27. Rzepka R, Ge Y, Araki K (2005) Naturalness of an utterance based on the automatically retrieved commonsense. In: Proceedings of IJCAI 2005—nineteenth international joint conference on artificial intelligence, Edinburg, Scotland, pp 996–998

28. Rzepka R, Dybala P, Sayama K, Araki K (2013) Semantic clues for novel metaphor generator. In: Proceedings of 2nd international workshop of computational creativity, concept invention, and general intelligence, C3GI
29. Schreiber M (1999) Potential of an interactive computer-based training in the rehabilitation of dementia: An initial study. Neuropsychol Rehabil 9(2):155–167
30. Shapiro DA, Firth JA (1985) Exploratory therapy manual for the sheffield psychotherapy project. Psychological Therapies Research Centre, University of Leeds, England
31. Shapiro DA, Barkham M, Rees A, Hardy GE, Reynolds S, Startup M (1996) Effects of treatment duration and severity of depression on the effectiveness of cognitive behavioral and psychodynamic-interpersonal psychotherapy. J Consult Clin Psychol 64:1079–1085
32. Skinner HA, Pakula A (1986) Challenge of computers in psychological assessment. Prof Psychol: Res Pract 17(1):44–50
33. Stevenson J, Meares R (1992) An outcome study of psychotherapy for patients with borderline personality disorder. Am J Psychiatry 149:358–362
34. Takagi K, Rzepka R, Araki K (2011) Just keep tweeting, dear:web-mining methods for helping a social robot understand user needs. In: Proceedings of AAAI spring symposium "help me help you: bridging the gaps in human-agent collaboration" (SS05)
35. Takizawa M, Rzepka R, Araki K (2013) Improvement of emotion recognition system considering the negative form and compound sentence. In: Proceedings of the 29th fuzzy system symposium (in Japanese), pp 774–777
36. Turkle S (2006) A nascent robotics culture: new complicities for companionship. In: Technical report, AAAI Technical Report
37. Tversky A, Kahneman D (1974) Judgment under uncertainty: Heuristics and biases. Science 185(4157):1124–1131
38. Urabe Y, Rzepka R, Araki K (2013) Emoticon recommendation for Japanese computer-mediated communication. In: The Proceedings of 2013 IEEE seventh international conference on semantic computing, pp 25–31
39. Wagman M, Kerber KW (1984) Computer-assisted counseling: problems and prospects. Counselor Edu Supervision 24(2):142–154
40. Weizenbaum J (1976) Computer power and human reason. W.H. Freeman and Co., New York
41. Wright JH, Wright AS, Albano AM, Basco MR, Goldsmith LJ, Raffield T, Otto MW (2005) Computer-assisted cognitive therapy for depression: maintaining efficacy while reducing therapist time. Am J Psychiatry 162(6):1158–1164
42. Yatsu M, Rzepka R, Araki K (2012) A domain analytic method in modular-designed reflexive agent. In: Linguistic and cognitive approaches to dialogue agents, AISB/IACAP, pp 25–30

Models of the Patient-Machine-Clinician Relationship in Closed-Loop Machine Neuromodulation

Eran Klein

Abstract Closed-loop neuromodulation represents an emerging area in clinical medicine. Neural devices capable of measuring brain function and using measurements to iteratively guide output, such as deep brain stimulation, will be a significant advance in neuromodulatory technology. The introduction of closed-loop devices, particularly "smart" machines, will require changes in clinical ethical practice. A model of the clinical relationship could be a useful tool for addressing ethical challenges arising in this new area. Traditional models of the clinical relationship, like Emanuel and Emanuel's "four models," are suited to current unidirectional forms of neuromodulation. An adequate model of the patient-machine-clinician relationship may need to move beyond traditional models. Thus, I explore three new models: the *design* model, the *customer service* model, and the *quality monitoring* model. The exploration of these models of the patient-machine-clinician relationship will benefit from keeping an Aristotelian ideal of friendship in mind.

1 Introduction

The use of implantable neural devices, like deep brain stimulators (DBS) in Parkinson's disease (PD), has added neuromodulation to the repertoire of clinician skills [17]. Specialist clinicians have learned to interrogate and modify settings of neural devices in response to patient symptoms and signs. The next generation of

E. Klein (✉)
Department of Neurology, Oregon Health and Science University and Portland VA Medical Center, Portland, OR, USA
e-mail: kleine@ohsu.edu

E. Klein
Department of Philosophy and Center for Sensorimotor Neural Engineering, University of Washington, Seattle, WA, USA

S.P. van Rysewyk and M. Pontier (eds.), *Machine Medical Ethics*, Intelligent Systems, Control and Automation: Science and Engineering 74, DOI 10.1007/978-3-319-08108-3_17

implantable neural devices aims to move beyond clinician modulation of neural devices, replacing standard programming visits with "intelligent" neural machines capable of detecting change in brain function and adapting stimulation output, creating "closed-loop" devices [32, 34]. Near-term closed loop devices will embody a rather modest intelligence (e.g., enhanced input–output algorithms), but the drive for more intelligent machines is clear: improved neuromodulation. More sophisticated and perhaps autonomous or semi-autonomous machines will be developed, which in turn will change the practice of medicine. The nature and extent of this change is uncertain, but implications for medicine, and specifically clinical relationships, are worth considering now.

Models of the patient-clinician relationship (or doctor-patient relationship) are common tools for understanding obligations and responsibilities in healthcare [10].[1] 20 years ago, Emanuel and Emanuel [15] distilled the clinician relationship into four principal ethical models: paternalistic, informative, interpretive, and deliberative. The models differ significantly and track the wide historical evolution of medical practice from physician-driven to patient-driven medical decision-making. The models offered by Emanuel and Emanuel provide a relatively straightforward approach to thinking through the normative features of currently available forms of neuromodulation, like unidirectional DBS.[2]

Advances in neuromodulation technology, however, raise questions about the adequacy of standard ethical models of the clinical relationship. The emergence of closed-loop neuromodulation complicates a straightforward application of traditional models. Traditional models are paradigmatically applied to discrete and temporally proximate clinical decisions.[3] Which chemotherapy drug should be chosen? Is undergoing surgery soon worth the risk? Does cardiac resuscitation fit with a patient's end of life preferences? Closed-loop modulation, on the other hand, will involve the clinician in decisions with outcomes *temporally distant* and *indistinct.* Algorithms governing neuromodulatory machines will be chosen in

---

[1] The term *models* can have many meanings. In the literature on the patient-clinician relationship, use of the term is almost always meant to be normative. That is, a model of the relationship aims to encapsulate the central normative features of the relationship. By applying the model, one hopes to at least understand how the relationship can be seen in ethical terms, and more ambitiously to be able to derive ethical guidance in particular instantiations of the relationship. Thus, the intuitive distinction between 'clinical models' of the relationship and 'ethical models' of the relationship typically collapses. For our purposes, discussion of models of the patient-clinician relationship will also follow this convention and will equate clinical models with ethical models of the relationship and use the terminology interchangeably.

[2] Unidirectional DBS refers to the currently predominant form of this technology in which a target area of the brain is stimulated but the neurophysiological (or other physiological) effect of this stimulation is not sensed or recorded by the device. In simplified terms, information travels in only one direction.

[3] There are, of course, exceptions. Advance directives are a kind of temporally extended decision-making in medicine. This kind of decision-making differs, though, in that advance directives are typically understood as a way to provide guidance to others (e.g., clinicians, surrogates) for future decision-making (e.g., when incapacity occurs) rather making future decisions *now*.

advance of the clinical outcomes they determine, perhaps well in advance. In addition, individual clinical outcomes, such as motor output of a given stimulatory amplitude and frequency, will become the immediate input for a next iteration of a sensory-stimulation loop. As such, the lines between individual clinical events—for example, a change in modulatory settings—will blur as modulatory settings are adjusted automatically and iteratively, outside of conscious decision-making and perhaps below the level of consciousness. Will the role of the clinician need to shift to help a patient *live with* a device?[4] It is worth asking: is there an appropriate clinical model for closed-loop neuromodulation?

The outline of the paper is as follows. Neuromodulation will be defined and described and Emanuel and Emanuel's four ethical models of the clinical relationship will be presented. The current practice of DBS programming will be used to illustrate how traditional ethical models, like the four models, can be used to explore the normative features of discrete decisions about neuromodulation. Then, it will be suggested that advances in neuromodulatory technology may raise new ethical questions for which a model could be a useful heuristic. One likely advance in this area, the development of closed-loop neuromodulation, will introduce the possibility of a new relationship—the patient-*machine*-clinician relationship—that the search for an appropriate clinical model may need to accommodate. Three potential candidate models of the patient-machine-clinician relationship will be offered: the *design* model, the *customer service* model, and the *quality monitoring* model.[5] It will be argued that each of these models is limited in what it offers normatively to an account of the clinical relationship. The search for a clinical model that can accommodate patient-machine-clinician relationships will be shown to benefit by incorporating an Aristotelian notion of friendship.

2 Neuromodulation and Models of the Clinical Relationship

Neuromodulation is a generic term for iterative intervention in the nervous system. It has been defined as "a form of therapy in which neurophysiological signals are initiated or influenced with the intention of achieving therapeutic effects by

[4] The ethos of rehabilitation medicine, in so far as it is centered on helping patients adjust to new functional abilities and devices, may be a useful analog. Neuromodulatory medicine, however, would seem to go beyond the traditional boundaries of rehabilitation medicine and involve life trajectories that are more open-ended.

[5] These three candidates need not be viewed as a wholesale replacement for more traditional models, such as the deliberative model. Rather, this exploration can be undertaken in parallel to debate over preferred models of the clinical relationship more broadly. For instance, it may be that a traditional model, like the deliberative model, may be sufficient to guide action in particular circumstances, such as whether to choose device A versus device B. The question being investigated here is whether traditional models might be inadequate or insufficiently developed to account for the *whole* of neuromodulatory medicine—and whether alternatives are worth considering.

altering the function and performance of the nervous system" [21]. The notion of neuromodulation is familiar enough, even if the word is not. Adjustment of an antidepressant dose based on medication response is neuromodulation, so too is reaching for coffee when concentration lags or a bite to eat at the onset of hunger or hypoglycemia. The use of devices to modulate function of the brain is an expanding scientific and medical domain. The rise of neural engineering has helped add brain and nerve-based devices to the repertoire of human neuromodulation.

Deep brain stimulation treatment for Parkinson's disease is the most common kind of device neuromodulation. Individuals with Parkinson's disease lose essential elements in dopaminergic brain circuitry responsible for motor activity and electrical stimulation to areas deep in the brain can be used to compensate, for a time, for this deficiency [37]. The optimum stimulatory frequency, amplitude, pulse width and wave form vary by individual and by stage of disease, and typically must be adjusted over time through reprogramming [8]. Individuals implanted with DBS currently require neurologic evaluations and re-programming in response to Parkinson's disease progression and stimulatory response. Periodic evaluations and adjustments are a kind of neuromodulation: adapting an intervention, e.g., a change in DBS stimulatory parameters, based on feedback response, clinical symptoms and signs.

The relationship between patient and physician has long been recognized as special [1].[6] Roles and responsibilities flow out of this relationship that are both familiar from other areas of life and yet clearly distinct. The search for a model (or group of models) to capture what is special about the relationship has been an often repeated exercise in bioethics. Models based on healing [31], negotiation [10], contract [35], covenant [26], shared-decision-making [6], narrative [9], and bureaucracy [16] have been offered. Models, when of appropriate fit, aim to bring issues arising in medical practice into sharper focus and provide a framework upon which to hang normative concepts implicated in medicine, such as confidentiality, trust, beneficence.

One of the most influential appeals to models has been that of Emanuel and Emanuel [15]. Emanuel and Emanuel argue that the clinical relationship can be encapsulated in four types of models: paternalistic, informative, interpretative, and deliberative. These four types are conceptually distinct but tend to map on to changes in western medicine in the last century.

[6] The literature in bioethics has traditionally referred to the clinical relationship as the 'doctor–patient relationship.' It has been pointed out that this terminology is less than ideal. The listing of 'doctor' first in this pair is seen to reinforce the privilege of the physician perspective. Further, by limiting the relationship to physicians this terminology hives off—and hides from view—the many other forms clinical relationships can take (nurses, psychologists, social workers, physician assistants, among others). As such, I will adopt the terms patient-clinician or patient-machine-clinician relationship and speak of clinicians, rather than physicians, except in presenting the views of others who use the 'physician' term.

In the paternalistic model, clinicians use clinical skills to diagnose and pursue therapeutic regimens that the clinician believes to be in the best interest of the patient. Others have called this kind of model the "priestly" [35] or "parental" model [10]. The paternalistic model spans a continuum along which the exercise of patient autonomy is inversely related. At one end of this spectrum, the clinician determines what is to be done and does not or only perfunctorily informs the patient. At the other end, the patient chooses among therapeutic options, but patient autonomy is restricted by the clinician narrowing or coloring possible options in line with clinician beliefs about what is in the patient's best interest [7].

The clinician in the informative model assembles and presents relevant information to the patient and executes patient decisions made in light of this information. The informative model embodies a robust division of labor with respect to facts and values. The clinician provides factual information about the diagnosis, therapeutic options, and prognoses, and the patient combines this factual information with his or her own values, and comes to a decision. Others have described this as an "engineering" [35] or a "technician" model [10]. The clinician is taken to be a kind of expert in facts relevant to medical decision-making, but makes no claim to expertise about what is best for the patient or what the patient values.

The clinician in the interpretive model not only provides the patient with factual information but works with the patient to clarify the patient's values and align them with available options. Brock [6] has referred to this kind of model as "values clarification." The clinician engages in a discussion of patient values to facilitate practical reasoning. A concrete example of which might be: "So, correct me if I am wrong, but from what you've said it seems that you place a very high value on X (independence, relationship with your family, dignity, etc.). If that is right, then therapy Y might be most consistent with that."

In the deliberative model, the clinician goes beyond clarifying the values a patient happens to have and suggests which health-related values *ought* to be pursued. The animating idea of the deliberative model is that the clinician is uniquely situated to help the patient strive toward certain moral ideals. The clinician's expertise is not merely technical. The deliberative model takes seriously that clinicians can be not just epistemically privileged with respect to scientific facts but also, at least in some meaningful respect, more experienced in how patient values generally cohere with medical options. Clinicians can develop an understanding of what values medicine can and cannot help promote. Combined with the professional obligation to care, this understanding puts clinicians in a position of not just laying out possible options, but of recommending options and offering persuasive arguments for these. The clinician is invested in the patient, not unlike that of a "teacher."

The current practice of clinician monitoring after DBS implantation can illustrate these models. Imagine the case of a woman with a DBS for Parkinson's disease tremor who visits her clinician for new symptoms of depressive thoughts. This patient has DBS placed in the subthalamic nucleus (STN), which is associated with increased risk of depression [37]. She has a remote history of severe post-partum depression. A clinician guided by each of the different models might approach the discussion of how to proceed with DBS programming changes differently. A paternalistic approach might involve her clinician strongly advocating for a reduction in stimulation, arguing that addressing her depressive symptoms, particularly given her history of depression, has priority over loss of motor benefit from her DBS. An informative approach might involve the clinician taking a more detached stance, presenting existing data on the relationship between STN settings and depression and methodically laying out her options. An interpretive approach might involve the clinician helping map her options onto her beliefs and values—how worried is she about depression or how happy is she with the motoric benefits of her DBS—and arrive at an option that is of best fit for her. Finally, a clinician within the deliberative model might encourage her to think about her options in terms of the type of person she could become. Perhaps a decision not to adjust her settings would affirm and strengthen certain valuable character traits she prizes, such as her success in overcoming depressive thoughts in the past.

As the foregoing example illustrates, models of the clinical relationship provide a reasonable starting point for thinking about ethical aspects of DBS programming in part because the programming visit embodies a paradigmatic *clinical decision*. The programming visit occurs at a discrete meeting between clinician and patient. It involves an evaluation of the current health state of the patient and a therapeutic decision made in light of this evaluation. The patient (and often family) give a subjective account of symptoms and changes in function and the clinician conducts tests, such as physical exams or timed motor tasks at varying device settings. A decision whether and how to change device settings is made by the patient and the choice is carried out (e.g., DBS settings are changed or not) by the clinician.

3 Advances in Neuromodulation

3.1 Near Future of DBS

The future of DBS programming will look quite different than it does today. Advances in technology already underway will make it possible for patients to change their own machine settings, for changes to machine settings to be made remotely, and for direct measurement of neural functioning to guide programming decisions. In light of these advances in the technology, the current practice of clinical neuromodulation in DBS will evolve.

The first change to programming practice will be in the ease with which stimulation settings can be changed. Programming devices that are portable and easily navigated will make it more common that patients are able to modify their own device settings. One such device currently allows patients to choose among groups of clinician-determined stimulation parameters [27]. The philosopher Dubiel [13] describes the value of this kind of self-modulatory technology in his autobiographical account of living with PD and DBS. Dubiel describes a trade-off he experienced with DBS:

> Depending on whether I need to walk for some distance or speak in public, I have to enter commands into my pacemaker much the way I would on my computer. If I want to enunciate clearly, I have to set the amplitude very low, which then regularly leads to relative immobility and depression. If I want to walk for more than half a kilometer, I have to set the level correspondingly high, which then makes me speak inaudibly and sound washed out [13, p. 120].

Dubiel adjusts his DBS settings to meet his needs. He could set his device amplitude high, allowing him to walk quickly across campus in time to deliver a lecture, or turn down the amplitude in order to make his voice comprehensible to his audience. This kind of self-modulation has an analog in pharmacologic treatment of Parkinson's disease. Patients adjust their dosage of medication to their current symptoms, taking an extra dose of levodopa if tremor or bradykinesia are worse, or conversely reducing or skipping a dose.

A second change likely to occur is the ability to adjust device settings remotely. The incorporation of wireless technology into neural devices will make remote modulation possible [14]. For instance, a patient who lives far from a DBS center will be able to discuss symptoms over the telephone with a clinician and changes to the devices will be made wirelessly (or with the aid of wireless technology). Coupled with other forms of remote data gathering, such as video conferencing or ambient monitoring, the quality of data needed for accurate device adjustment may come to rival that gathered in clinic visits. For instance, speed of gait is often used to help determine if a change in stimulation is warranted in DBS for PD [8] Wearable sensors can be used to give an accurate picture of a person's gait in environments of interest (e.g., the home) [30].

A third change likely to occur is the use of neural sensors to gather data relevant to neural device programming, such as DBS [24]. Neural stimulators are currently assessed in terms of subjective symptoms or observable effects on patient behavior. As we have seen, the DBS programming visit is an example where such data is collected. Neural devices capable of sensing brain function provide an alternative mode of assessing stimulation devices. A sensor that detects the electrical or chemical state of the brain, for example, a change in beta frequency oscillation near the STN, may provide useful information for adjusting neurostimulation. A patient with self-modulatory capabilities could use such information to guide modulation decisions, or a clinician could use such sensing capabilities to guide in-clinic or remote adjustment of device settings. Neural sensors promise to provide a new, and some might say more *direct*, source of information for neuromodulation decisions.

All of these changes to neural stimulation technology may put pressure on the traditional patient-clinician relationship. Self-modulation of neural devices by patients will change the locus of control over day to day modulatory activities. Remote modulation will change the form (the where and how) of the patient-clinician interaction. Sensing-enabled devices will add a new source of information together with, and perhaps at times in conflict with, traditional ways of assessing clinical benefit (e.g., subjective report, physical exam). These changes in technology will require adjustments in clinical practice.

The extent to which these changes in delivery of medical care will in turn require a rethinking of obligations and responsibilities in medicine are an important and open question. The ethical challenges occasioned by these technological developments at least have precedent. Many patients with Parkinson's disease, for instance, already have discretion over when and how to take symptomatic medications, in effect shifting the locus of daily neuromodulatory control from clinician to patient. In addition, we are becoming increasingly familiar and dependent upon devices over which others exert control by remote means. Networked computers can be turned on, off, reset, or reconfigured remotely. Furthermore, devices capable of recording health or health-related information, such as activity level or physical location, are becoming a part of modern daily life. While changes in neural engineering technology may put new demands on the patient-clinician relationship, and the models that can be used to elucidate its normative features, it is not obvious that they threaten, at a more fundamental level, the clinical relationship. The same may not be the case for another development in neural technology sitting on the horizon: closed-loop neuromodulation.

3.2 Closed Loop Neuromodulation in DBS

Closed-loop is a term for devices that use algorithms to adapt output to current needs. Closed-loop neuromodulatory devices, like a DBS capable of sensing changes in the neural milieu around the STN and adjusting stimulation accordingly, promise a significant improvement over current unidirectional devices like standard DBS. Many individuals with Parkinson's disease, for instance, currently experience fluctuations in symptoms based on time of day or activity level and could benefit symptomatically from more frequent changes to stimulatory settings [28]. As we have seen, self-modulation by patients or wireless technologies allowing patients and clinicians to make remote device adjustments are two ways in which to bridge this gap. A more direct way would be to cut out the patient and clinician as intermediaries, and close the neuromodulatory loop even tighter. Instead of patients and clinicians using clinical symptoms or information from neural sensors as a substrate for making stimulation decisions, neural sensors could provide information *directly* to a neurostimulatory device. Patients and clinicians would not have to make decisions about device adjustment. A closed-loop device, in its ideal form, will tie stimulation and the changing needs of the recipient so closely as to mimic normal physiologic function. The goal of closed-loop neuromodulation is for a device to become a seamless part of an individual's neural functioning.

Closed-loop devices can be more or less sophisticated based on the algorithms they embed. A machine can use a very simple algorithm to tie input and output. For instance, the frequency and amplitude of stimulation from a DBS can be set to increase if the activity in an area or circuit of interest (e.g., STN) falls belong a threshold level. Once the activity in the target area rises back above the threshold, stimulation settings can reset to baseline. Even this rather simple input–output loop would be an improvement over current DBS technology [32].

The promise and future of closed-loop neural devices lies in more sophisticated algorithms. In part, this is because the origination of most human behavior, even simple motor movements, is quite complex. Algorithms able to incorporate inputs from a broad range of neural circuits and brain regions influencing a particular behavior (e.g., gait) are more likely to allow for a fine-grained response (e.g., smooth and fast gait). Current neuromodulatory therapies, like DBS or pharmacologic treatment of depression, provide benefit through more coarse means (e.g., stimulating the entire STN, blocking serotonin reuptake across brain regions). Closed-loop devices that are able to respond to multiple neurophysiologic inputs will facilitate more targeted output.

The more significant advance in closed-loop technology will be the incorporation of a kind of *intelligence* into the input–output loop. Consider an extension of Dubiel's example. A DBS device based on a simple algorithm might be programmed to increase output during a sustained walk (e.g., across campus) and decrease output during sustained verbal activity (e.g., lecturing). A device programmed in such a simple way would support his goal of being a successful academic. Imagine, however, a "smarter" closed-loop machine, one that incorporated a more sophisticated algorithm in which the device *learned* about Dubiel's activities (and even priorities) over time, and facilitated their achievement in various ways. Perhaps a device could learn that a walk across campus of a certain duration occurring at certain time always preceded his lecture. The device might lower its amplitude automatically with his first step into the classroom, and save him the few awkward moments of unintelligible speech at the start of his lecture or the unwanted attention he draws in accessing a self-programming device. The closed-loop device might be taught to increase amplitude even higher in order to speed his gait if he is running late for the class. Conversely, the device might learn from experience that rushing across campus leads to worse lecturing performance, perhaps manifest by higher levels of activity in brain regions associated with stress or distress, and that slowing Dubiel down before arriving, even if making him a little late, is in his best interests, whether he is immediately aware of this or not. With a sophisticated algorithm and an expanded range of input data, a device might learn what Dubiel *really* values and help him achieve it.

It is not hard to see how sophisticated closed-loop neural devices of this kind—those that embody a kind of artificial intelligence (AI)—might raise ethical concerns. Is the device facilitating Dubiel's achievement of his goals or helping shape what those goals are? How is the algorithm in the device chosen in the first place and how informed ought this choice to be? Is it really possible to consider the various futures that one algorithm will usher in (versus another) without quickly surpassing the imaginative capacities of even the most sophisticated patient-clinician pair? How should the "intelligent" device be regarded? Does the machine's intelligence need to be *morally* respected?

Traditional models, like those of Emanuel and Emanuel, are not an easy fit for closed-loop neuromodulation. Interventions most typically the subject matter of such models do not involve intelligence in this way. A medication, a surgery, a unidirectional device (like current DBS, as we have seen) or even a counseling or educational intervention certainly involves or relies on the intelligence of practitioners. The intervention would not be possible or likely of value without this intelligence. A closed-loop device, on the other hand, not only involves the intelligence of practitioners but is *itself* intelligent. The possibility of *machine intelligence* has implications to which traditional dyadic models of the clinical relationship do not, and perhaps are not well-suited to, attend. Might there be alternative models that are a more natural fit with intelligent devices?[7]

4 Models of the Patient-Machine-Clinician Relationship

Consider three models that have resonance with engineering practice: clinician as *designer*, as *servicer*, or as *monitor*. I will present each of these briefly and then argue that none of them seems on first pass sufficient for guiding the patient-machine-clinician relationship.

The first model is that of the clinician as a *designer*. The designer's role is to help the patient pick a device and corresponding algorithm that best fits the patient's physiology, values, and preferences. The design model overlaps with interpretive or deliberative models discussed above, but is more clearly prospective in its orientation. The clinician works with the patient to "lock-in" a suitable algorithm for the device, one that encompasses the type of device intelligence and independence that is a best fit for how the patient sees him or herself now and projects into the future. The clinician assists the patient in making choices about the device. What inputs will the device be able to consider? Are there limits on the range of acceptable stimulations? The clinician helps the patient pick the optimum design for the closed-loop device. To do this effectively, the clinician will need an extensive knowledge of the patient and deep understanding of design possibilities.

The second model is the clinician as *servicer* of the device. After implantation, the patient will find the device to be meeting or not meeting expectations. In the service model, the clinician's role is primarily to address a possible mismatch. This may involve interrogating the device and making sure it is working as designed. It also may involve exploring what the patient wanted prior to implantation or currently wants out of the device. Finally, it may involve educating the patient on the current capabilities of the device or what would be possible if the device were changed, somewhat akin to the clinician in the informative model. Ultimately, the clinician works to make the device a satisfactory fit for the patient.

7 There is an interesting question about whether the machine itself is a party to the relationship. This gets to questions about machine agency and is interesting but beyond the scope of what is argued here. Nonetheless, even if we hold the status of the device an open question—and how it ought to be regarded—the relationship between the clinician and patient is still challenged in a fundamental way. This challenge is task enough.

The third model involves the clinician as a *monitor* of the device. Even closed-loop devices designed for a level of autonomy or independence must function within a set of acceptable parameters. Within the monitoring model, the clinician's role is to check whether the device is functioning within an acceptable range. This will involve interrogating the device to make sure it is working according to pre-established criteria (e.g., delivering the set stimulation). It will also involve the clinician "looking in on" the patient in the sense that a closed-loop device may result in unacceptable patient behavior. For instance, a closed-loop device may lead to personality changes (e.g., disinhibition) or destructive habits (e.g., gambling, overeating) that fall outside of what the patient originally deemed acceptable behavioral outcomes. How a clinician exercises this role if individuals do not want to be monitored or no longer view certain behaviors as unacceptable (e.g., gambling) raises important ethical questions about identity over time and the clinician's role in public health surveillance.

These three models have some appeal but each seems narrow in applicability. The design model, for instance, is principally focused on the initial decision of which device to adopt, with relative neglect of what happens after. While the design model does not preclude involvement of the clinician after the choice of device (designers can check in on their devices to see how they perform) the priority in this model is on designing the device correctly at the outset. The service model, on the other hand, is narrow in a different way. The patient's satisfaction with the device is the core feature of this model. It is what brings the patient and device to clinical attention and what motivates adjusting the device settings. Other considerations, though, such as what may be in the overall best interest of the patient or what might be in the interests of others (e.g., family), may also be important, but within the service model these would seem to take a subsidiary role. Finally, the quality-control monitor role is narrow in yet a different way. Such a monitor works by looking out for deviations from predetermined rules or outcomes. In the case of closed-loop devices, however, unacceptable performance is a moving target. Individuals will change over time as a function of the device in who they take themselves to be and what they find acceptable.[8]

While these limitations as sketched may not be sufficient to justify abandoning the search for an appropriate model amongst these candidates, they may give pause; none of these models, at least at first pass, seems sufficiently robust to do normative work across the full range of *living with* an intelligent neuromodulatory device. Perhaps the search for a model of the patient-machine-clinician relationship ought to travel in another direction, toward a model that is more straightforwardly normative in its ambitions. A friendship model of the clinical relationship would seem to fit such a mold.

---

[8] Monitoring, in the generic sense, is a part of every clinician's role, much as caring or diagnosing is. Clinicians monitor for changes in health as a core professional activity, and in so far as it is relevant to health or health decision-making, monitor for changes in the *person*. Whether monitoring ought to be taken as *the* orienting activity of a clinician is a different question.

5 Closed Loop Neuromodulation and Aristotelian Friendship

The notion that friendship might be a useful and relevant model for the patient-clinician relationship has a long history. Lain Entralgo [22] in his *Doctor and Patient* traces the history of friendship (*philia*) as a foundation for medical practice from the Hippocratic *Corpus* forward. Wadell [36] argues that a friendship model is a natural outgrowth of the three features shared by friendship and the clinical relationship: (1) benevolence, (2) mutuality, and (3) a shared good. Fried [18] suggests that physicians and patients can be thought of as "limited, special-purpose friends" in so far as one friend can assume the interests of the other in certain contexts.

The friendship model has its detractors. Loewy [25] argues that the ubiquitous and imprecise use of friendship poses an insurmountable barrier to taking seriously friendship as a model for the clinical relationship. More recently, Davis [12] points out two critical differences between friendships and clinical relationships. Moral responsibilities within friendships are mutual or reciprocal and friendships are freely entered into, whereas in the clinical relationship the extent of the physician's responsibilities are clearly asymmetric to that of the patient and the existential vulnerability of the patient precludes a relationship of truly free entry (and exit).

The normative potential of friendship is often traced to Aristotle [11]. Aristotle most explicitly addresses friendship in the *Nicomachean Ethics* [3]. He delineates three kinds of friendship in terms of the ends toward which they aim. The first two are limited species of friendship, friendship based on the achievement of a pleasurable outcome or on the usefulness of an association: "[T]hese friendships are only incidental; for it is not as being the man he is that the loved person is loved, but as providing some good or pleasure" (NE 1156a17–19). The third kind of friendship, one that Aristotle notes is capable of *perfection*, has the moral good of another as its end (NE 1156b7, 34):

> Now those who wish well to their friends for their sake are most truly friends; for they do this by reason of their own nature and not incidentally; therefore their friendship lasts as long as they are good—and excellence is an enduring thing (*NE* 1156b10–12).

Aristotle's third kind of friendship, what Cooper calls "character friendship" has as its end the moral goodness of another. Friendship involves recognizing the good qualities of character in another and acting in ways that support and promote these.

That Aristotle's, or any other, view of friendship would have relevance to the patient-machine-relationship might seem an odd, even quaint, suggestion. The way in which medical care is organized and delivered today can hardly be seen as hospitable to most notions of friendship. Economic and social forces drive patients and physicians together and then apart, with a lengthy or lifelong patient-clinician interaction becoming less the norm. The interactions of patient and clinician are dissected and "nudged" in one way or another in service to external ends (e.g., efficiency, satisfaction, resource utilization) [20]. Rather than a friendship, patients quite often want use of a technical expert and physicians want a job with unambiguous occupational boundaries. As such, the challenge of finding points of contact between friendship and the current instantiation of medicine has to be conceded. That said, the challenge does not vitiate the idea that friendship might be useful, only cautions against thinking the value will be self-evident or easily achieved.

Oakley and Cocking [29] argue that friendship is best viewed as a kind of regulative ideal. Like other ideals that can structure a good life, friendship involves

> ...internalizing a certain normative structure of excellence, in such a way that one is able to adjust one's motivation and conduct so that it conforms, or at least does not conflict, with that standard...For in exemplifying the good of friendship, one does not act for the sake of *friendship* per se, or even for the sake of *this* friendship, but rather for the sake of this *person*, who is one's friend. And one can properly be said to be acting for the sake of this person only if one has shaped one's perception in certain ways—for example, one must have developed some kind of understanding of what this person's well-being consists of, and of which ways of acting would promote it [29, pp. 28–29].

In this way, friendship is an ideal that structures the development and exercise of character, both in oneself and in one's friend, by contributing to a flourishing life.[9] Friendship as a regulative ideal proves useful in thinking about the clinical relationship. The clinical relationship too serves as a kind of regulative ideal. The development and exercise of certain character traits in clinicians and patients promote the achievement of certain morally valuable ends. The ends are nested and given meaning by the goals of medicine. Friendship, on this view, does not provide an ideal to which the clinical relationship should *conform*. Rather, both friendship and the clinical relationship serve as examples of regulative ideals for human flourishing.

Two important features of the closed loop neuromodulation relationship argue for exploring this friendship ideal further. The first is the centrality that identity plays in decisions about closed loop neuromodulation. The second is the way in which trust figures into the success of the relationship. Before ending, it is worth sketching very briefly how these features are prominent in closed loop neuromodulation and why the friendship ideal might be particularly suited to help illuminate them.

The initial decision to "lock in" a closed-loop device algorithm confronts questions of identity directly. As discussed above, a patient's choice of algorithm involves a patient and clinician in an exploration of patient values and preferences, followed by a mapping of these onto algorithm choices. At one level, such a decision involves identity just as any other consequential medical decision would [4]. Patients think about medical options in terms of who they are and who they want, or do not want, to become. The outcome of one medical decision eventually presents new options, affording an opportunity to reassess and reconfigure identity.

---

9 Oakley and Cocking note that this kind of view is not out of step with what Aristotle intended, though they acknowledge that traditional views of Aristotelian friendship often incorporate an explicitly moral disposition to friends (p. 72).

This is where closed-loop neuromodulation begins to take a different path.[10] The choice of algorithm is a choice of how *not* to be presented with discrete clinical decisions. Standard opportunities for adjusting identity are given over to an intelligent device that makes "decisions" hidden from view. Neuromodulation becomes akin to a neurophysiologic process rather than a series of patient-directed decisions.

This not only makes it that much more important to achieve an identity-congruent choice of algorithm at the outset, but it shifts the way in which identity figures into the life of the individual with the device. Notions of moral and legal responsibility, which in other contexts are distributed over a series of discrete decisions leading up to a given outcome, in the case of a closed-loop device trace back in full to the original choice of algorithm. The psychological and social burden of tying responsibility so tightly to the initial choice of algorithm will be substantial. Furthermore, closed-loop neuromodulation, by doing away with individual modulation decisions, will also do away with socially and medically acceptable decision-points for reconfiguring identity. In so doing, medicine may need to liberalize what counts as acceptable reasons for considering a change in device algorithm. Rather than requiring particular clinical symptoms, signs or diseases, clinicians may need to attend more carefully to considerations of identity. Comments like "I don't feel like myself anymore" [33] may need to be put on par with traditional clinical measures.

The other feature that promises to figure prominently in the closed-loop neuromodulation relationship is trust. Given how consequential the initial decision of algorithm is, trust between patient and clinician will be paramount. The patient will need to trust that the clinician knows the capabilities of the devices (and algorithms) and can accurately prognosticate about their benefits and harms. The patient will also need to trust that the clinician has the patient's best interests in mind when presenting and recommending device options. Similarly, the clinician will need to trust that the patient gives accurate personal information and engages in an honest and thorough self-exploration of values and preferences.

The importance of trust in closed-loop neuromodulation goes beyond the initial choice of algorithm. The nature of a closed-loop device is that it locks the clinician (and patient) out of iterative decision-making. This shifts the clinician into a role of following *alongside* neuromodulation. What it will mean to follow alongside a patient with a closed-loop device is an open question. It may involve technical monitoring of the device in some fashion, perhaps remotely, and intervening when

[10] This is not to say that identity is not a part of other clinical decisions made in medicine. For instance, identity is likely *the* most important consideration in decisions about advance directives. There is an important difference, however. In the case of advance directives, the decision is circumscribed and priority is given to trying to respect or reflect identity rather than influence it. Very rarely do people think about how *making* an advance directive will go on to influence the type of person one will become. Conversely, the decision about a closed loop device is conspicuously a choice about how to shape one's identity going forward, what features of the self to endorse and promote and what features to avoid.

certain objective parameters are met (e.g., battery failure, input–output mismatch). The parameters for intervening based on technical grounds will likely need to be outlined and agreed upon in advance, before implantation of the device or choice of device algorithm.

Intervention may also be needed for non-technical reasons. The closed-loop iterative process will lead some individuals to behave in ways that they or others find unacceptable. For example, Glannon [19] presents a case first described by Leentjens et al. [23] of a patient with PD and DBS who experienced megalomania and chaotic behavior as a result of unidirectional DBS stimulation. This behavior was felt to pose a danger to himself and prompted clinical intervention.[11] Intervention will similarly be needed in patients with closed-loop devices, complicated perhaps by more insidious development of behavioral change. Changes in behavior or personality that occur slowly over time in response to iterations of a feedback loop may be difficult to distinguish from changes due to non-device means. When the patient finds such changes worrisome or unacceptable, intervention may be easily justified, but when the patient does not, we can see why trust may be important.

The patient's initial choice of device algorithm will ideally involve a discussion with the clinician as to acceptable and unacceptable future outcomes of neuromodulation. For instance, an individual might find it unacceptable if the functioning closed-loop device led to behavior that was compulsive or disinhibited. "If I ever become a compulsive gambler, please intervene." Part of the process of deciding on a device algorithm will be deciding on acceptable parameters for future clinician intervention, even if a patient *in the future* finds such parameters unacceptable. The clinician may be party to a kind of Ulysses contract.[12] A patient may want assurance that the clinician will intervene if the functioning of a closed-loop device leads to changes that are *now* deemed unacceptable. Such assurance is grounded in trust. A patient will have to trust a clinician will be able and willing to intervene, and trust that the clinician will know the patient well enough to decipher when a change in behavior crosses a line. As Baier [2] notes, investing a friend with discretionary power is an ineliminable element of trust.

Aristotelian friendship embeds rich notions of identity and trust. Aristotle takes friendship to involve a commitment to the development of another's character [11]. Friends strive to make their friends better people. This involves as a prerequisite that friends delve into the character of their friends and use what they come to

[11] In this case, the intervention involved clarifying the patient's preferred state. When the patient was not stimulated—and presumably competent to make a decision comparing the two states—the patient chose to be stimulated and institutionalized, due to psychotic behavior, rather than have the stimulator turned off. Hence, an intervention may simply be initiating a discussion with a patient in order to clarify decision-making capacity. While even an 'intervention' of this type could be viewed as paternalistic from a synchronic perspective, it will fall far short of the more objectionable paternalism involved in changing device parameters without the patient's consent.

[12] Ulysses instructed his crew to bind him to the mast and ignore his exhortations to unbind so that he could safely enjoy the music of the Sirens.

know of each other as the basis for action. The commitment to character development carries throughout the friendship, but is particularly important for consequential decisions. Friends tend to wrap themselves up in such decisions because the implications for their friend's character going forward are so great. The notion of trust is also central to an Aristotelian notion of friendship. Friends trust one another to assist not just when asked, but also when needed. Friends notice things about their friends that others miss; they often have a longer and deeper perspective, in Aristotle's terms "time and familiarity" (NE 1156b, 25–29), from which to view the behavior of their friends. Friends have a unique ability to *perceive* what a friend needs [5]. This sometimes justifies, even requires, intervening in the life of a friend even when the friend does not initially desire the intervention.[13] Thus, the friendship ideal provides a glimpse of how to steer a course between unpalatable extremes of paternalism and bureaucratic impartiality in the clinical relationship.

We can start to see the outlines of how an Aristotelian ideal of friendship would be useful for thinking about the patient-machine-clinician relationship in closed-loop neuromodulation. The role of the clinician extends beyond the patient's initial decision to get or not get a device; it reaches into helping a patient make sense of his or her identity and what it means to *live with* a device. At the same time, the patient and clinician are bound by notions of trust; the depth of trust required for success of this relationship extends beyond limited notions of reliability or contract and into entrusting one's future to the hands of a caring other. The friendship ideal provides support for thinking that merely designing, monitoring or servicing a patient's device will not capture the richness required of the clinician's role in neuromodulation.

A robust model of the patient-machine-clinician relationship would be valuable, but it may be that we have to settle for something more preliminary. An Aristotelian ideal of friendship, conceived as a regulative ideal, may be a useful tool for understanding the role of the clinician (and patient) in the emergence of neuromodulatory technologies. While not itself a model, the ideal of friendship provides an anchor for thinking about the patient-machine-clinician relationship and exploring the adequacy of clinical models.

Acknowledgments This work was supported by a grant from the National Science Foundation (NSF Award #EEC-1028725).

[13] The "initially" is important here, because there are limits to the extent of paternalism that any friendship can accommodate. The exercise of paternalism can also violate other features of friendship (e.g., mutual respect) and so must be exercised judiciously. How much paternalism any *particular* friendship can bear is particular to the history and nature of *that* friendship. Exercised too often or without a sense that friends are both oriented to the same good, this paternalism can harm or even destroy a friendship. Friendship is not a blank check to co-opt another's path toward the good.

References

1. Amundsen DW, Ferngren GB (1983) Evolution of the patient-physician Relationship: antiquity through the renaissance. In: Shelp EE (ed) The clinical encounter. D. Reidel Publishing Company, Dordrecht, pp 3–46
2. Baier AC (1994) Moral prejudices: essays on ethics. Harvard University Press, Cambridge
3. Barnes J (1984) The complete works of Aristotle. Princeton University Press, Princeton
4. Baylis F (2011) I am who I am: on the perceived threats to personal identity from deep brain stimulation. Neuroethics 6(3):513–526. doi:10.1007/s12152-011-9137-1 (Published Online First: 14 Sept 2011)
5. Blum L (1980) Friendship, altruism and morality. Routledge & Kegan Paul, London
6. Brock DW (1991) The ideal of shared decision making between physicians and patients. Kennedy Inst Ethics J 1(1):28–47
7. Brody H (1987) The physician-patient relationship: models and criticism. Theoret Med 8:205–220
8. Bronstein JM, Tagliati M, Alterman RL et al (2011) Deep brain stimulation for Parkinson disease: an expert consensus and review of key issues. Arch Neurol 68:165–171
9. Charon R (2001) Narrative medicine. JAMA 286(15):1897–1902
10. Childress JF, Siegler M (1984) Metaphors and models of doctor-patient relationships: their implications for autonomy. Theor Med Bioeth 5(1):17–30
11. Cooper JM (1980) Aristotle on friendship. In: Rorty AO (ed) Essays on Aristotle's ethics. University of California Press, Berkeley, pp 301–340
12. Davis FD (2000) Friendship as an ideal for the patient-physician relationship: a critique and an alternative. In: Pellegrino ED, Thomasma DC, Kissell JL (eds) The healthcare professional as friend and healer. Georgetown University Press, Washington, DC
13. Dubiel H (2009) Deep within the brain: living with Parkinson's Disease. Europa editions, New York
14. Eberle W, Penders J, Yazicioglu RF (2011) Closing the loop for deep brain stimulation implants enables personalized healthcare for Parkinson's disease patients. In: Engineering in medicine and biology society, EMBC, 2011 Annual international conference of the IEEE, p 1556–1558
15. Emanuel EJ, Emanuel LL (1992) Four models of the physician-patient relationship. JAMA 267:2221–2226
16. Engelhardt HT (1983) The physician–patient relationship in secular, pluralist society. In: Shelp EE (ed) The clinical encounter. D. Reidel Publishing Company, Dordrecht, pp 253–266
17. Ford PJ, Henderson JM (2006) The Clinical and Research Ethics of Neuromodulation. Neuromodulation Technol Neural Interface 9(4):250–252
18. Fried C (1974) Medical experimentation: personal integrity and social policy. North-Holland Pub. Co., New York
19. Glannon W (2009) Stimulating brains, altering minds. J Med Ethics 35(5):289–292
20. Klein E (2012) Redefining the clinical relationship in the era of incentives. Am J Bioeth 12(2):26–27
21. Krames ES, Peckham PH, Rezai AR (2009) What is neuromodulation. In: Krames ES, Peckham PH, Rezai AR (eds) Neuromodulation. Elsevier-Academic Press, London, pp 3–8
22. Lain Entralgo P (1969) Doctor and patient. McGraw-Hill Book Company, New York
23. Leentjens AF, Visser-Vandewalle V, Temel Y et al (2004) Manipulation of mental competence: an ethical problem in case of electrical stimulation of the subthalamic nucleus for severe Parkinson's disease. Ned Tijdschr Geneeskd 148(28):1394–1398
24. Little S, Brown P (2012) What brain signals are suitable for feedback control of deep brain stimulation in Parkinson's disease? Ann NY Acad Sci 1265(1):9–24
25. Loewy EH (1994) Physicians, friendship, and moral strangers: an examination of a relationship. Camb Q Healthc Ethics 3(1):52–59
26. May WF (1975) Code, covenant, contract, or philanthropy. Hastings Cent Rep 5(6):29–38

27. Medtronic Inc. http://professional.medtronic.com/pt/neuro/dbs-md/prod/dbs-patient-programmer-model-37642/index.htm#.UmGLW3bn_IU. Accessed 31 Oct 2013
28. Moro E, Poon Y, Lozano AM et al (2006) Subthalamic nucleus stimulation: improvements in outcome with reprogramming. Arch Neurol 63:1266–1272
29. Oakley J, Cocking D (2001) Virtue ethics and professional roles. Cambridge University Press, New York
30. Patel S, Lorincz K, Hughes R et al (2009) Monitoring motor fluctuations in patients with Parkinson's disease using wearable sensors. IEEE Trans Inf Technol Biomed 13(6):864–873
31. Pellegrino ED (1979) Toward a reconstruction of medical morality: the primacy of the act of profession and the fact of illness. J Med Philos 4(1):32–56
32. Priori A, Foffani G, Rossi L et al (2013) Adaptive deep brain stimulation (aDBS) controlled by local field potential oscillations. Exp Neurol 245:77–86
33. Schüpbach M, Gargiulo M, Welter ML et al (2006) Neurosurgery in Parkinson disease a distressed mind in a repaired body? Neurology 66(12):1811–1816
34. Stanslaski S, Afshar P, Cong P et al (2012) Design and validation of a fully implantable, chronic, closed-loop neuromodulation device with concurrent sensing and stimulation. IEEE Trans Neural Syst Rehabil Eng 20:410–421
35. Veatch RM (1972) Models for ethical medicine in a revolutionary age. Hastings Cent Rep 2(3):5–7
36. Wadell P (1995) Friendship. In: Reich WT (ed) Encyclopedia of bioethics. Simon and Schuster, New York, pp 888–891
37. Yu H, Neimat JS (2008) The treatment of movement disorders by deep brain stimulation. Neurotherapeutics 5(1):26–36

Modelling Consciousness-Dependent Expertise in Machine Medical Moral Agents

Steve Torrance and Ron Chrisley

Abstract It is suggested that some limitations of current designs for medical AI systems (be they autonomous or advisory) stem from the failure of those designs to address issues of artificial (or machine) consciousness. Consciousness would appear to play a key role in the expertise, particularly the moral expertise, of human medical agents, including, for example, autonomous weighting of options in (e.g.,) diagnosis; planning treatment; use of imaginative creativity to generate courses of action; sensorimotor flexibility and sensitivity; empathetic and morally appropriate responsiveness; and so on. Thus, it is argued, a plausible design constraint for a successful ethical machine medical or care agent is for it to at least model, if not reproduce, relevant aspects of consciousness and associated abilities. In order to provide theoretical grounding for such an enterprise we examine some key philosophical issues that concern the machine modelling of consciousness and ethics, and we show how questions relating to the first research goal are relevant to medical machine ethics. We believe that this will overcome a blanket skepticism concerning the relevance of understanding consciousness, to the design and construction of artificial ethical agents for medical or care contexts. It is thus argued that it would be prudent for designers of MME agents to reflect on issues to do with consciousness and medical (moral) expertise; to become more aware of relevant research in the field of machine consciousness; and to incorporate insights gained from these efforts into their designs.

S. Torrance (✉)
Centre for Research Cognitive Science (COGS), School of Engineering and Informatics, University of Sussex, Falmer, Brighton, UK
e-mail: stevet@sussex.ac.uk

R. Chrisley
Departmentof Informatics, Centre for Cognitive Science (COGS),
Sackler Centre for Consciousness Science, University of Sussex, Falmer, Brighton, UK
e-mail: ronc@sussex.ac.uk

S.P. van Rysewyk and M. Pontier (eds.), *Machine Medical Ethics*,
Intelligent Systems, Control and Automation: Science and Engineering 74,
DOI 10.1007/978-3-319-08108-3_18

1 Introduction

Since the advent of artificial intelligence, and more specifically, of AI-based artificial ethics systems, technology has threatened to acquire moral agency, or moral status, in a direct sense: not merely in mediating the moral activities and attitudes of humans in the way writers such as Latour [28] and Verbeek [49] have outlined, but also in being designed to explicitly mimic, or reproduce, in as rich and detailed a way as possible, the activities and attitudes of human moral participants. The goal of AI has been to develop artificial agents that are increasingly interactive, autonomous and adaptable (using the criteria of agenthood deployed by Floridi and Sanders [13]). From the earliest days of AI, many have asserted that it is a necessity to ensure that such increasingly autonomous agents operate according to goals or codes that conform to the moral expectations of the people they interact with; in a word, that such artificial agents are, as far as possible, *moral* agents [3, 29, 50].

Thus the field of machine ethics (ME) appears on the scene. ME can be divided into two endeavors: the investigation of the ethical issues that arise around intelligent or autonomous artificial moral agents; and the attempt to design and build such moral agents—*theoretical* and *constructive* ME, as we shall call them. The autonomous agents that are being developed in constructive ME are like human moral agents in that they can comfortably be described in terms of interactivity, autonomy and adaptability, and, moreover, that their actions are assessable in terms of moral good or ill [13].

However, such agents are unlike human moral agents—to date at least—in that they do not have, except in a vestigial or metaphorical sense, any kind of consciousness. Moreover, consciousness seems to be one feature that resists being built into AI agents, at least in any convincing way. Since many take consciousness to be central to human agency, and therefore to human moral agency, this presents a *prima facie* challenge to the very idea of constructive ME. By contrast, the emerging field of machine consciousness (MC) sees the recalcitrance of consciousness as a contingent matter, and attempts to import some, at least, of the human features of consciousness into our technologies. If MC can succeed in this an apparent obstacle to the very idea of constructive ME can be removed.

In the following discussion, we will examine how far the field of MC may impact on the parallel project of ME; both in general as two offshoot studies of Artificial Intelligence, and in particular, within the application areas of medicine and care. As we will see, there are several analogies between the two fields of ME and MC; thus, those working in machine ethics, particularly in the medical and care application areas, may benefit from looking at developments in machine consciousness.

2 Consciousness and Ethics in the Medical/Care Domain

Consciousness, and several properties related to it, is involved in a vast number of human activities. In health and care settings, conscious processes are enlisted when non-routine or challenging tasks need to be performed, like sifting through alternative possibilities to select the best diagnosis, performing a challenging surgical operation, counseling a bereaved relative, or providing care to a physically frail patient. The term "consciousness" denotes a broad cluster of attributes; and many of these attributes seem to play a key role in the expertise—particularly the moral expertise—of human medical agents. Attributes closely associated with consciousness such as focused attention, mindful exercise of bodily skills, use of imagination in scenario-evaluation, empathetic projection to the states of others, etc. are clearly relevant in a variety of different ways to health and care practice, and to delivering medical and social care performance in professionally sound, and morally responsible, ways. Here are some specific examples of how different aspects of consciousness appear to be strongly implicated in the exercise of ethical skills or expertise within medical and care practice:

- autonomous balancing of options in judgment and decision; for example in diagnosis, or the planning of treatment, etc., particularly where moral considerations play a prominent role in factors that require consideration in the weighting.
- use of imagination and creativity to generate possible courses of action that satisfy situational constraints; for example, when considering moral choices in clinical contexts.
- the exercise of sensorimotor flexibility and sensitivity; for example, when physically interacting with human patients, performing medical procedures, and so on, in ways that respect moral constraints, such as avoiding discomfort, invasiveness, etc.
- empathetic (and morally appropriate) interpersonal responsiveness; for example, in face-to-face communication with patients and their carers; in dealing with fear and distress, etc.

Other such examples abound. In describing the various personal attributes that are demanded of medical and care practitioners—such as surgeons, consultants, nurses, clinical advisors, care support workers, and so on—when operating in such situations, a number of facets of consciousness come to the fore, such as attentiveness, imagination, awareness, and so on. What is true of human medical agents would presumably also carry over into the domain of artificial medical agents. So, when considering the field of machine medical ethics (MME)—and indeed machine ethics more generally—it would be wise to consider work in the neighboring field of machine consciousness. Any artificial agent that fulfilled functions such as those above, and others, might need to be designed in a way that took account of developments in MC.[1]

[1] We use the term "Machine" Ethics/Consciousness here rather than "Artificial" following many (but not all) in the respective fields. There is a lot of fuzziness about what counts as a "machine" here, but the de facto emphasis in this discussion is on computational systems.

3 The Scope of Machine Consciousness

ME and MC might both be considered as fringe areas, or extensions, of AI, alongside artificial life, artificial social modelling, artificial emotion, computational creativity, etc. It is worth differentiating each from their parent field, AI. As a first approximation, it can be said that, while AI (in the main) is the attempt to create machines (using computational and robotic techniques) that are *intelligent*, or *cognitive*, agents, (constructive) MC is the attempt to create machines that are *conscious* agents, and (constructive) ME is the attempt to create machines that are *ethical* agents.

There are a number of ways in which those thumbnail descriptions require qualification to be even broadly useful. Let us start with a difficulty specific to MC, one that concerns the relation between intelligence and consciousness. AI is often thought to cover more than the narrowly intellectual or cognitive aspects of mind. In this way AI strongly overlaps with, or even subsumes, those mental features that are central to being conscious. Thus many would say that MC must be fully subsumed under AI, or nearly so. However, others would argue that there are key aspects of consciousness which are not able to be captured by traditional AI or computational techniques, for example, because they concern, not so much the publicly-accessible aspects of consciousness, as the "subjective feel" or "first-person phenomenology". Such aspects of consciousness—the core aspects, to many eyes—are often thought to present a particular barrier in terms of computational modelling or replication. So many questions are raised, by critics and friends of MC alike, as to whether MC, taking it as a primarily computational endeavor, could ever really "capture" core aspects of consciousness as easily as AI might seem to "capture" core aspects of intelligence or cognition.

Some researchers draw a distinction between "functional" consciousness, on the one hand, and "phenomenal" consciousness, on the other (e.g., [8, 14, 21, 47] see also [7], for a broadly comparable distinction between "access" and "phenomenal" consciousness). Intuitively, there seems to be a distinction between what might be called "consciousness as consciousness does"; that is, consciousness considered in terms of the roles it plays within the exercises of different mental powers and "consciousness as consciousness seems (or feels)" (see [19]). There is considerable controversy over whether these are two distinct kinds of consciousness, or whether phenomenal consciousness can be explained without remainder in terms of functional consciousness (e.g., [38, 39]).

For the purposes of this discussion we keep an open mind about the phenomenal-functional distinction. Upholding the distinction would lead to viewing MC as having at least two different research targets. The first branch—trying to create computational models that enable us to understand or reproduce[2] the

[2] For the moment we are eliding over the distinction between artificially modeling some target area, one the one hand, and reproducing, or replicating that target area. This distinction is usually marked in the area of AI, by the phraseology "Strong AI"/"Weak AI". A similar division between Strong and Weak MC has been suggested (refs); and between Strong and Weak ME. We will discuss the Strong/Weak antithesis later on in the paper.

functionality of consciousness—is perhaps much more amenable to the spectrum of methods in AI that MC for the most part relies on. The second target—the experiential or phenomenal aspects of consciousness—may still be amenable to AI methods, at least as far as modelling is concerned. However, proponents of a robust functional/phenomenal distinction are likely to say that MC cannot reproduce phenomenal consciousness using AI or computational methods—although they may allow that other approaches, for example via synthetic biology, may at some point succeed in developing artificial organisms which possess phenomenal consciousness. (This possibility would suggest that it is best to avoid making "Machine Consciousness" and "Artificial Consciousness" interchangeable terms.) Rejecting the distinction (seeing phenomenal consciousness as explicable without remainder in terms of functional consciousness) would lead to the view that the second research goal is in principle fully achievable by AI methods—including that of reproducing all of consciousness's "first person" aspects.

4 Machine Ethics: Some Core Distinctions

We have suggested that according to a popular conception of consciousness, the field of MC can be thought of as having two distinct targets: functional consciousness and phenomenal consciousness. The former target seems to be more comfortably within the scope of MC, when understood as using core AI methods. In a similar way, one might think it is possible to distinguish between *functional* ethical status and *genuine* ethical status as targets for constructive ME (e.g., [50, 51]; see also [31] for a more nuanced set of distinctions).

Making this intuitive distinction more precise is decidedly tricky, but a proper analysis of previous attempts to do so would take us too far afield. So, we simply offer this clarification: Let the ethical evaluation of a behaviour b be denoted by $E(b)$. Then an agent X is *functionally* ethically equivalent to agent Y iff, for each of Y's behaviors b that are ethically evaluable (i.e., that are such that $E(b)$ exists), the corresponding behaviour a of X is causally indistinguishable from some possible or actual behaviour b' of Y such that $E(b') = E(b)$. Note that according to this definition, X can be functionally ethically equivalent to agent Y while still being behaviorally distinct from Y, in two ways: (1) there need be no correspondence between X's behaviour and Y's non-ethically evaluable behaviour; and (2) X's behaviour can be distinct from Y's ethically evaluable behaviour as long as it is causally indistinguishable from an actual or possible behaviour of Y's that has the same ethical evaluation as Y's actual (ethically evaluable) behaviour. Note that this definition is silent concerning the ethical evaluability of X's behaviour, including X's behaviors that correspond to Y's ethically evaluable behaviour. Note also that the definition appeals to the notion of causal indistinguishability. One gloss on that notion is this: b and b' are causally indistinguishable iff the (non-relationally individuated) effects of b are the same as the (non-relationally individuated) effects of b'.[3]

By contrast, we can stipulate that an agent X is *genuinely* ethically equivalent to agent Y iff, for each of Y's behaviors b that are ethically evaluable (i.e., that are such that $E(b)$ exists), the corresponding behaviour a of X is such that $E(a) = E(b)$. It follows that if X is genuinely ethically equivalent to Y then X is functionally identical to Y, but not vice versa. Most relevantly, the distinction between functional and genuine ethical identity would allow the following possibility: an agent X is functionally ethically equivalent to agent Y, even though none of X's behaviour is ethically evaluable. Presumably, the functional aspects of moral action and decision are amenable to replication by the AI methods which are currently most accessible to constructive ME researchers. As AI systems are increasingly able to generate, through their own autonomous decision-making, actions that are (functionally) morally sensitive, in their potential to cause good or harm to people, it is thought desirable to be able to build moral constraints into the functionality of such systems. This functional approach to ethical modelling leaves wide open (and thus side-steps) the question of whether such AI agents, or indeed any agents developed using exclusively AI techniques, could ever instantiate genuine moral agency, or be regarded as having genuine moral status.

At this point an important clarification should be made. There are at least two senses in which people talk of a moral or ethical agent: a classificatory sense and a normative sense. The classificatory sense is: something that is, or engages in behaviour that is, morally *evaluable*; something to which moral norms apply. The normative sense is: something that is *morally good*, in that it (on balance) does what is right. The second sense presupposes the first sense, but, notoriously, not vice versa. An example of the first usage is found in an often-cited article by Moor [31], which details four ways (impact/implicit/explicit/full) in which an artificial agent might produce behaviour that is morally evaluable. This is in contrast with the notion of "ethical" that the public, lawmakers, and the laity in general typically have in mind when asking "Can we build an ethical robot?" That a robot is ethical in the first sense would be of cold comfort to its wronged victims. Similarly, saying under oath "Bloggs has always exhibited ethical behaviour", while true in the first, classificatory, sense of "ethical",[4] will likely score you a perjury charge if you say it while knowing Bloggs to be a liar, cheat and thief. Of course, these two senses are related, and Moor's idea is that some ways of making robots that are

---

[3] The "non-relationally individuated" qualifier is meant to block certain trivializations of the notion of causal indistinguishability. For example, suppose, *per impossible*, that b and b' have the same effect, that of light L coming on. Further suppose that they both have no other effects on any other objects. Intuitively, b and b' are causally indistinguishable. However b, but not b', has the (relationally individuated) effect of *light* L *coming on as a result of* b. This would render b and b' (and all other pairs of behaviors) causally distinguishable, making any notion defined in terms of it (such as functional ethical equivalence) vacuous. Adding the restriction prevents this.

[4] Or perhaps not, strictly speaking: plausibly, only a proper subset of our (and Bloggs') behavior is ethically evaluable.

ethical in the first sense will be of greater use than others in making robots that are ethical in the second, normative, sense. But the distinction (and potential confusion) remains. In this article, we will be using the first, classificatory, sense, except where noted otherwise.

However, this first sense of "moral agent" can itself be understood in two importantly different ways. When considering what it is for an artificial agent, or a human being or other creature, to be a moral agent (in the classificatory sense), or to have moral status, one may be focusing on at least two different kinds of property. (a) One may be thinking of a being that is capable of acting in a way that is open to moral appraisal. (b) Alternatively, one may be thinking of a being *to whom* morally appraisable actions can be done; that is, a subject with moral interests. Agents of the first sort are sometimes called "moral agents", in a restricted sense; those of the latter sort are then by contrast called "moral patients" (e.g., [13, 17, 18]). For rough and ready illustrations, consider, as moral agents, someone staffing a helpline for potential suicides, or a tax evader (both would be moral agents in this restricted sense, examples with contrasting moral polarities have deliberately been used here). For examples of moral patients, consider a burglary victim or someone receiving financial assistance to pay for expensive medical treatment. (The last-mentioned individual will be both a moral patient and a medical one.) An alternative terminology that, we suggest, is less confusing (especially when discussing ME in the context of medical care) will be used here: we will talk of "moral producers" and "moral consumers", respectively.[5]

What are the conditions that determine what counts as a moral agent/producer and how do these differ from those that determine what counts as a moral patient/consumer? It is doubtful that both types of role require an identical set of conditions. For example, to count as an ethically responsible moral producer one would arguably need to be capable of engaging in intelligent moral deliberation, with due regard for the relevant facts of the case, to the extent to which these are accessible, and so on. Being a moral consumer, on the other hand, may, arguably, involve a smaller set of conditions: according to many commentators, infants, and even fetuses, as well as many kinds of non-human animals are taken to qualify for moral consumerhood, even while lacking the cognitive features (such as the potentiality to engage in rational deliberation about courses of action) which mark out a genuine or fully-fledged moral producer.

---

[5] The terms "moral agent/patient" are used more frequently than "moral producer/consumer" in the literature, but there are pitfalls to such a usage. First, when one talks about an "artificial agent", or indeed of a "moral agent", one is often using the term "agent" in a wider sense than when one is using "agent" to distinguish between moral agency and patiency. For example, there is a very real question of whether an artificial agent—in a wide sense of "agent"—could be a genuine moral patient (moral consumer) (see [46, 48]). Second, as suggested above, the term "moral patient" is particularly awkward in a medical context, as in the present discussion. Medical patients may invariably be moral patients (or moral consumers) as well, but the class of moral patients (consumers) whose interests may be affected by a particular medical intervention may be larger than the actual patient receiving that intervention (e.g., there may be family members whose interests would be severely affected if the intervention went wrong, and so on).

It seems that there is a strong link between qualifying as a moral consumer and certain kinds of consciousness or sentient awareness. For example, in an often-quoted passage from Bentham [6, 283], the capacity to "suffer" is singled out as a qualification for moral (consumer) consideration. Providing "suffering" is taken so as to imply ability to have sentient experience of pain, terror, physical discomfort or curtailment of movement, to be aware of being deprived, and so on (rather than, e.g., to suffer a financial loss in a certain trading period—which could apply to a more abstract entity such as a corporation), this would widen the class of moral consumers to a large population of animal species. Perhaps it is not just any kind of sentience that is needed, but rather a type of self-awareness, or, in Regan's phrase [33], the fact of being "the subject of a life".

Moral producers, standardly, are also conscious—but that may not be what makes them qualify for that designation; or if it is, it need not be the only route to qualification. For one thing, many recognize the moral responsibility/producerhood of corporations; that this purported status may derive from the consciousness of the people that a corporation comprises does not derail the general point.[6] For many people working in ME, that certain classes of artificial agents may at some future time be considered as fully-fledged moral producers is not at all to do with their being conscious or sentient—in anything like Bentham's sense. A completely non-conscious or non-sentient machine might be a moral producer, on one version of such a view, even while it does not possess any kind of conscious awareness (and is thus, in this respect at least, quite unlike human moral producers, presumably). Being a moral producer would rather be based on such capacities as its ability to deliberate on its behaviour in a specifically moral way, to justify such behaviour in moral terms, to engage in particular forms of interaction with humans (and other machines), and so on. Another version of this view might allow for the existence of conscious artificial agents, even to the point of denying that one could have the above capacities without also being conscious, yet still assert that it is the capacities, and not the concomitant conscious states, that qualify the artificial agent for moral producer status. In terms of the notion of functional (as opposed to genuine)

[6] A similar point can be made concerning moral consumerhood. Many see corporations as possessing rights, as entities toward which we can have obligations; for an example, one only need look at the recent (2010) US Supreme Court ruling in the case of Citizens United v. Federal Election Commission, which held that the first amendment in the Bill of Rights applies to corporations. If corporations have rights, then they can be wronged. And being something which can be wronged is, plausibly, sufficient for moral consumerhood, e.g., "An entity has moral status if and only if it or its interests morally matter to some degree for the entity's own sake, such that it can be wronged" [26]. Again, that corporations currently comprise conscious beings does not in itself undermine the general point; further, it is not *prima facie* absurd to suppose that a corporation could persist, and retain its rights, even if the number of humans constituting it shrank to zero. There is also the much-discussed issue of potentialism. Presumably we retain our full moral status even when we are unconscious; this (and the moral consumerhood of, say, fetuses) is usually explained in terms of a potential for being conscious, rather than consciousness itself. Perhaps, then, some artificial agents might be properly seen as moral consumers, as long as they could be understood in some way to be potentially, even if not actually, conscious.

morality, introduced earlier, it would be argued on (either version of) this view that, once one has in an artificial agent a sufficiently rich instantiation of functional moral productivity, one has all that is necessary for admitting that such an agent is a genuine moral agent, in the productive sense, even if (phenomenal) consciousness is absent. On an opposing view, it would be argued that such machines, as long as they are considered to be non-conscious, could not really be taken as genuine moral producers, even though they might be considered as functional moral producers (see [46] for further discussion). For example, one might take the capacity for empathy to be central to moral producerhood in a way that implies that only moral consumers can be genuine moral producers (on the ground that consciousness is a precondition of genuine moral consumerhood); see Sect. 5. So, on this latter view, an agent *X* might engage in behaviors that are causally indistinguishable from those of a genuine moral producer, therefore making *X* functionally (ethically equivalent to) a moral producer, without *X* being genuinely (ethically equivalent to) a moral producer.

Linking this discussion to the earlier distinction between functional and phenomenal consciousness, it is worth pointing out the following parallel. The development of artificial moral *consumers* will be thought by many to be resistant to traditional methods in AI/robotics in a way that parallels the way the quest for artificial phenomenal consciousness is resistant to such methods. This parallel may not be a coincidence. Some people would indeed insist that having phenomenal consciousness is a fundamental precondition to genuine moral consumerhood. So the perceived intractability of constructive ME which targets moral consumerhood may be a consequence of the perceived intractability of MC when the latter is concerned with targeting phenomenal consciousness. Perhaps one could avoid this intractability by developing merely functional moral consumers—but it is not clear that there is much to gain by this.[7] It is not clear if a merely functional moral consumer would be one towards which we (humans) would need to regard ourselves as having any moral obligations. Even if the possibility of merely functional moral consumers may be seen as an advantage from the perspective of constructive ME, it is easy to imagine (especially if aided by some of the many science fiction scenarios investigating this possibility) the moral hazards of living in proximity to agents that behave exactly as if they were moral consumers, but toward which we have (or believe we have) no ethical responsibilities.

In the case of developing artificial moral *producers*, on the other hand, it is easier to imagine that there could be a point in developing functional moral producers which were capable of apparently morally good behaviour (that is, behaviour that, were it performed by a genuine moral producer, would be considered morally good), but which were not to be taken as having full-blooded moral producer status (e.g., because they lacked real consciousness, or other key properties). Indeed,

[7] It is also unclear what (non-nefarious) motives one could have for creating "genuine" moral consumerhood in an artificial agent, unless one takes the view that it is a requirement for, e.g., moral producerhood.

this is, broadly speaking, the working goal of constructive machine ethics, in its current incarnation.

However, some might question the ability to produce (replicate) functionally morally good behaviour (of humans) without reproducing what they take to be a central cause of that behaviour in humans: genuine moral producerhood, and/or consciousness. On such a view, certain issues arise, of which the issue to do with occasional moral mis-performance seems to be one of the most pressing: any attempt to create agents that are *functionally* ethically equivalent to us, but not *genuinely* ethically equivalent to us, must fail to reproduce our ethical performance, possibly quite spectacularly or tragically. The worry is that any *attempt* to achieve functional ethical equivalence without genuine ethical equivalence would necessarily (or very likely) fail, with possibly disastrous results. So attempting to achieve the first where the second is unattainable, is fraught with unacceptable risks. Supporters of the quest for functional morality will, however, attempt to assuage such worries by pointing out that good constructive ME engineering design will ensure that such cases are kept to a minimum, and that, in any case, the risks may well fall below the risks that attach to human moral agents ("failing to reproduce our ethical performance" is neutral on whether the failure is due to their being less, or more, ethically good than we are). We further consider the view that genuine consumerhood is necessary for genuine or functional producerhood in the next section.

Perhaps this would be sufficient to make ME a theoretically, and indeed practically, useful field of investigation—particularly in the medical and care fields. The hope is that artificial agents could be constructed which could take moral decisions which conformed to best moral practice in a variety of medical or personal-care contexts (for example as robotic nurses or computational diagnosticians)—even while not having the right attributes in terms of consciousness (say) to qualify as genuine moral consumers, and possibly even while not being genuine moral producers. This would give a lot of mileage to Machine Medical Ethics as a potential field for development in order, hopefully, to greatly enhance medical practice and policy.

5 Medical Moral Producerhood and Conscious Empathy

We have just considered the view that, in order to be classed as a genuine moral producer—or even as a workable artificial surrogate for a moral producer—an artificial agent would also have to fulfill the conditions to fit it out to be a moral consumer. It could be said, for example, that an artificial agent could not be expected or guaranteed to make decisions in a given medical setting that would *even approximate* the ones that responsible human medical practitioners would make in that setting, *unless* it were capable of imagining or understanding, at first hand, *what it was like* to experience pain, distress, bereavement, and so on—or to be capable of doing so at least in outline. Arguably, such first-hand understanding requires being able oneself to experience such states; that is, to be phenomenally conscious. And few would deny that this would render such an agent a moral consumer; hence, on this view, performance functionally equivalent to human moral producerhood requires moral consumerhood. (This would seem to be a particularly apt argument to consider in a domain of activity such as clinical medicine, where acute life-or-death issues are involved all the time; the same goes for social or personal care contexts.) Skeptical expectations concerning merely functional or apparent moral agents could be theoretical (they are conceptually or nomologically impossible) or practical (they are theoretically possible, but prohibitively difficult to build).

Such capabilities, it might be said, would require any such artificial agent to have a far more intimate set of similarities to a real human medical or care practitioner than it would ever be possible to achieve through standard AI techniques. For example, consider the ability, when the situation calls for it, to put oneself imaginatively in the place of various people likely to be affected by particular courses of action. This would seem to be essential to human ethical behaviour in general and human medical ethical behaviour in particular. Yet, some might think that such a capacity requires phenomenal consciousness of sorts. And some of those might deem such phenomenal consciousness to be irreducible to functional MC-tractable, aspects of consciousness.

Hence, just as problems were raised in relation to an MC that soft-focused on phenomenality, so analogous problems might be raised about an ME that soft-focused on phenomenality (and thus which concentrated on moral producerhood without attending to the capabilities that also constitute moral consumerhood).[8] But the issue cuts two ways: some MC researchers, agreeing that imaginative abilities require phenomenal consciousness have used this to argue from the fact that their systems seem to have imaginative abilities to the conclusion that they therefore can be said to have phenomenal states [1, 20, 22–24].

Further, one might question the assumption that the only way to produce morally good behaviour is to reproduce human moral behaviour, including the methods (e.g., empathic imagination) from which it derives. To counter the claims of the aforementioned MC researchers by claiming that such MC systems achieve their imaginative functionality without the use of true (empathy or) consciousness would undermine the argument that consciousness is required for apparently morally ethical behaviour, medical or otherwise.

Nevertheless, some may still doubt the feasibility of a partial machine ethics that concentrates on the producer aspects of moral subjecthood, while avoiding the consumer aspects, and on producing functional morality as opposed to some more fully-fledged version of moral agency or moral productivity. Such doubt may

[8] It could be argued, in a related way, that an artificial agent which was unable to experience pain or affective suffering could not be a subject of moral praise or blame, since the latter requires a capability of experiencing the positive or negative sanctions, whatever they might be, that attaches to any such praise or blame.

derive from a holistic ethics (with respect to the moral producer-consumer distinction) that would severely hamper any attempt to develop a partial constructive ME which focused on moral producerhood and ignored moral consumerhood. But it looks as though, because of the central role played by phenomenal consciousness in moral consumerhood, and because of the supposed resistance of phenomenal consciousness to computational techniques, a *complete* holistic ME conceived as a purely AI project; that is, a joint descriptive/constructive machine ethics project that fully implemented, in a machine, all (major) aspects of our moral thinking, experience and action may not be possible.

However, while there may be some truth in these holistic arguments, they might be resisted. Although some might think that accepting the conclusions of such arguments would have the effect of annulling ME (at least using currently-known AI techniques) as a viable topic of inquiry, we have seen that at least some MC researchers believe that currently known techniques *are* adequate, at least in principle, for conferring the imaginative/empathic abilities on which sophisticated medical moral producerhood likely relies, independently of whether such abilities require phenomenal consciousness (and thus moral consumerhood) themselves. Therefore, the holistic arguments, while deserving a wider airing than they have had in the past, and than what we can give them here, would be challenged by a significant number of people within the field.

Nevertheless there is another, possibly deeper, kind of concern surrounding the notion of empathy that may be thought to affect the viability of constructive ME using current AI and computational techniques. There is a tradition of thinking about empathy that stems from phenomenologists such as Husserl and his pupil Edith Stein, but also taken up by Merleau-Ponty, and more recently Evan Thompson and others, that sees empathy as indeed closely related to consciousness, but to a strongly embodied notion of consciousness [41]. It is claimed that this embodied conception of empathy is particularly relevant to the way illness is experienced, both in oneself as an ill person, and in others. This claim was explored in a paper by Toombs [43, 44] on the relation between empathy and the experience and treatment of illness.

A key to understanding this approach is the recognition of a duality in our conceptions of the body. This duality runs through the phenomenological tradition from Husserl on [25, 30, 42]. We can consider the human body first as an objective, biological entity, with its own physical structures, laws of functioning and principles of breakdown or pathology: this is the body as a subject of science and of medicine as the latter is conventionally understood. However, a second conception of the body is of the body-as-lived, the way one lives through one's body as one goes through one's life. Embodiment conceived in this second way is often described as involving a *pre-reflective* immediacy or awareness, since we normally take our embodiment for granted as we engage in our projects in the world. For example, one is typically unaware of the movement of one's feet as one hurries to catch a train. Often, one becomes explicitly aware of aspects of one's body-as-lived: for example when one trips and falls while running, or through certain forms of meditation. Illness of different sorts, as Toombs explains, brings one's awareness of the body-as-lived into sharp focus, and she suggests that the (typical) difference between a doctor's and a patient's conception of the patient's body is essentially characterized by the difference between the body as an objective entity and the body as lived.

Toombs then discusses how these different conceptions of the body affect clinical practice, and how the empathetic grasp of the patient's body-as-lived can be screened out in many current forms of medical practice because of an excessive concentration on the objective mode of conceiving the patient's bodily characteristics. Such an objective approach is, of course, indispensible to medical science and practice; however Toombs points out that an empathetic grasping of the patient's illness-as-lived is also crucial to developing a properly "humanly-grounded" approach to medicine [44].

These observations can be used to provide a philosophical underpinning to proposals for improving medical training and practice, e.g., by more extensively teaching skills in recognizing the lived experience of illness in order to show how such practice can be ethically deepened. This is indeed a central aim for Toombs in her paper. However it is not hard to see how they might also need be borne in mind by those embarked on the constructive ME project, and how they might engender a certain pessimism about that project. In order to deepen the possibilities for empathetic bodily transference between doctor (or carer) and patient one can, of course, draw upon the fact that the patient shares with the doctor or carer a common human embodiment. Nevertheless, artificial medical agents, physically constructed via technologies of the foreseeable future, will have little physically in common with the ill humans whose treatment they will be designed to assist in. So the possibilities for the deep kinds of embodied empathy that writers like Toombs regards as necessary for "humanly grounded" medical care may seem to be severely limited in medical machines as currently conceived.

The designers of ethical medical or care agents thus need to be aware of these difficulties when engaging in their R&D. It suggests that medical agents may be far from ideal in reproducing the empathetic skills of our most ideally trained human practitioners. On the other hand, this point is not an in-principle objection to the constructive medical ME program as such. To say of a technology that it is far from ideal is not to say that it cannot be developed to be of selective benefit or to be continually improved; and all technologies we have ever had have been less than ideal.

Moreover, empathy, as many have pointed out, often requires building a bridge across differences between individuals as much as drawing upon what is common between those individuals. Suppose I am seeking to empathetically understand the lived experience of a Parkinson's patient. Clearly there will be important differences in embodiment, yet the hope is that those differences can be bridged to a degree that allows for suitable treatment and care. Similarly for differences between genders, ages, races, ectomorphs/endomorphs, disabled/abled, and so on. Why, then, can't the gap between an anthropomorphic robot and a human also be bridged, to allow empathetic kinds of responses from one to the other in ways that may be acceptable or helpful at least in selected situations of use?

In order to progress the discussion further, we propose a pragmatic strategy. Despite the problematic issues raised here, we suggest that there is a lot of mileage in exploring how AI-based techniques may be used to develop ethically-sensitive machine medical or care agents. It will be important for those working on such AI techniques to be aware of the problems raised in this and earlier sections, concerning the relation between functional and genuine ethical status, concerning the duality between moral consumer and moral producer status, of the role of phenomenal consciousness and empathy in ethical productivity, and on the bifurcation between objective and lived conceptions of embodiment. Nevertheless it is understandable. Proper cognizance of all these issues is desirable when seeking to achieve an adequate understanding of what is involved in machine medical ethics. All the same it is surely acceptable, in proceeding with such a project, to concentrate more on the functional aspects of machine consciousness than on the phenomenal aspects, and more on the producer aspects of machine ethics than on the consumer aspects. Nevertheless we stress that the developer of medical ME agents must be sensitive to the role played by phenomenal consciousness, and on the issues concerning empathy, lived embodiment, etc. in the moral aspects of the practice of people working in the medical and social-care fields.

6 Strong Versus Weak in MC and ME

Earlier we gave a quick characterization of constructive ME as "the attempt to create machines that are moral." and of constructive MC as "the attempt to create machines that are conscious". In the case of ME, it has become clear that the phrase "machines that are moral" is ambiguous as between "machines that have status as moral producers" and "machines that have status as moral consumers". In addition there is the "morally evaluable/morally good" ambiguity, also discussed earlier. However there is a further dimension of ambiguity, which centers round the distinction, also discussed above, between merely functional moral status and genuine or intrinsic moral status. One might understand talk of "machines that have genuine moral status" in terms of any of the various senses referred to above (moral producer/consumer; morally evaluable/morally good).

Another way to characterize the space of possible approaches within ME is to differentiate between "strong (constructive) ME" and "weak (constructive) ME", by analogy with strong and weak AI. The debate over the merits of strong AI, which dominated the philosophy of AI in the 1980s and beyond, has spawned satellite debates in various of the fringe areas of AI, of which ME is one flourishing example. Although originally (in [34]) a different distinction, the strong AI/weak AI distinction has sometimes been used to distinguish between the goal of producing artificial agents that are genuinely intelligent on the one hand, vs. producing artificial agents that are merely functionally intelligent (that behave as if they are intelligent) on the other. Similarly, one could characterize weak constructive ME as seeking to create agents with merely functional moral status, and strong constructive ME as seeking to create agents that have genuine moral status.

One might understand talk of "machines that have genuine moral status" in a number of different ways. For example, there is the distinction between moral producers and moral consumers, referred to above. Then there is the distinction, also discussed above, between moral agency in the sense of displaying behaviour that is morally evaluable, and in the sense of displaying behaviour that is morally good. Clearly these different distinctions allow a variety of different ways of understanding the strong/weak contrast in the field of ME.

Within the field of machine consciousness a similar contrast has been drawn between strong and weak MC [12, 45].[9] Strong constructive MC would have the goal of creating artificial agents that are genuinely conscious (where "genuine" may be understood in terms of phenomenality, or in some other way). Weak constructive MC would rather seek to create artificial agents that behave *as if* conscious, but which may in fact not be so. This corresponds to the distinction between genuine ethical equivalence and functional ethical equivalence, made in Sect. 4. So, in MC, as in ME, one can support a distinction between strong and weak positions or goals, depending on whether one's research concerns do, or do not, aspire to replicating "the real thing".

However, these contrasts between strong and weak ME, and between strong and weak MC, are in need of still further qualification and refinement. It has been suggested [11] that a more subtle distinction needs to be made in the case of MC.[10] We believe it fruitful to examine this proposal and then to go on to discuss its implication for ME, and in particular for medical ME.

The discussion of the strong/weak MC distinction in [11], in that it more closely follows Searle's original strong-weak distinction, pursues a path somewhat different from, and more complex than, that just given. Analogously to strong AI, A supporter of strong MC would be one who believes that the (computational. robotic, etc.) technology[11] involved in an MC system could, in principle, be *sufficient* to generate "real" consciousness in (certain) such systems. In parallel with weak AI, a supporter of weak AC would not claim that computational technology would be either necessary or sufficient to generate consciousness: rather their support for MC would stress the value of building computational models in helping us to understand the nature of consciousness (rather as a computational model of a hurricane could be of high predictive or explanatory value, even while neither it, nor any more sophisticated development from it, might ever actually instantiate a hurricane).

However, [11] also suggests that there is an important mid-way position, in a zone of neglected possibility between strong and weak MC, as defined above. This is called

[9] Normally the strong/weak distinction is thought of as primarily applying to intelligence. In fact, one can see the context of Searle's original use of the distinction, the Chinese room argument, as applying most directly to consciousness, and only indirectly to intelligence, via the assumption that genuine intelligence requires understanding accessible from a first-person perspective.

[10] Chrisley there talks of Artificial Consciousness (AC) rather than Machine Consciousness.

[11] Here we are generalizing Searle's strong-weak distinction in another way: his distinction dealt specifically with the technology of digital computer programs, whereas we are open to considering any computational/robotic technology.

"lagom" MC (*lagom* is a Swedish word that we understand, perhaps incorrectly, to mean something like "perfection through moderation" or "just right"—perhaps, taking a leaf from the book of exo-planetary science, one might talk of "Goldilocks MC"). To a first approximation, someone who supports *lagom* MC would say that certain computational/robotic systems might have features that, while not sufficient, are necessary for consciousness. But that needs some refinement. A key idea behind *lagom* MC is that there may be significantly different ways in which consciousness is instantiated within different kinds of system. One way in which consciousness is realized is found in the realm of biological systems, of which humans, and no doubt various mammalian and non-mammalian species, are instances. However the "space of possible consciousnesses" [37] may be vast, and may include creatures or systems—including, potentially, artificial ones—with very few properties in common with those terrestrial biological systems that are recognized to be conscious. Such genuinely conscious systems, it might be said, may include certain examples that could, at least in principle, be developed from current computer-based technologies.

There are all sorts of ways in which computer-based technologies may converge on biological systems. For example, various directions in which current artificial life research may be going. However current computational technologies may also develop in ways that take one further from the space of biological systems; and certain developments in far-from-natural-biological systems may end up in a region of the space of possible consciousnesses. So a supporter of *lagom* MC may claim that certain decidedly non-biological systems may have certain features that are necessary (given the non-biological medium at hand) for consciousness—at least for certain kinds of consciousness—perhaps even ones which are morally significant in the sense that beings instantiating them would deserve our moral concern if they experienced certain forms of pain or distress; or would deserve our moral commendation if they made decisions to act in such-and-such a way, or which at least are capable of deliberation or action (concerning what would normally be considered moral issues) in ways which may seem to make them capable of being members of a moral community.

However, it is also possible to adopt a position exemplifying *lagom* MC that involves investigating possible MC systems that are much closer to the space of natural biological consciousnesses. So it might be claimed that there are certain features which are necessary to consciousness, where those features can be described (at a certain level of generality or abstractness) in a way which applies, *albeit in markedly divergent manners,* to human and other animal systems on the one hand, and to certain artificial computational/robotic systems, on the other. Such a claim would involve the suggestion that there is a set of features F such that the members of F are (a) necessary to consciousness (or at least certain types of conscious state) and (b) exemplified in certain biological systems and in certain other computational/robotic systems in rather different ways, even if (c) they are not (even jointly) sufficient for (any types of state of) consciousness in either class of systems.

What would be interesting about such a claim, if validated, is that feature-set F would provide a partial explanation of what it was to be conscious (or, at least to have that particular type of conscious state), by explicating certain fundamental features necessary to (those sorts of state of) consciousness. For example, even if one were to concede (because of being persuaded by the Chinese room argument, say) that it is never the case that running some program P is *sufficient* for consciousness, one might at the same time hold that running P is *necessary* for (certain kinds of) consciousness, that is, that one could not be considered to have those kinds of consciousness if one did not have the abstract causal structure involved in running P.[12]

One important point made by this explanation of *lagom* MC (two variants of it have been described here) is that it shows that the conventional debates between strong and weak MC tend to ignore an important middle territory of positions in that debate (and in this respect they mirror the far more extensive literature concerning strong and weak AI). But why should this be of interest to investigators of machine *ethics* (and in particular of medical ME)? What, it might be asked, could be gained by trying to develop consciousness in artificial medical agents? First, this discussion should help to make it clear that simply seeing consciousness as of no interest to people working in the area of artificial moral systems modelling (and more specifically the modelling of the ethics of artificial medical agents) is to take a far too oversimplified view of the matter. Second, we will argue that the strong/weak/*lagom* trichotomy can be directly applied in the area of ME, in a way that would be of value for developers of artificial medical agents to take into account.

7 Lagom ME?

Is there a correlate to *lagom* MC in the domain of ME? The answer seems to be assuredly, yes. It is possible to imagine that there might be certain features (call them set G) such that features G are (a) necessary to some being's counting as a

[12] Some might be unimpressed by a shift of focus from sufficient to necessary conditions. Necessary conditions, it might be thought, are ubiquitous to the point of being non-explanatory. Compare a similar move in chemistry: "I don't know what the sufficient conditions for a sample *X* being water are", a scientist might have said in the time before the structure of water was known, "but I do know a lot of the necessary conditions: *X* must be a substance, must be located in space-time, must be composed of atoms, must either be an element or a compound, must be capable of being a liquid at room temperature…" and so on. While true, all of these conditions, even taken together, fall short of explaining what water is, unlike the sufficient condition "*X* is H_2O". Two things can be said to quell this worry. First, the example is rather anachronistic. From our current, H_2O-informed perspective it is easy to underestimate the explanatory and heuristic value of the necessary conditions just cited. Second, the necessary conditions can be ordered, from the most widely applicable, to the very-narrow-but-still-not-narrow-enough-to guarantee-waterhood, such as "has, at sea level, a boiling point of 100 °C and a freezing point of 0 °C", which may be very useful in the determination of sufficient conditions. The suggestion, then, is that synthetic investigation of the "narrow" end of the corresponding hierarchy of necessary conditions for consciousness can play an important role in explaining consciousness.

moral agent in either or both the producer sense or the consumer sense; (b) exemplified, on the one hand, in certain biological systems, and on the other hand, in certain other computational/robotic systems, *but in rather different ways;* yet which are (c) not sufficient to allow us to attribute genuine moral status (in either the producer sense or the consumer sense) to any system displaying features G. Since we are primarily concerned with moral producerhood here, we will confine our discussion to that aspect of moral agency henceforth.

In other words, a computational/robotic system could possess certain characteristics that would be regarded as necessary to such a system counting as a moral producer, where those features may be instantiated in somewhat contrasting ways in such artificial agents as compared to their human counterparts, but also where we may not have sufficient grounds to say that possession of those characteristics are in themselves sufficient for such artificial agents to be 'genuine' moral agents (producers). Two questions that arise are: (i) what might such features be? (ii) what are the features that are missing such that were they present as well in the artificial system, then we would think that we had a set of properties that were sufficient to justify us in attributing genuine moral producerhood to that system?

As for (i) there seem to be a number of different ways to supply examples of what such necessary features might be. One such way is suggested by the work of Stan Franklin and colleagues, on the development of a general-purpose decision-making system, based on the LIDA cognitive architecture, where such a system is envisaged as being able to make morally significant decisions in various contexts (for example in medical or care contexts). In brief, Franklin and colleagues have developed LIDA (previously IDA) systems as attempts to instantiate Bernard Baars' Global Workspace model of consciousness [5, 14, 15]. So a LIDA system is a general purpose problem-solving agent whose cognitive organization involves a number of sub-agents, specialized for achieving certain tasks within the problem-structure, where these sub-agents form a succession of temporary alliances, and such that these alliances of sub-agents compete for control of a centralized information-broadcast system which is used by successive coalitions to post information for use by other sub-agents within the overall cognitive system, in order to tackle the ongoing problem-solving task which is the system's top-level goal.

The LIDA example is particularly apposite in the present discussion about machine ethics, since it is intended, not just as a model of generalized problem-solving (AGI) but also as a functional model of consciousness (and hence is a line of research fitting into the research program of MC). However, it has been argued by some of its proponents that it is highly relevant to ethical modelling. So, they claim, it exhibits one important way in which the modelling of consciousness may be useful or required in order to provide good candidate templates for effective machine ethics agent design [51].

Note that the point being discussed here about LIDA systems is not that all human moral agents must instantiate a LIDA-style (or GW-style) cognitive architecture, although that might be plausible at some very broad level of generality. Rather, the claim would be that the LIDA architecture provides one way of realizing, in an artificial agent, a certain kind of global cognitive organization, *some variant of which* is necessary for any agent (natural or artificial) to count as a moral producer.

As for question (ii), what additional features might need to be added to turn an agent containing these *lagom ME* features into a genuine, full-blooded, moral producer? One such feature that might be supplied, one that chimes in with our previous discussion is (some form of) *phenomenal* consciousness. Possibly this is the sole such additional feature which is required to provide us with genuine moral producerhood, or maybe it is one among a handful, or possibly a rather long list, of features that, when added to the mix, would jointly suffice to make the agent possessing those features a genuine moral producer. As remarked, the LIDA system has been offered as a model of consciousness; more specifically, of functional consciousness. Its supporters appear to believe that the LIDA architecture, if developed in a sufficiently rich and detailed way, could actually allow us to conclude one day that an artificial agent exemplifying it was also *phenomenally* conscious (W. Wallach, in conversation.) Some supporters might go so far as to say that *any* realization of that rich and detailed development of LIDA would *ipso facto* be conscious (that is, would be conscious solely by virtue of realizing that architecture), effectively embracing strong MC and the idea of future-LIDA as specifying sufficient causes for consciousness. But one does not have to embrace such a position in order for the example to do the work it is doing here; namely, to illustrate how a certain AI model (the product of years of collaborative work in its development and refinement) could be used to bolster a *lagom* position in the domain of machine ethics.

We are not here trying to argue that either the LIDA-style cognitive architecture, or the property of phenomenal consciousness (however that might be implemented in a computational system, even assuming that it could be) *must* play the roles that we have given them in the above discussion; rather they are meant to provide illustrations of how the *lagom* position could be fleshed out in the case of machine ethics. However, it may well be that something like a Global Workspace model of which LIDA is one variant (for others, see [21, 35, 36]) may provide a powerful development pathway for rich ME systems, particularly ones for use in the medical/care domains.

8 Human (or Biological) Versus General

Another distinction discussed in [11] is worth mentioning. The study of consciousness, whether natural or artificial, can have as its explanatory or replicatory target either consciousness as an Earthly phenomenon—as found in humans and other animals on our planet—or consciousness understood more broadly, to include beings which might inhabit non-terrestrial biospheres, as well as potential non-biological beings (be they artificial or natural) that possess consciousness. These different research emphases may of course reflect doctrinal differences about the nature of consciousness. In a similar way, there has been a historical tension within moral theory between those who see ethics as a specifically or exclusively human phenomenon, and those who see morality in broader, more ideal terms, abstracting away, as far as possible, from human or animal contingencies. The moral philosophies of Aristotle and Hume are representative of the first approach; those of Plato and Kant are examples of the second. Again these two different research foci reflect theoretical differences over the nature of morality and moral thinking (or perhaps what it is to be human).

This tension also provides a choice of focal emphasis within machine ethics. Should constructive ME researchers develop machine simulations which track as closely as possible human moral thought/choice/action? Or should ME (constructive or theoretical) be taking a more abstract stance, especially given the ways in which computational systems differ from, and indeed could, in principle, improve on human ethical decision-making, in virtue of their more powerful processing capabilities? (Compare the deliberations of a would-be aeronautical engineers at the end of the 19th century: To achieve machine flight, should we track as closely as possible the features of natural flyers (birds, bees, etc.) by including flapping wings made of feathers or gossamer? Or should machine flight be taking a more abstract stance, given the differing materials at our practical disposal? [2, 10, 11, 24–25, 40, 52] There has been considerable debate within the ME community about the best approach to take: several papers have been written taking different historical models within the history of ethical theory as departure-points for building prototype ME systems (e.g., [4, 16, 32]). Perhaps different approaches are necessary within different application domains, or even within the same domain. As far as medical and care systems are concerned, each approach might be necessary for different purposes. In situations where agents are required to interact closely with patients it may be necessary for machine systems to instantiate an extended array of human skills of physical and emotional engagement; where agents are required to incorporate large, complex medical knowledge-bases, a more abstract, rational approach may be required. What is important is that designers of medical ethics machines should be aware of this debate and their designs should take cognizance of the options available.

9 Conclusion

The above discussion has been intended to provide a philosophical backdrop to the development of ethically-sensitive artificial agents or systems in the medical or care spheres. A main concern here has been to stress the parallels between theoretical issues in machine ethics, and those in the field of machine consciousness, another fringe area of AI. But there are not merely strong parallels: consciousness and ethics are strongly intertwined in various ways, both as mental or mind-related phenomena in humans (and other biological creatures), and in relation to AI developments.

However, we started our discussion in this paper by pointing out a contrast between theoretical and constructive ME and of MC as research fields. The possibility of producing machine models of various aspects of ethical thinking, decision or action seems quite inviting. So machine ethics might be thought to be, on the face of it, quite a flourishing area for development: if the aim of ME (at least in its constructive mode) is to develop artificial agents that can behave in ethically acceptable ways, this seems to promise a quite feasible set of research objectives via AI-style methodologies. On the other hand the term "consciousness" appears to cover a range of mental features that may seem—some of them at least—quite recalcitrant to being modeled, let alone reproduced, using AI-based, computational methods. While the cognitive or behavioral aspects of conscious activity may respond to AI methods, the phenomenal or felt aspects of consciousness do not seem likely to do so in as straightforward a way. Also, some may be skeptical about how relevant issues concerning modelling consciousness are to the machine ethics researcher.

So it might be thought that, in developing practical ethical models for use in the medical or care spheres, it would be better to avoid questions of consciousness. Deep models of ethical thinking or emotion may be needed to take account of work in AI research on consciousness, it might be said. But, there is a practical urgency to design autonomous robotic and AI agents to be 'ethically safe'; that is, to operate in such a way that we can be assured that they observe appropriate ethical rules or values. This is especially so in sensitive areas such as medicine and personal care. So, it may be argued, there just won't be the development time for machine ethicists to get too embroiled in issues to do with consciousness, even if the latter does have some bearing on ethics.

Our response to this skeptical challenge has been a complex one. In brief, we have differentiated between theoretical and constructive machine consciousness. Further, we have noted that humans may be attributed moral status in at least two senses, both as moral "producers" and as moral "consumers". With respect to theoretical MC, we have shown that there are deep interrelations between various key notions in (meta-) ethics and consciousness, both in relation to moral production and moral consumption. For example, despite the considerations to the contrary which we have also presented, it might be insisted that consciousness is a precondition of being an ethical consumer, on the grounds that the ability to *experience* the positive or negative aspects of the actions of others is essential to having ethical consumer status. But consciousness may also be a precondition of being an ethical producer, although the connections are less direct and even more subject to challenge in the latter case, perhaps. Certainly, if being an ethical producer involves having the capacity for some kind of empathetic identification with the possible recipients of one's actions, then some kind of conscious, reflective entry into the potential experiences of others seems necessary for moral producerhood as well as consumerhood.

Additionally, we have shown that, according to certain models of decision-making (for example the LIDA/Global Workspace model), especially ethical decision-making, one needs to be mindful of the core cognitive functionality of consciousness. So attempts to build ethically-sensitive agents (especially medical ones) that fail to take proper account of the theoretical issues in machine consciousness may be too simplistic to work robustly and may therefore not succeed in achieving a chief practical goal of constructive ME, namely achieving autonomous action that was ethically dependable.

A question that was considered in our discussion concerned the possible role played by empathy in the range of moral actions and responses of medical and care practitioners, and how such empathetic actions and responses might be realized in a constructive ME machine. Two kinds of arguments were considered: the first suggested that empathetic grasping of the experiences of others may involve an understanding of "what it is like" to be in that state, in a way that implies phenomenal consciousness. Such phenomenal consciousness, it might be urged, could not be realized in an artificial agent whose design was based on current AI technologies. The second argument, drawing upon phenomenological conceptions of embodiment stemming from Husserl, provides a picture of empathy that stresses, not our imaginative grasping of the state of the other, but the common lived embodiment of one human facing another. Again, any non-embodied AI agent, or an agent created with current and foreseeable AI technology, would be unlikely to realize the kind of commonality of embodiment that underlies this kind of human-to-human empathic response.

Such critical points raise deep issues that need to be taken into account in the program of constructive ME, particularly within medical and care domains. Nevertheless, we suggest that they are not fatal to such a program. It is always important to recognize the limitations of the current state of the art in any research field. So the progress of constructive ME will be challenged, for a considerable time, by the need to take proper account of the limitations in the imaginative capacity available to machines, relative to humans, and by the very different forms of embodiment exemplified by these two types of agent. But, as noted earlier, there are also many deep differences between humans, and such inter-human differences also mean that empathetic responses of human carers or healers are often far from easy to achieve, or ideal. So we recommend a strategy which takes care not to be too ambitious, based on small results, and geared towards selected situations of use. This way, the kinds of differences between human and machine ethical agents alluded to by these arguments will have less of an impact on the program's success.

More broadly, we have suggested that theoretical tools used in discussion of some general issues in MC or ME, or in AI in general need to be deployed with more subtlety and nuance. For example the frequent allusions to strong versus weak AI, and the corresponding notions for ME, may need to be enhanced by our recognizing a middle zone, which we have referred to as *lagom* ME (by analogy with an earlier treatment of a similar notion in MC). We have here discussed how this relates to both the ME and MC fields. While we referred specifically to the LIDA model in this connection, it seems clear that other cognitive approaches to consciousness may also have important applications with the field of ethics.

The issues at stake can be summarized in three claims/lines of reasoning (which we do not purport to have conclusively established), listed here in order of decreasing strength:

- For an artificial agent to perform well[13] in the medical domain, it will have to be a genuine moral producer, and therefore a genuine moral consumer. Since one

[13] "Well" is a relative term that we are intentionally leaving unspecified to allow us to cover a broader range of claims. But as an example, one (strong) gloss on it would be: "at a level sufficient to allow replacement, or at least significant supplementation, of human medical personnel".

or both of these require being phenomenally conscious, artificial medical agents, to perform well in the medical domain, will have to be phenomenally conscious.

- Even if *genuine* moral status is not required, for an artificial agent to perform well in the medical domain, it will have to at least be a *functional* moral producer (and possibly a functional moral consumer). Although functional ethical status may not require being phenomenally conscious, it is likely to require meeting some of the (non-trivial) necessary conditions for such, which, plausibly, can be achieved by techniques continuous with traditional or current AI approaches, even if sufficient conditions for consciousness cannot. In the limit, functional ethical status (being ethically functionally equivalent to a genuinely ethically competent medical practitioner) may require functional consciousness status.
- Aside from the above claims, which concern constructive ME, an important component of theoretical ME (e.g., delineation of responsibility in situations involving informationally sophisticated computational/robotic medical technology) will benefit from understanding the ways in which a particular piece of technology fails to meet the sufficient conditions for phenomenal consciousness and (therefore) moral producerhood/consumerhood. Understanding the shortcomings of such technology will put the medical professionals involved in a better position to know which aspects of their expertise are most likely to be needed. Further, the tools developed for such understanding will also be of use in understanding the (phenomenally or ethically relevant) strengths and weaknesses of the humans involved in such a technology, possibly suggesting areas for training or development to better reach desired medical outcomes.

There are thus a number of ways in which concerns relating to machine consciousness impinge upon machine ethics in general and in the medical and care domains in particular. Probably the most important issue concerning consciousness in relation to AI is how far various aspects of consciousness, and in particular the *phenomenality* of consciousness, can be captured within a computational framework, at least in principle. Few would deny that this must be an open question at this stage of theoretical and technological development. But, even if one remains optimistic about eventually producing a computational agent that is phenomenally conscious in a rich sense, this is not likely to be done without a lot more time and/or effort. (However, "singularity" theorists are very optimistic indeed on this score; see [9, 27].) It may also be possible to create deep models of agency that incorporate many of the grosser functional or behavioral features of consciousness without achieving a rich "inner feel".

We have also discussed whether full or genuine ethical status (at least ethical-agent or ethical-producer status) could be realizable within a computational agent model. Again the answer may be yes in principle, though not without a great deal more development effort (and thus time). But, as with the consciousness case, many aspects of human ethical action or decision-making may be reproduced in AI models. Our discussion suggests that a pragmatic approach is desirable in both the MC and ME spheres. And moreover, we have argued that the task of building useful and appropriate machine ethics agents will best be done by incorporating insights from machine consciousness. The increasing operational autonomy of robots and AI agents will have repercussions for many fields. But the fruits of work in MC are likely to be of particular use within the range of application areas that come under medicine and care. These areas will require, of the artificial agents we build, the exercise of physical emotional and imaginative sensitivity, complex interpersonal skills and reflective decision-making. These are all areas that have been much targeted by philosophers and psychologists interested in both consciousness and ethics. So progress in medical machine ethics is likely to be greatly enhanced by attention to developments in machine consciousness.

Acknowledgements The authors would like to thank Robert Clowes, Mark Coeckelbergh, Madeline Drake, DavidGunkel, Jenny Prince-Chrisley, Wendell Wallach, and Blay Whitby. We would also like to thankmembers of the audience of the following institutions/meetings for their helpful comments:University of Sussex (COGS, E-intentionality); Twente (SBT); AISB meetings on MachineConsciousness (Universities of Hertfordshire, Bristol, and York) and on Machine Ethics(Birmingham and Goldsmiths Universities). Thanks also to the EU Cognition network for theirsupport.

References

1. Aleksander I (2000) How to build a mind: toward machines with imagination. Weidenfeld & Nicolson, London
2. Armer P (2000) Attitudes toward intelligent machines. In: Chrisley RL (ed) Artificial intelligence: critical concepts, pp 325–342. Routledge, London. (Originally appeared In: Feigenbaum E, Feldman J (eds) Computers and thought. McGraw-Hill, NY, pp 389–405
3. Anderson M, Anderson SL (eds) (2011a) Machine ethics. Cambridge University Press, Cambridge
4. Anderson SL, Anderson M (2011b) A prima facie duty approach to machine ethics: machine learning of features of ethical dilemmas, prima facie duties, and decision principles through a dialogue with ethicists. In: Anderson M, Anderson SL (eds) 2011a, pp 476–492
5. Baars BJ (1988) A cognitive theory of consciousness. Cambridge University Press, Cambridge
6. Bentham J (1989/2005) An introduction to the principles of morals and legislation. In: Burns JH, Hart HL (eds) Oxford University Press, Oxford
7. Block N (1995) On a confusion about a function of consciousness. Behav Brain Sci 18(2):227–247
8. Boltuc P (2009) The philosophical issue in machine consciousness. Int J Mach Conscious 1(1):155–176
9. Chalmers DJ (2010) The singularity: a philosophical analysis. J Conscious Stud 17(9–10):7–65
10. Chrisley RL (2003) Embodied artificial intelligence. Artif Intell 49:3–50
11. Chrisley RL (2007) Philosophical foundations of artificial consciousness. Artif Intell Med 44:119–137
12. Clowes RW, Seth AK (2008) Axioms, properties and criteria: roles for synthesis in the science of consciousness. Artif Intell Med 44(2):91–104
13. Floridi L, Sanders JW (2004) On the morality of artificial agents. Mind Mach 14(3):349–379
14. Franklin S (2003) IDA: a conscious artefact? J Conscious Stud 10(4–5):47–66
15. Franklin S, Patterson FGJ (2006) The LIDA architecture: adding new modes of learning to an intelligent, autonomous, software agent. In: IDPT-2006 proceedings, integrated design and process technology: society for design and process science

16. Grau C (2011) There is no "I" in "Robot": robots and utilitarianism. In: Anderson M, Anderson SL (eds) 2011a, pp 451–463
17. Gunkel D (2012) The machine question: critical perspectives on AI, robots and ethics. MIT Press, Cambridge
18. Gunkel D (2013) A vindication of the rights of machines. Philos Technol 27(1):113–132. doi:10.1007/s13347-013-0121-z
19. Guzeldere G (1997) The many faces of consciousness: a field guide. In: Block N, Flanagan O, Guzeldere G (eds) The nature of consciousness: philosophical debates. MIT Press, Cambridge, pp 1–67
20. Haikonen PO (2005) You only live twice: imagination in conscious machines. In: Chrisley RL, Clowes RC, Torrance SB (eds) Proceedings of the AISB05 symposium on machine consciousness. AISB Press, Hatfield, pp 19–25
21. Haikonen PO (2012) Consciousness and robot sentience. World Scientific, Singapore
22. Hesslow G (2002) Conscious thought as simulation of behaviour and perception. Trends Cogn Sci 6:242–247
23. Hesslow G, Jirenhed DA (2007) The inner world of a simple robot. J Conscious Stud 14:85–96
24. Holland O, Goodman R (2003) Robots with internal models: a route to machine consciousness? J Conscious Stud 10(4–5):77–109
25. Husserl E (1952/1989) Ideas pertaining to a pure phenomenology and to a phenomenological philosophy: second book: studies in the phenomenology of constitution (trans. Rojcewicz R, Schuwer A). Kluwer Academic Publishers, Dordrecht, The Netherlands
26. Jaworska A, Tannenbaum J (2013) The grounds of moral status. In: Zalta EN (ed) The Stanford encyclopedia of philosophy (summer 2013 edn). URL: http://plato.stanford.edu/archives/sum2013/entries/grounds-moral-status/
27. Kurzweil R (2005) The singularity is near: when humans transcend biology. Viking Press, NY
28. Latour B (2002) Morality and technology: the end of the means. Theor Cult Soc 19(5–6):247–260
29. Lin PA, Abney K, Bekey GA (eds) (2012) Robot ethics: the ethical and social implications of robotics. MIT Press, Cambridge
30. Merleau-Ponty M (1945/1962) The phenomenology of perception (trans Smith C). Routledge and Kegan Paul, London
31. Moor JH (2006) The nature, importance and difficulty of machine ethics. IEEE Intell Syst 21(4):18–21
32. Powers TM (2011) Prospects for a Kantian machine. In: Anderson M, Anderson SL (eds) 2011a, pp 464–475
33. Regan T (1983) The case for animal rights. The University of California Press, Berkeley
34. Searle JR (1980) Minds, brains, andprograms. Behav Brain Sci 3:417–424. doi:10.1017/S0140525X00005756
35. Shanahan M (2006) A cognitive architecture that combines internal simulation with a global workspace. Conscious Cogn 15:433–449
36. Shanahan M (2010) Embodiment and the inner life: cognition and consciousness in the space of possible minds. Oxford University Press, Oxford
37. Sloman A (1984) The structure of the space of possible minds. In: Torrance SB (ed) The mind and the machine: philosophical aspects of artificial intelligence. Ellis Horwood, Chichester, Sussex, pp 35–42
38. Sloman A (2010) Phenomenal and access consciousness and the "Hard" problem: a view from the designer stance. Int J Mach Conscious 2(1):117–169
39. Sloman A, Chrisley RL (2003) Virtual machines and consciousness. J Conscious Stud 10(4–5):133–172
40. Sloman A, Chrisley RL (2005) More things than are dreamt of in your biology: information processing in biologically-inspired robots. Cogn Syst Res 6(2):45–74
41. Thompson E (2001) Empathy and consciousness. J Conscious Stud 8(5–7):1–32

42. Thompson E (2007) Mind in life: biology, phenomenology, and the sciences of mind. Harvard University Press, Cambridge
43. Toombs SK (1992) The meaning of illness: a phenomenological account of the different perspectives of physician and patient. Kluwer Academic Publishers, Norwell
44. Toombs SK (2001) The role of empathy in clinical practice. J Conscious Stud 8(5–7):247–258
45. Torrance SB (2007) Two conceptions of machine phenomenality. J Conscious Stud 14(7):154–166
46. Torrance SB (2008) Ethics and consciousness in artificial agents. Artif Intell Soc 22(4):495–521
47. Torrance SB (2012) Super-intelligence and (super-) consciousness. Int J Mach Conscious 4(2):483–501
48. Torrance SB (2013) Artificial consciousness and artificial ethics: between realism and social-relationism. Philos Technol (Special issue on 'Machines as moral agents and moral patients') 27(1):9–29. doi:10.1007/s13347-013-0136-5
49. Verbeek PP (2011) Moralizing technology: understanding and designing the morality of things. Chicago University Press, Chicago
50. Wallach W, Allen C (2009) Moral machines: teaching robots right from wrong. Oxford University Press, Oxford
51. Wallach W, Allen C, Franklin S (2011) Consciousness and ethics: artificially conscious moral agents. Int J machine Conscious 3(1):177–192
52. Whitby B, Yazdani M (1987) Artificial intelligence: building birds out of beer cans. Robotica 5:89–92

Emotion and Disposition Detection in Medical Machines: Chances and Challenges

Kim Hartmann, Ingo Siegert and Dmytro Prylipko

Abstract Machines designed for medical applications beyond usual data acquisition and processing need to cooperate with and adapt to humans in order to fulfill their supportive tasks. Technically, medical machines are therefore considered as affective systems, capable of detecting, assessing and adapting to emotional states and dispositional changes in users. One of the upcoming applications of affective systems is the use as supportive machines involved in the psychiatric disorder diagnose and therapy process. These machines have the additional requirement of being capable to control persuasive dialogues in order to obtain relevant patient data despite disadvantageous set-ups. These automated abilities of technical systems combined with enhanced processing, storage and observational capabilities raise both chances and challenges in medical applications. We focus on analyzing the objectivity, reliability and validity of current techniques used to determine the emotional states of speakers from speech and the arising implications. We discuss the underlying technical and psychological models and analyze recent machine assessment results of emotional states obtained through dialogues. Conclusively we discuss the involvement of affective systems as medical machines in the psychiatric diagnostics process and therapy sessions with respect to the technical and ethical circumstances.

K. Hartmann (✉) · I. Siegert · D. Prylipko
Cognitive Systems Group, Otto von Guericke University Magdeburg,
Magdeburg, Germany
e-mail: kim.hartmann@ovgu.de

I. Siegert
e-mail: ingo.siegert@ovgu.de

D. Prylipko
e-mail: dmytro.prylipko@ovgu.de

S.P. van Rysewyk and M. Pontier (eds.), *Machine Medical Ethics*,
Intelligent Systems, Control and Automation: Science and Engineering 74,
DOI 10.1007/978-3-319-08108-3_19

1 Introduction

Machines designed for medical applications beyond usual data acquisition and processing need to cooperate with and adapt to humans in order to fulfill their supportive tasks. Hence, from a technical point of view, modern medical machines are specific human computer interaction (HCI) systems. More precisely, due to the particular type of interaction within the usual physician-patient-relationship, medical machines interacting with patients must be designed as affective systems. This fact holds regardless of whether the system interacts autonomously or supervised.

The chapter demonstrates why the introduction of affective systems in medical application scenarios is premature and discusses the ethical issues raised. While affective systems in medical scenarios raise general ethical questions such as human indispensability in healthcare, trustworthiness and the inviolability of human dignity, certain applications additionally imply negative consequences for the individual patient. We focus our discussion on the use of dialogue-based automated emotion detection in psychiatric applications. The issues raised are due to technical deficiencies of affective systems applicable to all HCI-applications, but are especially evident in psychiatric application scenarios. One application that raises serious ethical questions is the use of emotion recognition methods in order to assist in the diagnostic process of psychiatric disorders and the use of affective systems in therapy sessions [24, 59].

Although machines have the advantage of proficient storage, processing and sensory equipment, it is questionable whether machines would (and should) be able to recognize emotions and react truly affectively to patients. This would imply the technical integration of enhanced human abilities which are often only imperfectly developed in humans. Due to several concerns, such as privacy and misuse issues as well as objectivity, reliability and validity of the outcome, any type of intervention of physician-patient-interactions by "non-physicians" have led to moral debates in the past. While concerns of privacy and misuse issues may be addressed through the utopia of secure systems, machines used to assess and modify patient behavior still impose non-neglectable ethical concerns. We will investigate the validity, reliability and objectivity of dialogue-based automated emotion and disposition analyses. We will show that validity, reliability and objectivity are not assumable and that hence, the unreflected incorporation of automated emotion detection from speech in medical applications inherently imposes risk of both misuse and abuse, contradicting the supportive purpose of medical machines.

The chapter is structured into four further sections. Section 2 introduces the psychological background for affective HCI and discusses current emotion theories and emotion categorization/annotation issues. Section 3 describes the technical state-of-the-art in speech-based emotion detection and the technical deficiencies associated with the approaches explained. Section 4 reports on published results related to the dependability of emotion detection from speech in different applications. Based on the previous sections, the justifiability of the integration of automated emotion detection from speech in medical application scenarios is discussed in the concluding Sect. 5.

2 Emotion Models and Theories in HCI

Emotions are widely accepted as essential to human-human interaction and inherent in physician-patient-interactions. Psychiatric disorders are often expressed through irregularities in emotional patterns and observable in the patients' behavior and speech. Although the *American Psychiatric Association* (APA) and the *World Health Organization* (WHO) published international classification tools (DSM-5 and ICD-10 respectively) used for the diagnosis of psychiatric disorders, diagnoses are often debatable and partly subjective. The difficulty lies in the term and understanding of human emotion and disposition. Debates on emotion theories are therefore indispensable for the design and understanding of affective systems.

In medical applications, affective systems may be faced with three types of difficulties:

1. The reliable, valid and universally reproducible recognition of emotional states in users,
2. the correct classification of the emotional states and
3. generating the appropriate affective response to the emotional input.

While (1) implies the need for universal, cooperative emotion models in both psychology and engineering, (2) addresses the technical difficulties of identifying sufficient, measurable observables associated with emotional expressions and the appropriate categorization of emotions, and (3) attends the difficulties associated with abstract emotion understanding neither inherent to humans nor to technical systems. The following discussion focuses on the issues associated with (1) and (2), as solving these may be considered a prerequisite to accomplishing (3).

2.1 Terminology

The emotion models incorporated in affective systems designed for medical applications must be psychologically founded. The terms *emotion*, *mood*, *personality traits* and *disposition* should be cooperatively defined for both psychological and technical investigations to allow the correct implementation in technical systems. However, precise definitions of these terms are already subject to disputes within the initiating discipline. The most common definitions used within the HCI research community are given in the following passages.

Affect A subjective experience of external events which may be characterized by specific bodily response patterns.

Emotion As stated in [4], emotions reflect short-term affects, usually bound to a specific event, action, or object. Following this definition, an observed emotion reflects a distinct user assessment directly related to a specific event and is of short duration. Hence, an affective system may not conclude the overall emotional state of an individual due to a single observation.

Mood In contrast to an emotion, a mood reflects a medium-term affect, not necessarily related to a unique event [47]. Moods last longer and are considered as rather stable affective states that influence the user's cognitive functions.

Personality reflects a long-term affect and specifies individual behavioral characteristics. The *NEO-FFI* is a common scheme used to categorize personalities and is based on the *Five Factor Model* [40, 44]. The NEO-FFI uses five traits to define the associated typical behavior. A prediction of the user's mood in combination with knowledge about the personality can be used to draw conclusions about the conversation course and the validity of the collected data [14].

While the research communities were able to partially agree on "best practice" definitions for the above named terms, disagreement regarding the choice and validity of emotion theories is persistent. However, the emotion models incorporated in affective systems must be founded on an approved emotion theory in order to be adaptable in medical applications. Due to potentially incompatible emotion theories, the selection of an appropriate emotion model has far reaching consequences for the affective system and its applications. To highlight these implications, a short introduction to some of the most common emotion theory approaches is given in the following.

2.2 Emotion Theories

In 1919 McDougall presented the concept of *primary emotions* as psychologically primitive building blocks [41]. He described functional behavior patterns with descriptive labels, such as anger or fear. The utilization of labels, the amount needed and the correct assignment of labels to emotions are still being debated [11, 58].

In the 1970s, Ekman introduced the concept of *basic emotions* following his observations on non-verbal behavior. Ekman and Wallace V. Friesen observed that the expressions of certain emotions seemed culturally, ethnically, gender and age independent. They generated a list of emotions expressed and recognized universally through similar facial expressions, yielding six basic emotions (anger, disgust, fear, happiness, sadness, surprise) [18].

However, this type of categorical emotion classification does not support the investigation of relationships between emotions. To address this issue, Plutchik [49] proposed the wheel of emotions in the 1980s with eight bipolar emotions and several combinations of these. This classification allowed to describe the intensity of an emotion and introduced so called *complex emotions* [49] (Fig. 1).

Another approach was to introduce a "*total-feeling*" by Wundt [74]. The idea is to represent a *feeling* as a mixture of potentially elementary feelings, each described as a single point in a three dimensional emotion space (pleasure $\leftrightarrow$ displeasure, excitement $\leftrightarrow$ inhibition, tension $\leftrightarrow$ relaxation) (Fig. 1).

Wundt's [74] concept was that an external event triggers a sequence of total-feelings over time. This results in a specific, continuous course within the described

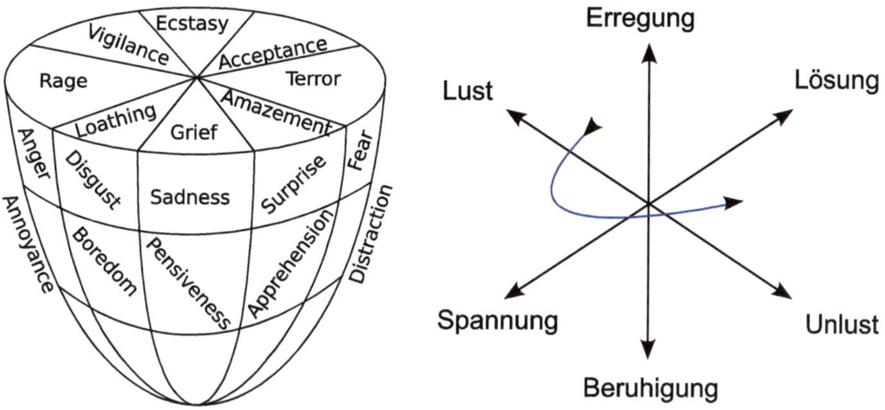

Fig. 1 *Left hand side* Plutchik's [49] three dimensional structural model of emotions [49], 157]; *right hand side* Wundt's three axes of an orthogonal emotion space, with emotion trajectory (following Wundt [74]). The axes correspond to: x-axis ("inclination-disinclination"), y-axis ("tension-relaxation"), z-axis ("excitement-sedation") [66, 67]

three-dimensional space. The resulting trajectory describes the specific course within this space. The description of emotions independent of labels and the model-inherent transitions are an advantage of this approach. Unfortunately, Wundt's theory neither locates emotions within the emotional space, nor does it provide a solution for the integration of intensity levels.

The exact configuration and dimension of the human emotion space is still being discussed within the research community. Prominent results are those of Schlosberg [60], Plutchik [49], and Russell and Mehrabian [52, 53].

A particularly relevant detail was found through the investigations of Russell and Mehrabian [52], who were able to emphasize that three dimensions are needed to distinguish emotional states [52]. Furthermore, Russell and Mehrabian located 151 emotional terms in the "Pleasure-Arousal-Dominance-space" (PAD-space).

However, the reliability of Russell and Mehrabian findings has been questioned repeatedly. Unfortunately, their results were not reproducible within a comprehensive study performed by Gehm and Scherer [23], nor in similar studies. A possible explanation of the irreproducibility was that the ability of the involved subjects to emotionally rate relevant words or pictures was not comprehensible [57].

3 How Machines Learn Emotions

To enable medical machines to automatically detect and recognize emotions, the methods used for affective systems must be applied. Figure 2 provides an overview of the methods involved and the general emotion learning process in technical systems.

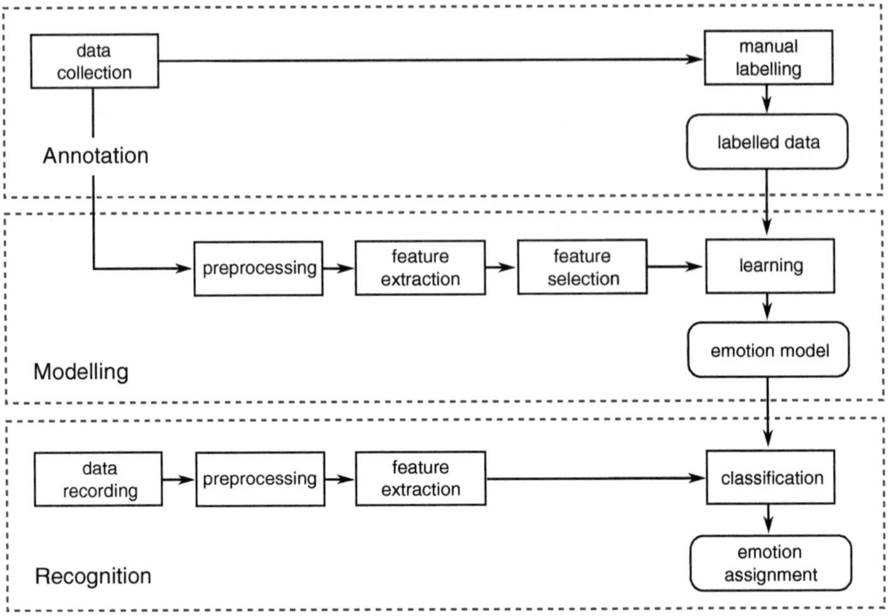

Fig. 2 How a machine learns

In the first step, the data (i.e., audio and video featuring emotional content displayed by humans) is recorded and saved. This data is then annotated by human experts with respect to the emotional labels provided. Due to the inter-rater reliability issues discussed in Sect. 4.1, several annotators should be involved in this task. The emotional label for one specific sequence is assigned based on the result of (weighted or unweighted) majority voting. This inherently implies that the emotional labels reflect the emotional awareness and understanding of the annotators, not necessarily representing the truth. See Sect. 5 for further details on this discussion.

The second step to allow machines to recognize emotions automatically involves preprocessing of the recorded sequence. We focus on speech data; however, any other modality may be chosen as well. Within the second step, the signal is represented through *features*: informative descriptors that represent the relevant information of the speech signal. Hence, the feature extraction process prepares the source data for further automated processing by machines.

The machines learn to match emotional labels to signals that have a specific configuration of the extracted features through pattern recognition methods. To enable machines to learn the association of an annotated label with the corresponding feature configuration, the annotated data (consisting of labels and features) is given to the machine. As an outcome of this learning process, the machine understands a certain feature configuration and emotion label as being connected, generating models of "emotions" (or more general: learned classes).

The generated emotion models may then be applied to new recordings to recognize emotions/classes automatically. The quality of the performance is highly dependent on the model applied.

The following sections give a brief description of each of the three steps involved in the emotion learning process in machines.

3.1 Annotation Methods

In healthy humans, the most reliable method to obtain a valid assignment of an expressed emotion may be a self-assessment by the observed subject [25]. Unfortunately, this is not always feasible due to numerous reasons (experimental design, the subjects' ability to reflect upon itself, health status of the subject). Hence, designers of affective systems must rely on third-parties to annotate the emotional sequences recorded. In order to achieve a minimum of homogeneity and reliability in the annotation, annotators are provided with annotation schemes. These schemes add to the uniformity of the annotation process and increase the likelihood of similar assessments by the annotators despite age, gender and socio-cultural differences. These annotation schemes are often based on the emotion theories discussed in Sect. 2.2.

The categorical emotion theories have the advantage of common terms, i.e., the annotators are provided with a selection of terms out of which they may choose to annotate the observed emotional sequence. However, the categorical emotion theories do not include the relations between the emotions observed. Furthermore, the assignment of relations between the terms used is highly subjective and differs depending on age, gender and socio-cultural background of the annotator. This may influence the annotation outcome.

The dimensional theories inherently incorporate emotional transitions, but the annotation process is more complex and the annotators need further training. The annotation through dimensions is not intuitive and the uncertainty of (dimensional) emotion theories inherently threatens the approach. There is little consent on the axes labels; however, the terms assigned to the axes may have an unforeseeable influence on the annotators' assessment. Regardless of the choice of annotation scheme, to heighten the chances of obtaining reliable labels, several annotators (mostly >6) are employed.

In the following, a short overview of common emotion annotation methods is given. Unfortunately, these methods do not necessarily reflect the emotion truly felt by a subject. Fragopanagos and Taylor analyzed a number of "input- and output-specific issues" that influence the assessment, such as display rules and cognitive effects [22]. Alarmingly, a study by Truong found that felt emotions are not always perceivable by observers [70] and that the average agreement between self-rater and annotators was lower than the inter-rater agreement [69].

Word Lists A common method used to assign emotional labels are *Word Lists* (WLs). In this method, descriptive labels, usually not more than ten, are selected

Table 1 Common word lists and related corpora

Affective labels	Used in
Negative, non-negative	ACC [35]
Angry, bored, doubtful, neutral	UAH corpus [10]
Fear, anger, stress	BSS 04 [16]
Positive, neutral, negative	ISL Meeting Corpus [8]
6 Basic emotions	DES [19], EmoDB [9]
Joyful, empathic, surprised, ironic, helpless, touchy, angry, bored, motherese, reprimanding, rest	AIBO database [3]

to describe the emotional state of an observed subject (Table 1). These labels can be formed from counter-parts, such as positive versus negative, or designed for a specific task (e.g., aggression detection) [16, 35, 38]. Several databases comprise a set of emotions, comparable to Ekman [18] or Plutchik [49]. The missing relationship between labels is one of the disadvantages of WLs, making it difficult for the annotator to give a reliable assessment [54] of the emotional sequence. The quantification of affective changes is especially difficult using WLs.

Geneva Emotion Wheel A prominent solution to overcome problems associated with WLs is the *Geneva Emotion Wheel* (GEW) (Fig. 3) by Scherer [56]. This assessment tool is related to Scherer's appraisal theory [55]. It is a theoretically derived and empirically tested instrument used to assess emotional reactions to objects, events, and situations. The GEW consist of 16 emotion families, each with five degrees of intensity, arranged in a wheel shape on the two axes "control" and "pleasantness". Additionally, no emotion and neutral options are provided.

This arrangement supports the annotator in assessing a single emotion family with a specific intensity by guiding him through the axes and quadrants. The labeling effort using GEW is quite high. However, the GEW allows quantification of the relation between distinct emotions, as it combines both the categorical and the dimensional approach. However, in dimensional approaches, the dimensions "pleasure" and "arousal" are more common terms used to describe human affective states [25, 51]. The dimension "control" is an object of ongoing discussions, mostly because studies investigating the axes to be used may rely on different methods and interpretations. Their outcome for the axis referred to as "control"/"dominance"/"coping" varies from being regarded unnecessary [52, 75] to immanent [23] in order to distinguish emotions.

Picture-Oriented Assessment This verbal description of emotions causes difficulties in assessment, as the application of these terms to a different language requires translation and validation of the used categories [7]. Furthermore, the relation between each literal description can differ from annotator to annotator, so that the relations differ not only from the subjective observation, but also from the subjective interpretation of the verbal description [45]. To address these issues, Lang invented a picture-oriented instrument to assess the pleasure, arousal, and dominance dimension directly [34].

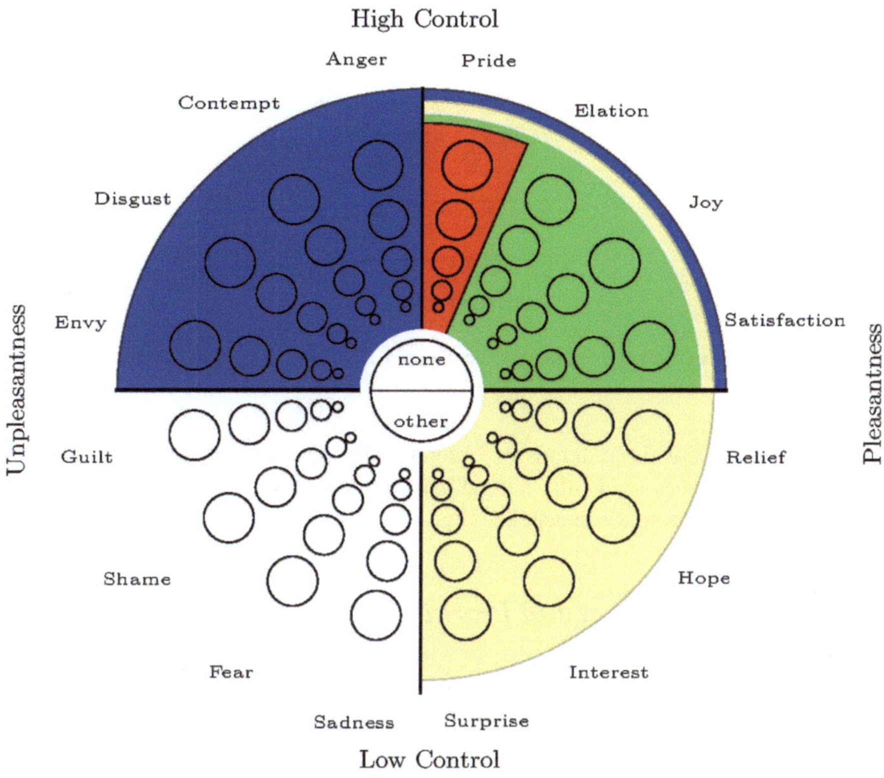

Fig. 3 Geneva emotion wheel. An example to illustrate the labelling process: An annotator assesses an emotional sequence as having "high control" (*blue semicircle*), and "pleasant" (*yellow semicircle*). The resulting assessment is the *green quadrant*. The choice "pride" is marked in *red*

Here, due to a dimensional representation, the annotators are guided to judge the relation between observed emotions. This presentation illustrates the relation between affective states much better than a literal transcription. Additionally, it reduces the evaluation effort. The *Semantic Differential Scale* uses 18 bipolar adjective pairs to generate judgments along three axes, whereas the *Self-Assessment Manikins* (SAM) depict the same dimensions through figures. The granularity is adjustable and spans from a 5 figure-scale for each dimension [46] to a 9 point-scale using intermediate steps between the figures [29]. However, due to the dimensional description and the limiting granularity, the ability to evaluate distinct or blended emotions is missing.

Feeltrace An entirely different approach to assess emotions is utilized with Feeltrace [13]. To examine the emotional evolvement, this instrument tracks the assessed affect over time. Feeltrace is based on the activation-evaluation space and arranged in a circular manner. The evaluation of the resulting labels is difficult, as each annotator produces a constant track with a step width of 0.02 on time and 0.0003

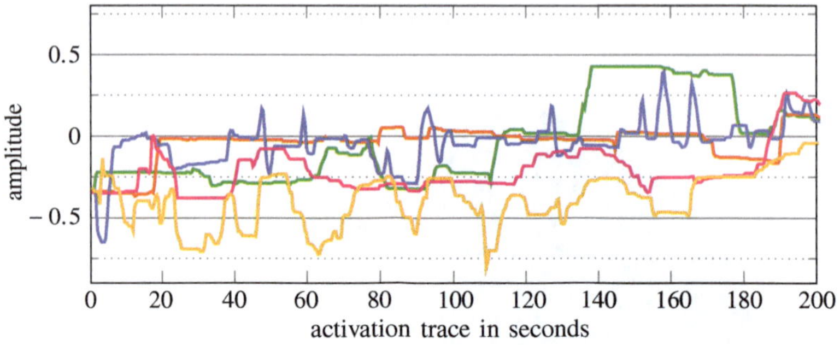

Fig. 4 Example trace plot for female speaker 1, trace 29 from SAL annotated by 5 annotators [50]

on value axis, where the relations of observed changes are very individual. Hence, Feeltrace is rather used to identify a trend in the emotional evolvement than to define a specific point in the emotion space. See Fig. 4 for an example of Feeltrace traces.

3.2 Feature Extraction and Learning

To automatically recognize emotions from utterances, technical systems heavily rely on pattern recognition methods. These methods describe typical configurations of selected descriptors to identify and describe a common state. In the emotion detection from speech, these descriptors are called features. The configuration of speech features is matched to specific classes or emotions based on the annotated training data given. The machines learn to abstract from the training data to a common model and to assign emotional labels similarly to the annotators' assignments.

A majority of the approaches to classify utterances use acoustic features. The most popular features are prosodic (e.g., pitch-related features, energy-related features etc.) and spectral features (e.g., MFCC). Prosodic features are more widely accepted as these are associated with variations in vocal tract configuration and hence directly relate to sympathetic and parasympathetic activation, known to be associated with affect.

The selection of features used to describe the observed signal influences the classification performance greatly. First approaches to the automated emotion detection from speech utilized the same spectral features as used for speech recognition, resulting in rather weak classifications. However, more recent approaches incorporate more advanced prosodic and voice quality features [31]. Recently, the automatic construction of feature vectors that are optimal for emotion recognition is being investigated [28].

Depending on the unit of analysis, classification approaches can be separated into two major groups: static and dynamic approaches. Within the static approach, the whole input utterance is regarded as a single unit. Hence, only a single feature vector is extracted for the complete input utterance (also referred to as "turn"). Since the value of each single feature varies over time, the characteristics of the input signal (so called *low level descriptors*: pitch- and energy-related features, spectral content etc.) are represented using *functionals* such as mean, standard deviation, extremes, ranges, and quartiles [72, 73]. Typical classification methods for static (or turn-level) analysis are support vector machines (SVM) [20], multilayer perceptrons (or other types of neural networks) [61], and Gaussian mixture models (GMMs) [39]. Also, Bayes classifier [71], Bayesian networks [20], random forests [30], decision trees [37], k-nearest neighbor classifiers [15] are widely applied.

Within the dynamic approach to classification, the unit of analysis is a short-time (around 25 ms) frame window, shifting through the signal. Features are extracted for every single frame, thus providing a sequence of feature vectors. Among dynamic (also referred to as "frame-level") analysis, hidden Markov models (HMMs) are predominantly used [36, 72, 73]. Each emotion is modeled with a corresponding HMM. The combination of turn- and frame-level analysis is also possible, e.g., Vlasenko et al. [72, 73]. For an overview of advanced emotion classification techniques see Palm and Glodek [48]. However, a detailed understanding of these methods may be considered negligible for the discussion of the dependability of automated emotion detection/recognition methods. The accuracy rates of some of the most popular classification methods for emotional speech were investigated in Ayadi et al. [2]. These findings are given and discussed in the Sect. 4.2.

3.3 Automatic Emotion Recognition

Speech-based emotion classification systems typically consist of two stages: (1) front-end signal processing ("feature extraction"), and (2) classification that evaluates incoming features and assigns the class labels. Having finished training, the classifier defines a model associated with each label, which may be applied to new recordings. These recordings do not need to be part of the training set nor do the speakers need to be known. If the training of the classifier was successful, the recorded data does not need to undergo an annotation process. The annotation of classes is done automatically by the classifier. In order to do so, the same pre-processing and feature extraction methods are applied [5]. Depending on the learning methods used, the training data and the annotation given, the outcome of these classifiers may vary greatly. Some of the most recent emotion recognition results based on automated classification are given and discussed in the Sect. 4.2.

4 Dependability of Emotion Detection and Recognition

As stated previously, the assessment and labeling of emotions is a subjective task. Annotation schemes are used to determine the emotional information of data and assign specific labels to emotional sequences in order to allow machine learning methods to be applied. These learning methods are applied on larger datasets, trying to extract and generalize feature changes according to the annotation given. This procedure implies that the outliers are smoothened. However, if the annotation for emotional sequences is varying strongly, the learning outcome will result in highly simplified and erroneous classifications. Sequences will only be identified on a "best fit" and "least common denominator" level. These considerations highlight the necessity for high inter-rater reliabilities in order to design solid emotion detection classifiers.

This section discusses the dependability issues for the annotation, modelling and recognition phase raised in the previous sections. To allow a solid discussion of the chances and challenges for affective systems in medical applications, the three steps identified to allow machines to learn emotions are analyzed with respect to their dependability. The Sect. 4.1 discusses the reliability of the annotation process and the elements involved. Section 4.2 investigates issues that may influence the dependability of the emotion detection based on the feature extraction and learning methods. The final Sect. 5 discusses the reported results of different emotion recognition challenges to provide concrete numbers for specific applications of speech based emotion detection.

4.1 Annotation Reliabilities

There are only few studies available that investigate the specific effects of the annotation process on emotion detection [17, 38, 69]. The major issue in the annotation of emotional sequences is "inter-rater reliability" (IRR), a measure of the reliability of the given annotation by two different annotators for one specific emotional sequence labeled.

The IRR pays tribute to the fact that the assessment of an emotional sequence is subjective. Annotation schemes try to minimize the effects of age, gender and socio-cultural background on the annotation process; however, these effects cannot be eradicated. Despite the knowledge of the unreliability of human annotators, little research has been done to improve the IRR [17]. A detailed discussion of these issues can be found in Siegert et al. [66, 67].

Krippendorff's Alpha A measure often used to determine the IRR for a given speech corpus is *Krippendorff's alpha*, α, [32]. Krippendorff's alpha measures the disagreement of the annotated labels for one specific emotional sequence. To obtain a measure for the agreement of the annotated labels the value calculated for the annotations disagreement is subtracted from 1. Furthermore, several distance metrics for nominal, ordinal, interval, and ratio labels are available. An advantage of α compared to other IRR measures is its greater reliability utilizing two or more annotators and its robustness when dealing with incomplete data [27].

The obtainable values for α range from 0 ("low agreement") to 1 ("high agreement"). There are several interpretation schemes available for the calculated values of α. The most common classification used was suggested by Landis and Koch [33]. Other schemes are provided by Altman [1], Fleiss et al. [21] and

Table 2 Calculated IRR for VAM, distinguishing nominal (nom) and ordinal (ord) metric for each dimension and part

Part	Valence nom/ord	IRR	Dominance nom/ord
		Arousal nom/ord	
VAM I	0.106/0.189	0.180/0.485	0.176/0.443
VAM II	0.086/0.187	0.210/0.431	0.137/0.337
VAM (I + II)	0.108/0.199	0.194/0.478	0.175/0.433

Krippendorff [32] respectively. For a comparison of the named schemes refer to Fig. 6. The differences in the interpretations and labels of Krippendorff's alpha make it hard to compare results unless the exact values for α are provided additionally.

Reported Inter-Rater Reliabilities The IRR assessment reported by Callejas [10] was based on the annotation of the UAH emotional speech corpus. The UAH emotional speech corpus contains 85 dialogues from a telephone-based information system spoken in Andalusian dialect from 60 different users. They used a WL to distinguish the four emotional terms (angry, bored, doubtful and neutral). The annotation process was conducted by 9 labelers and performed on complete utterances. To infer a relation between the emotional categories, the authors arranged the emotional terms on a 2-D activation-evaluation space and defined angular distances, in the range of 0°–180°. The authors reported an α of 0.338 for their "angle metric distance" [10].

A corpus using the dimensional labeling approach is the "Vera am Mittag" audio-visual emotional speech data-base (VAM). It contains spontaneous and unscripted discussions between two to five persons from a German talk show [26]. Selected sentences were labeled with SAM (see Sect. 3.1) using a 5-point scale. The resulting assessments were translated into the interval of $[-1, 1]$ and a weighted average assessment was calculated.

The VAM database contains 499 items derived from very good quality speakers (denoted as VAM I), evaluated by 17 human listeners and 519 items from good quality speakers evaluated by 6 human listeners (denoted as VAM II). The calculated α with nominal and ordinal distance is given in Table 2.

Table 2 shows that the achieved inter-rater reliabilities were quite low. The values for a nominal distance metric ranged from poor to slight, while the ordinal distance achieved fair to moderate alpha values. However, with a smallest value of 0.086 and a highest value of 0.478, the IRR of annotations done with ordinal labels must still be considered unreliable. The training level of the annotators had only an effect on the IRR for the assessment along the arousal and dominance dimension.

The SEMAINE corpus contains the recordings of emotionally colored conversations. Utilizing four different operator behaviors, the scenario is designed to evoke emotional reactions. To obtain annotations, trace style continuous ratings were made on five core dimensions (Valence, Activation, Power, Expectation, Overall Emotional Intensity) utilizing Feeltrace [42]. A subset, the SEMAINE

Table 3 IRRs for selected functionals of SAL comparing α and $\alpha_{0.05}$ with given Cronbach's Alpha α_{Cr} values [43] for the Traces Intensity (I), Valence (V), Activation (A), Power (P), and Expectation (E)

		I	V	A	P	E
Median	α	0.14	0.12	0.12	0.11	0.09
Sd	α	0.14	0.14	0.12	0.11	0.09

Solid-SAL corpus (SAL), was made available to the research community. The number of labelers varied between 2 and 6, and the experiments were divided into segments with a fixed length of about 5 min. An example trace is given in Fig. 4. The resulting IRRs, using Krippendorff's alpha with ordinal metric distances, are given in Table 3.

The discussed results support our hypothesis that the IRR for the annotation of emotional or affective sequences is poor. Due to the reported values for alpha, the annotations must be considered unreliable. It must be noted that our inference also holds for popular and widely used speech corpora such as VAM and SEMAINE. Many results used for the general design of affective systems rely on research results obtained from investigations of these corpora.

All reported alpha-values are given in Figs. 5 and 6 for comparison and arranged on the interpretation scheme by Landis and Koch [33].

4.2 Feature Extraction and Learning

While the feature extraction methods should not influence the emotion detection's dependability negatively, feature selection does have an impact on the validity and significance of the results obtained.

As mentioned earlier, some features, especially those categorized as "prosodic features" are closely related to the vocal tract configuration and hence associated with the activation of the sympathetic and parasympathetic systems. Some features are currently being discussed as well suited for the detection of psychological disorders [59].

The learning process is closely related to the annotation process. The learning process is therefore dependent on both the annotation quality as well as the quality of the preprocessed data provided. Classifiers may be considered as the automated, technical correspondent to a human annotator. The classifier generates the emotion models that match a given recording to a specific feature configuration in order to automatically assess the emotional label associated with that specific feature configuration. Some of the most common classifiers used were briefly introduced in Sect. 3.2. The average accuracies that may be achieved with these classifiers have been investigated by Ayadi et al. [2] (Table 4).

The average accuracies given for some of the popular classification methods allow us to deduce that the accuracy of the classification—i.e., the correctness of

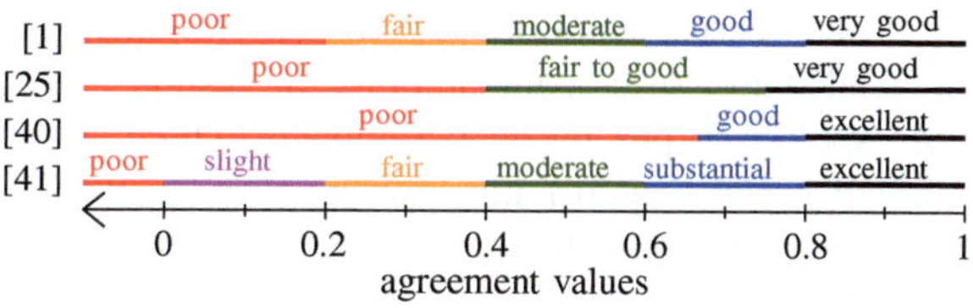

Fig. 5 Comparison of different interpretation schemes for α. Line *[1]* refers to the scheme according to Altman [1]; *[2]* Fleiss et al. [21]; *[3]* Krippendorff [32]; *[4]* [33]

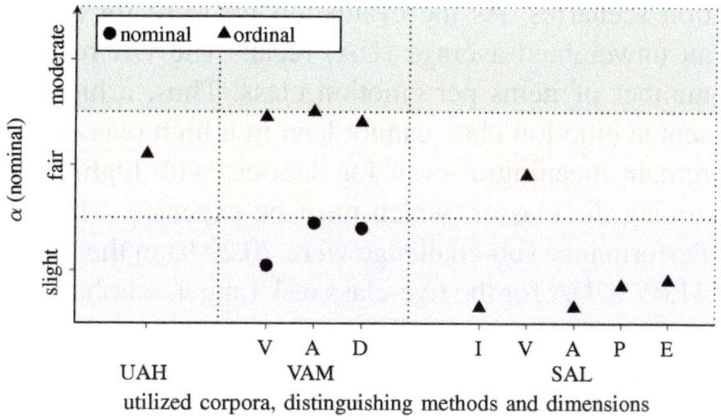

Fig. 6 Compilation of reported IRRs, plotted against the agreement interpretation Landis and Koch [33]. Both nominal and ordinal values are given (Table 2)

Table 4 Classification performance of popular classifiers, employed for the task of speech emotion recognition

Classifier	HMM (%)	GMM (%)	ANN (%)	SVM (%)
Average accuracy	75.5–78.5	74.83–81.94	51.19–70	75.45–81.29

HMM hidden Markov model, *GMM* Gaussian mixed model, *ANN* artificial neural network, *SVM* support vector machine

the assignment performed through automated emotion detection—varies with the classifier chosen and ranges between 51.19–81.94 % on average. Better performances may be possible due to good training data and wise selection of features, depending on the specific application and task.

5 Emotion Detection Reliability: INTERSPEECH Challenges

The rapid growth of the affective computing field resulted in several methods being applied to various corpora using different evaluation strategies and features. This diversity led to a low comparability of the methods used and the results obtained. Aiming to bridge the gap between the reported studies and their poor compatibility, INTERSPEECH 2009 Emotion Challenge [62] was the first open public evaluation of speech-based emotion recognition systems. The INTERSPEECH Challenge demanded the strict comparability of results through the utilization of one specific speech corpus by all participants. Three sub-challenges (Open Performance, Classifier, and Feature) aimed at the classification of five non-prototypical emotion classes (anger, emphatic, neutral, positive, remainder) or two emotion classes (negative, idle).

A corpus was built using spontaneous speech data (FAU AIBO, [68]) in order to estimate the performance of state-of-the-art methods in real-world application scenarios. As the evaluation measure for all challenges, the organizers chose an unweighted average (UA) recall. The UA recall does not take into account the number of items per emotion class. Thus, a high recall for just one widely represented emotion class cannot lead to a high overall recall. This allows the measure to remain meaningful even for datasets with highly unbalanced distributions of items among the classes, which must be expected. The best achieved results in the Open Performance sub-challenge were 70.29 % in the two-class task (negative vs. idle) and 41.65 % UA for the five-class task (anger, emphatic, neutral, positive, remainder).

The Emotion Challenge was followed by annual INTERSPEECH events devoted to the detection and classification of various aspects of human traits. The 2010 Paralinguistic Challenge focused on the classification of age, gender and affect [63]. The sub-challenges of age (four different age groups) and gender (female/male/children) were performed on the aGender corpus and resulted in 53.86 % UA and 84.3 % UA, respectively. For the recognition of affect (determination of speakers' interest) the TUM AVIC corpus was employed. The winner of this sub-challenge managed to achieve a cross-correlation measure of 0.146 (cross-correlation between the annotators' mean "Level of Interest" annotation ground truth and the prediction).

The Speaker State Challenge in 2011 aimed to recognize the intoxication and sleepiness of speakers [65]. For this purpose, organizers provided the Alcohol Language Corpus and Sleepy Language Corpus of genuine intoxicated and sleepy speech. Within the Intoxication sub-challenge speaker intoxication was determined in terms of a two-class classification problem: intoxication for a blood alcohol concentration exceeding 0.5 or non-intoxicated. The best achieved UA was 70.54 %. In the Sleepiness sub-challenge, sleepiness of speakers had to be determined by a suited algorithm and acoustic features. While the annotation provides sleepiness in ten levels, only two classes were accordingly recognized: sleepiness for a level exceeding level seven or no sleepiness. The winners of the Sleepiness sub-challenge achieved 71.69 % UA.

In 2012, the topics of the Speaker Trait Challenge were personality, likability, and pathology. For these tasks, the Speaker Personality Corpus, the Speaker Likability Database, and the NKI Concomitant Chemo Radiation Treatment (CCRT) Speech Corpus (NCSC) with high diversity of speakers of different personality and likability and genuine pathologies were provided. The speaker personality traits in the first corpus were measured with the popular "Big Five" OCEAN dimensions (openness, conscientiousness, extraversion, agreeableness, and neuroticism). The NCSC corpus provides 3 h of speech from 40 speakers with head and neck cancer (tumors located in the vocal tract and larynx) recorded before and at various times after treatment.

The winners of the Personality sub-challenge were able to achieve 69 % mean UA over 5 OCEAN dimensions. In the Likability sub-challenge, the likability of a speaker's voice had to be determined by a suitable learning algorithm and acoustic features. While the annotation provides likability in multiple levels, only two classes have to be recognized accordingly: likability above or below average. The best result achieved within the sub-challenge was 65.8 % UA. In the Pathology sub-challenge, the intelligibility of a speaker had to be determined by a suited classification algorithm and acoustic features. The winners result was 76.8 % UA.

In 2013, the INTERSPEECH Computational Paralinguistics Challenge covered a number of aspects, among others social signals, conflict, emotion, and autism [64]. From these, the last two are the most interesting within medical applications. Despite the recent trend towards naturalistic data in affective computing, for the 2013 Emotion sub-challenge, the organizers introduced (for the first time) a corpus of acted data in order to "fuel the ever on-going discussion on differences between naturalistic and acted material and hope to highlight the differences". The winners of the Emotion sub-challenge managed to provide 73.87 % UA for binary (positive/negative) arousal classification task, 63.26 % UA for valence classification and 42.29 % UA for the classification among 12 emotional categories.

In the Autism sub-challenge, the type of pathology of a speaker had to be determined from speech using acoustic features. This task comprises two subtasks: detecting children with autism spectrum disorder (ASD) and classifying them into four subtypes. For this task the Child Pathological Speech Database was provided. The best results reported were 93.58 % UA for binary classification (ASD vs. non-pathological development), and 69.42 % for the classification of the speakers into 4 subtypes of ASD.

From these results, we conclude that while the state-of-the-art methods for affect and trait recognition are promising, they are not mature enough for wide usage in medical applications.

6 Summary

This chapter described the state of the art in machine emotion detection from speech and demonstrated some currently discussed applications in psychiatry and medicine. The different emotion theories and models were described and disagreement between researchers concerning key terms was highlighted. However, without a valid emotion theory, machines cannot be taught to detect and recognize real emotions. Technical deficiencies of automated emotion detection were outlined. The annotation process yields special difficulties. However, feature extraction and learning methods may also induce erroneous output. The lack of dependability due to different aspects in each of the three steps identified in the automated emotion detection from speech was also discussed.

It is evident that ethical issues associated with the application of automated emotion detection methods are ignored by a majority of technical researchers. Reflecting on the previous sections, the propagation of the technology as mature for the use in medical applications, especially in the treatment and diagnosis of psychiatric patients, appears quite premature. We feel that several questions need to be addressed prior to incorporating automated emotion recognition mechanisms in medical applications. Although some technical researchers may consider these questions as negligible, ethical issues arising may have far reaching effects on our society and the individual. Some of the questions are:

- Privacy considerations
- Responsibility for patients and machines
- Objectivity of machines

The following summarizes our discussion, focusing on each of the named questions separately.

Privacy. Modern technical systems are capable of storing a tremendous amount of data. The recordings made are typically collected as so called "speech corpora", to be used for medical or research purposes. While the individuals contributing to a speech corpus will have signed a contentment form and are aware of their data being recorded, psychiatric patients may not have this option or information.

The content of therapy sessions must be considered highly vulnerable; however, discussions on how to safely store these recordings are marginal. The use of "clouds" [12] to permanently store data outside of the machine in use and to allow the access of data through networks alarmingly contributes to the raised privacy issues.

To automatically detect emotions in speech, large datasets are needed, both of individuals and groups. Affective machines may even learn to recognize and adapt to speakers, however, again needing large datasets of the individual. Due to the nature of the methods involved, a conflict between system developers and patients is unavoidable.

Apart from the issue of confidentiality and access, integrity must also be guaranteed. If the data stored should be adjusted, exchanged or destroyed, the consequences for the individual patient will be unforeseeable. It must be considered that the data stored may contain very detailed and confidential information about the individual, the individual's family and work.

These three aspects, integrity, availability (in terms of selected availability to authenticated users) and confidentiality of data are generally referred to as "the three aspects of IT security" (for further details on computer security, see [6]). The design of systems guaranteeing to fulfill all three aspects, under any circumstances, remains an unsolved problem.

Responsibility. Under normal conditions, the patient's physician or therapist is responsible for the patient's well-being during ambulant or stationary treatment. If a physician decides to externalize the diagnostic or therapeutic process partially or completely, the responsibility for the patient would normally be transferred accordingly. However, currently, technical systems are not considered capable of bearing responsibility for human individuals due to ethical, legal and technical considerations. Hence, technical (affective) systems will need to be supervised to be used in medical applications.

However, important questions remain: who is responsible for machine failures and who for invalid results? How are these two cases to be distinguished and who should decide upon this? While clear, technical system failures should be corrected by technicians, invalid results may be tracked back to poor classifiers, incomplete training data, bad annotators or less obvious failures in the technical set-up. It is questionable whether the physician can be made responsible for invalid results of a system he neither designed nor maintained. One could argue that the responsible physician must monitor the system's assessments and review each decision made by the system; however, doing so would complicate the physician's tasks, and hence contradict the idea of utilizing affective systems to assist in the diagnostic process and/or treatment of patients.

Objectivity of machines. Machines are presumed to be objective. While tragic events in the past have led to greater awareness that objective results may be objective but false, it is intuitively presumed that machines do not produce subjective output. Subjectivity is implied as a characteristic specific to human consciousness. Regardless, we neglect that high level applications as affective computing are designed by humans, based on human annotations and therefore are subjective.

As discussed in Sect. 4, the annotation process is subjective, as the annotator's assessment is subjective. Hence, classifiers trained on the annotators' assessment are subjective too. Interestingly, the classifiers learn the annotators' subjectivity. Hence, classifiers trained with a specific socio-cultural group of annotators will reflect this group's subjectivity. Although this fact may be easily derived, it is astonishing. *Automated emotion detection through machines is no less subjective than the emotion detection through humans.* In fact, machines may even be *more* subjective as they represent an abstraction of the subjective assessment of several individuals (see Sect. 3). Currently, there are no methods available to avoid this effect apart from the employment of carefully chosen and balanced annotator groups.

Due to the above considerations, one is confronted with the unavoidable truth that automated emotion detection is not objective. Hence, results achieved through medical machines assessing the emotional states of patients are not reliable. This fact leads to the conclusion that at least certain diagnoses made by automated emotion detection machines will be invalid. This result is widely known in the emotion detection research community, and the community is justifiably proud due to the good performances of modern classifiers (see Sect. 4.2). However, in medical applications, one should not consider to develop machines that categorize humans as "normal" or "abnormal" correctly in 70–80 % of the cases investigated. Especially due to the stigmatization of psychiatric disorders, a diagnostics tool with 30–20 % failure rate is not acceptable. The assumed objectivity of machine medical assessment raises genuine doubts that should be associated with the use of automated emotion detection methods for use in psychiatric applications.

Due to considerations of privacy, responsibility and objectivity, as well as investigations of current emotion detection and classification results (see Sect. 4.2) it seems untimely to incorporate machines in diagnosis and therapy of psychiatric patients. Thus, allowing medical machines to categorize human personalities and behaviors, and to interact with humans in fragile patient-physician contexts, is premature.

Refer ences

1. Altman DG (1991) Practical statistics for medical research. Chapman & Hall, London
2. Ayadi ME, Kamel MS, Karray F (2011) Survey on speech emotion recognition: Features, classification schemes, and databases. Pattern Recogn 3(44):572–587
3. Batliner A, Hacker C, Steidl S, Nöth E, Russell M, Wong M (2004) "You stupid tin box"-children interacting with the AIBO robot: a cross-linguistic emotional speech corpus. Proc. of LREC. LREC, Lisbon, Portugal, pp 865–868
4. Becker P (2001) Structural and relational analyses of emotions and personality traits. Zeitschrift für Differentielle und Diagnostische Psychologie 3(22):155–172
5. Bishop CM, Nasrabadi NM (2006) Pattern recognition and machine learning. Springer, New York
6. Bishop M (2004) Introduction to computer security. Addison-Wesley Professional, USA
7. Bradley MM, Lang PJ (1994) Measuring emotion: the self-assessment manikin and the semantic differential. J Behav Ther Exp Psy 25:49–59
8. Burger S, MacLaren V, Yu H (2002) The ISL meeting corpus: the impact of meeting type on speech style. ICSLP, Colorado, pp. 301-304
9. Burkhardt F, Paeschke A, Rolfes M, Sendlmeier W, Weiss B (2005) A database of German emotional speech. Proc. of Interspeech. ISCA, Portugal, pp 1517–1520
10. Callejas Z, López-Cózar R (2008) Influence of contextual information in emotion annotation for spoken dialogue systems. Speech Commun 50:416–433
11. Cambria E, Livingstone A, Hussain A (2012) The hourglass of emotions. In: Cognitive behavioural systems. Springer, Berlin Heidelberg, pp 144–157
12. Chao L (2013) Cloud database development and management. Auerbach Publications, USA
13. Cowie R, Douglas-Cowie E, Savvidou S, McMahon E, Sawey M, Schröder M (2000) FEELTRACE: an instrument for recording perceived emotion in real time. Proceedings of ISCA tutorial and research workshop (ITRW) on speech and emotion. ISCA, France, pp 19–24
14. Davidson R (1994) On emotion, mood, and related affective constructs. In: Ekman P (ed) The nature of emotion: fundamental questions. Oxford University Press, Oxford, pp 51–56
15. Dellaert F, Polzin T, Waibel A (1996) Recognizing emotions in speech. Proc. ICSLP 1996. ICSLP/ISCA, Philadelphia
16. Devillers L, Vasilescu I (2004) Reliability of lexical and prosodic cues in two real-life spoken dialog corpora. Proceedings of LREC. European Language Resources Association, Lisbon
17. Devillers L, Vidrascu L, Lamel L (2005) Challenges in real-life emotion annotation and machine learning based detection. Neural Netw 4(18):407–422
18. Ekman P (1992) Are there basic emotions? Psychol Rev 99:550–553
19. Engberg IS, Hansen AV (1996) Documentation of the Danish emotional speech database (DES). Aalborg University, Aalborg
20. Fernandez R, Picard RW (2003) Modeling drivers' speech under stress. Speech Commun 40:145–159
21. Fleiss JL, Levin B, Paik MC (2003) Statistical methods for rates and proportions, 3rd edn. Wiley, USA
22. Fragopanagos NF, Taylor JG (2005) Emotion recognition in human-computer interaction. Neural Netw pp 389–405

23. Gehm T, Scherer KR (1988) Factors determining the dimensions of subjective emotional space. In: Scherer KR (ed) Facets of emotion. Lawrence Erlbaum Associates, USA, pp 99–113
24. Gratch J, Morency L-P, Scherer S, Stratou G, Boberg J, Koenig S, et al (2013) User-state sensing for virtual health agents and telehealth applications. Medicine meets virtual reality 20—NextMed, MMVR. IOS Press, Shanghai, pp 151–157
25. Grimm M, Kroschel K (2005) Evaluation of natural emotions using self assessment manikins. IEEE workshop on automatic speech recognition and understanding. IEEE, San Juan, pp 381–385
26. Grimm M, Kroschel K, Narayanan S (2008) The Vera am Mittag German audio-visual emotional speech database. Proceedings of ICME. ICME, Monterry, pp 865–868
27. Hayes AF, Krippendorff K (2007) Answering the call for a standard reliability measure for coding data. Commun Methods Meas 1:77–89
28. Hübner D, Vlasenko B, Grosser T, Wendemuth A (2010) Determining optimal features for emotion recognition from speech by applying an evolutionary algorithm. Proceedings of Interspeech. ISCA, Makuhari, pp 2358–2361
29. Ibáñez J (2011) Showing emotions through movement and symmetry. Comput Hum Behav 1(27):561–567
30. Iliou T, Anagnostopoulos C-N (2009) Comparison of different classifiers for emotion recognition. Proceedings of the 13th panhellenic conference on informatics. IEEE Computer Society, Los Alamitos, pp 102–106
31. Kane J, Scherer S, Aylett M, Morency L-P, Gobl C (2013) Speaker and language independent voice quality classification applied to unlabelled corpora of expressive speech. Proceedings of international conference on acoustics, speech, and signal processing (ICASSP). IEEE, Vancouver, pp 7982–7986
32. Krippendorff K (2012) Content analysis: an introduction to its methodology, 3rd edn. SAGE Publications, Thousand Oaks
33. Landis JR, Koch GG (1977) The measurement of observer agreement for categorical data. Biometrics (33):159–174
34. Lang PJ (1980) Behavioral treatment and bio-behavioral assessment: computer applications. In: Sidowski JB, Johnson JH, Williams TA (eds) Technology in mental health care delivery systems. Ablex Pub Corp, New York, pp 119–137
35. Lee CM, Narayanan S (2005) Toward detecting emotions in spoken dialogs. IEEE Trans Speech Audio Process 2(13):293–303
36. Lee CM, Yildirim S, Bulut M, Kazemzadeh A, Busso C, Deng Z et al (2004) Emotion recognition based on phoneme classes. Proceedings of Interspeech 2004. ICSLIP, Jeju Island
37. Lee C, Mower E, Busso C, Lee S, Narayanan S (2009) Emotion recognition using a hierarchical binary decision tree approach. In: ISCA (ed) Proceedings of interspeech'2009. IEEE, Brighton, pp 320–323
38. Lefter I, Rothkrantz LJ, Burghouts GJ (2012) Aggression detection in speech using sensor and semantic information. In: Sojka P, Horak A, Kopecek I, Pala K (eds) Text, speech and dialogue, vol LNCS 7499. Springer, Berlin Heidelberg, pp 665–672
39. Lugger M, Yang B (2007) An incremental analysis of different feature groups in speaker independent emotion recognition. Proceedings of the 16th international congress of phonetic sciences. ICPhS, Saarbrücken, pp 2149–2152
40. McCree RR, John OP (1992) An introduction to the five-factor model and its applications. J Pers 2(60):175–215
41. McDougall W (1908) An introduction to social psychology [Dover edition (2003)]. Dover Publications Inc, London
42. McKeown G, Valstar M, Cowie R, Pantic M (2010) The SEMAINE corpus of emotionally coloured character interactions. Proceedings of ICME. ICME, Singapore, pp 1079–1084
43. McKeown G, Valstar M, Cowie R, Pantic M, Schröder M (2012) The SEMAINE database: annotated multimodal records of emotionally coloured conversations between a person and a limited agent. IEEE Trans Affect Comput 3:5–17

44. Mehrabian A (1996) Pleasure-arousal-dominance: a general framework for describing and measuring individual differences in temperament. Curr Psychol 4(14):261–292
45. Morris JD (1995) SAM: the self-assessment manikin an efficient cross-cultural measurement of emotional response. J Advertising Res 35:63–68
46. Morris JD, McMullen JS (1994) Measuring multiple emotional responses to a single television commercial. Adv Consum Res 21:175–180
47. Morris WN (1989) Mood: the frame of mind. Springer, New York
48. Palm G, Glodek M (2013) Towards emotion recognition in human computer interaction. In: Apolloni B, Bassis SE, Morabito FC (eds) Smart innovation, systems and technologies. Neural nets and surroundings, vol 19. Springer, Heidelberg, pp 323–336
49. Plutchik R (1980) Emotion, a psychoevolutionary synthesis. Harper & Row, New York
50. Prylipko D, Rösner D, Siegert I, Günther S, Friesen R, Haase M, Vlasenko B, Wendemuth A (2014) Analysis of significant dialog events in realistic human-computer interaction. J Multimodal User Interfaces 8(1):75–86
51. Russel J (1980) Three dimensions of emotion. J Pers Soc Psychol 9(39):1161–1178
52. Russel J, Mehrabian A (1974) Distinguishing anger and anxiety in terms of emotional response factors. J Consult Clin Psych 42:79–83
53. Russell JA, Mehrabian A (1977) Evidence for a three-factor theory of emotions. J Res in Pers 273–294
54. Sacharin V, Schlegel K, Scherer KR (2012) Geneva emotion wheel rating study. Center for Person, Kommunikation, Aalborg University, NCCR Affective Sciences. Aalborg University, Aalborg
55. Scherer KR (2001) Appraisal considered as a process of multilevel sequential checking. In: Scherer KR, Schorr A, Johnstone T (eds) Appraisal processes in emotion: theory, methods, research. Oxford University Press, Oxford, pp 92–120
56. Scherer KR (2005) What are emotions? And how can they be measured? Soc Sci Inform 4(44):695–729
57. Scherer KR, Dan E, Flykt A (2006) What determines a feeling's position in affective space? A case for appraisal. Cogn Emot 1(20):92–113
58. Scherer S, Schels M, Palm G (2011) How low level observations can help to reveal the user's state in HCI In: D'Mello S, Graesser A, Schuller B, Martin J-C (eds) Proceedings of the 4th international conference on affective computing and intelligent interaction (ACII'11). Springer, Memphis, pp 81–90
59. Scherer S, Stratou G, Mahmoud M, Boberg J, Gratch J, Rizzo A et al (2013) Automatic behavior descriptors for psychological disorder analysis. IEEE conference on automatic face and gesture recognition. IEEE, Shanghai
60. Schlosberg H (1954) Three dimensions of emotion. Psychol Rev 2(61):81–88
61. Schuller B, Rigoll G, LangM(2004) Speech emotion recognition combining acoustic features and linguistic information in a hybrid support vector machine—belief network architecture. Proceedings of IEEE international conference on acoustic, signal, and speech processing (ICASSP'2004). IEEE, Montreal, pp 577–580
62. Schuller B, Steidl S, Batliner A (2009) The INTERSPEECH 2009 emotion challenge. Proceedings of INTERSPEECH'2009. ISCA, Brighton, pp 312–315
63. Schuller B, Steidl S, Batliner A, Burkhardt F, Devillers L, Müller CA, et al (2010) The INTERSPEECH 2010 paralinguistic challenge. Proceedings of INTERSPEECH'2010. ISCA, Makuhari, pp 2794–2797
64. Schuller B, Steidl S, Batliner A, Vinciarelli A, Scherer K, Ringeval F, et al (2013) The INTERSPEECH 2013 computational paralinguistics challenge: social signals, conflict, emotion, autism. Proceedings of INTERSPEECH'2013. ISCA, Lyon
65. Schuller B, Steidl S, Batlinger A, Schiel F, Krajewski J (2011) The INTERSPEECH 2011 Speaker State Challenge. Proceedings of INTERSPEECH'2011. ISCA, Florence, pp 3201–3204
66. Siegert I, Böck R, Wendemuth A (2014) Inter-rater reliability for emotion annotation in human-computer interaction: comparison and methodological improvements. J Multimodal User Interfaces 8(1):17–28

67. Siegert I, Hartmann K, Glüge S, Wendemuth A (2013) Modelling of emotional development within human-computer-interaction. Kognitive Systeme
68. Steidl S (2009) Automatic classification of emotion related user states in spontaneous children's speech. University of Erlangen-Nuremberg
69. Truong KP, Neerincx MA, van Leeuwen DA (2008) Assessing agreement of observer- and self-annotations in spontaneous multimodal emotion data. Proceedings of INTERSPEECH'2008. ISCA, Brisbane, pp 318–321
70. Truong KP, van Leeuwen DA, de Jong FM (2012) Speech-based recognition of self-reported and observed emotion in a dimensional space. Speech Commun 9(54):1049–1063
71. Ververidis D, Kotropoulos C (2004) Automatic speech classification to five emotional states based on gender information. Proceedings of the 12th European signal processing conference (EUSIPCO'2004). EUSIPCO'2004, Austria, pp 341–344
72. Vlasenko B, Schuller B, Wendemuth A, Rigoll G (2007) Combining frame and turn-level information for robust recognition of emotions within speech. Proceedings of INTERSPEECH'2007. ISCA, Antwerp, pp 2249–2252
73. Vlasenko B, Schuller B, Wendemuth A, Rigoll G (2007) Frame versus turn-level: emotion recognition from speech considering static and dynamic processing. In Paiva A, Prada R, Picard RW (eds) Affective computing and intelligent interaction, vol LNCS 4738. Springer, Berlin Heidelberg, pp 139–147
74. Wundt WM (1922/1863) Vorlesungen über die Menschen- und Tierseele. L.Voss, Leipzig
75. Yang Y-H, Lin Y-C, Su Y-F, Chen H (2007) Music emotion classification: a regression approach. Proceedings of IEEE international conference on multimedia and expo (ICME'2007). IEEE, Beijing, pp 208–211

Ethical and Technical Aspects of Emotions to Create Empathy in Medical Machines

Jordi Vallverdú and David Casacuberta

Abstract This chapter analyzes the ethical challenges in healthcare when introducing medical machines able to understand and mimic human emotions. Artificial emotions is still an emergent field in artificial intelligence, so we devote some space in this paper in order to explain what they are and how we can have an machine able to recognize and mimic basic emotions. We argue that empathy is the key emotion in healthcare contexts. We discuss what empathy is and how it can be modeled to include it in a medical machine. We consider types of medical machines (telemedicine, care robots and mobile apps), and describe the main machines that are in use and offer some predictions about what the near future may bring. The main ethical problems we consider in machine medical ethics are: privacy violations (due to online patient databases), how to deal with error and responsibility concerning machine decisions and actions, social inequality (as a result of people being removed from an e-healthcare system), and how to build trust between machines, patients, and medical professionals.

1 Ethical Issues in Medical Machines

1.1 Data Privacy

Privacy is one of our key rights, and probably the one that is facing more problems under our digital revolution. Privacy is different from *secretism*: we shouldn't think that people worried about third party access to their medical records have

J. Vallverdú (✉) · D. Casacuberta
Philosophy Department, Universitat Autònoma de Barcelona, Barcelona, Spain
e-mail: jordi.vallverdu@uab.cat

D. Casacuberta
e-mail: david.casacuberta@uab.cat

S.P. van Rysewyk and M. Pontier (eds.), *Machine Medical Ethics*,
Intelligent Systems, Control and Automation: Science and Engineering 74,
DOI 10.1007/978-3-319-08108-3_20

anything to hide; they are just exercising their right to privacy. When considering how artificial intelligence (AI) based medical machines can affect patients' privacy, there are three main situations to analyze: the development of the machine, the training period, and the performance.

When developing a software machine, in order to have a system which is reliable and based on real situations and real people, it is fair to suppose that the design will use data about medical conditions from hundreds, or even thousands, of patients. This is an example of the Big Data paradigm. Big Data involves the collection of sensitive information from real people and gifting it to third parties that are not directly related to the health system.

During the training and performance period, it is reasonable to expect that the AI machine will collect data about the people under its care. In training, the central learning task is to understand more about the specific needs and even the personality of the patient it has to supervise. In the performance phase, the system collects health information both for a human physician that might track how the illness is developing in the patient and in order to improve its own performance.

Is this type of data collection safe from an ethical point of view? The answer is tricky. Let's consider first one-on-one processing of private information in the training and performance period. The first problem to consider is how the user might feel when knowing he or she is under surveillance. Besides the intrinsic negative feeling of being watched, there is also the problem of how the patient may start to consider what to do and how, feeling apprehensive about what the people behind the project might think [1]. Besides, as stated in Reeves and Nass (1996), people tend to interact with software and computers as if they were real people, even when they know they are not. Depending on the personality of the patient, this may be helpful in order to comply with a doctor's orders, but it can also backfire.

When considering how the AI machine is developed, we enter the territory of Big Data. According to the McKinsey Report on Big Data [2], we can loosely define "big data" as a series of related technologies that can analyze datasets "whose size is beyond the ability of typical database software tools to capture, store, manage, and analyze." The idea is to put together dozens, even hundreds of different variables, and look for correlations and patterns in an automatic way. Big Data can have many different applications, from national security to stock markets and, of course, to health issues as well.

When considering relationships between Big Data and health, the question of privacy inevitably rises. A fundamental question is: Who is going to accept a third-party database which can store highly personal and distressing information about mental disorders, child abuse or sexually transmitted diseases? The usual answer is to point out that third-party researchers are not interested in the user name or address. But the problem remains.

Let's say that a person's username is: "11100AAAK". How can we assure that no-one will be able to check that database and find out the real identity of 11100AAAK? The conventional solution is to "anonymize" the data, removing identifying characteristics such as name, residential address and telephone number, from the medical database. That way the researchers have all the relevant information, but no personal information which could help to re-identify the patient. But this protective measure does not resolve the fundamental ethical problem described above. In order to make the research relevant, one needs some background information about the people in the database, like the neighborhood they live, their race, whether they have a university degree, and so on.

In the mid-nineties, the state of Massachusetts decided to release "anonymized" data of all hospital visits of public workers. The then Governor William Weld stated in a press conference that the privacy of the data was assured. Privacy researcher Latanya Sweeney got a copy of that data, filtered it using the voting census and was able, among other things, to locate records of the hospital visits of Governor Weld himself [3]. Later, Sweeney was able to demonstrate [4] that up to 87 % of Americans can be associated with their hospital records even if name, address, and other identifying information are removed from the registry, but date of birth, zip code, and census are not.

Another approach is called "differential privacy" in which researchers make queries against a "statistical database". Between the user and the original database there is a randomized algorithm which is supposed to assure that private information is no longer available. However, recent research points out that this system is unreliable. According to Bambauer et al. [5], either you obtain erroneous research results or unnecessary privacy protections (because there is too much complexity).

There is also the question of disclosure: What if computer criminals or unscrupulous detectives attack a research database to find sensitive information about possible new victims for blackmailing or scams? Related to disclosure we should also consider the lack of transparency problem: we know that hundreds of thousands of data points about us are being taken every day, which are used to make all sorts of decisions on health plans, and maybe, if they leaked to third parties, about getting life insurance or a new job. However, we don't know exactly what these data are, or who has access to them. Wu [1] also indicates the problem of discrimination: how people might be treated differently depending on the type of data we have about them. We'll discuss these problems in the section devoted to social inequality.

1.2 Error and Responsibility

Machines fail: early morning, the toaster burns the slice of bread, the Arianne 5 rocket explodes 40 s after initial flight sequence (due to software error caused during execution of data conversion from 64-bit floating point to 16-bit signed integer value), your car isn't starting, missiles hit the wrong place,[1] mobile phones

[1] For a list of important big software bugs, see WIRED paper at: http://www.wired.com/software/coolapps/news/2005/11/69355?currentPage=all. For engineering errors, see Henry Petrosky: To Engineer Is Human: The Role of Failure in Successful Design (1992); Design Paradigms: Case Histories of Error and Judgment in Engineering (1994); To Forgive Design: Understanding Failure (2012).

collapse, and so on. Such machine malfunctions are very common, because attrition and complexity affects machines. In some special cases, like planes, trains or weapons, these errors are highly supervised, and precise protocols have been designed to control or minimize them.

Until very recently, the only robots or machines that could be involved in healthcare procedures were those in science fiction movies or books, and consequently this question was not a priority. Certainly, nowadays there are some classic technologies, like artificial organs ("classic" artificial hearts like Syncardia or more modern artificial hearts without pulse as made by doctors Bud Frazier and Billy Cohn), cochlear implants, advanced prostheses, and many others artificial devices that have a direct impact on human health. But we intend to discuss a different technology: *intelligent autonomous machines* (IAMs) that make independent decisions concerning human health. By the virtue of this capacity, they can also inflict harm. Three serious problems arise from this technology: first, can we trust them? Secondly, who is responsible for their functioning? Finally, due to the complexity of medical decisions, can they show moral responsibility?

Two questions should be immediately asked about such intelligent autonomous robots and responsibility:

- Are robots autonomous agents responsible for their acts? (in the same way that a human is, neglecting the fact that human actions occur in social environments and that responsibility is distributed among a hierarchy of subjects)
- Is the company that produced the robot responsible for its actions?

Clearly, robots belong to the second situation; they are *quasi*-agents, and their producers are responsible for their correct functioning, under normal conditions. The level of malfunctioning (e.g., software problems, mechanical disturbances, incorrectly implemented modeling processes) can be traced and ascribed to several sources, always external (from an identity perspective).

1.3 Social Inequality (Rich vs. Poor in E-Health Access)

When considering the ethical implications of IAMs, another key issue is the digital divide. Most research concerned with the digital divide tends to focus on the issue of access: Whether you can get your hands on a computer or not. Unfortunately, the ultimate beneficiary being targeted is all too often ill-defined. Generally, they are of two types: either disabled individuals, or individuals without digital literacy skills [6]. The blanket approach taken for all these groups is to organize practical courses wherein these diverse individuals are taught how to surf the net, e-mail, and so on.

However, mental barriers are as important as simple access. Giving people a computer and an internet connection is not enough to help them to use digital technologies in a profitable way, especially in health contexts. Major causal factors of marginalization from the information society are those such as mistrust often felt towards new technologies and the lack of any content attractive or useful to either the socially excluded or those at risk of being so. Instead, we should consider an approach based on empowerment [6, 7].

We propose that what is important is not merely knowing how to use, for instance, a web browser or email, but rather the educational and liberating potential of new technologies. Slowly, we are advancing, but still, we can see big differences regarding who is using e-health and who isn't. For example, in a recent paper, [8] showed that when using a portal to help people with diabetes, having access to computer was not the main relevant element, but mental barriers on how to use computers for empowerment. According to the authors: "those most at risk for poor diabetes outcomes may fall further behind as health systems increasingly rely on the internet and limit current modes of access and communication" [8, p. 321].

When we consider the intersection between digital exclusion and Big Data, the divide grows thicker. IAMs are going to be developed and trained using databases obtained with big data methods. So, only data from computer savvy users will be considered, thereby excluding people that do not use social networks. According to Lerman [9], "Billions of people worldwide remain on big data's periphery. Their information is not regularly collected or analyzed, because they do not routinely engage in activities that big data is designed to capture. Consequently, their preferences and needs risk being routinely ignored" [9, p. 60]. The exclusion problem also links to an epistemological problem: if relevant people are left out of the study, then the analysis will be based on data which could well be partial, or in the worst case scenario, plain wrong.

What should be done? The benefits obtainable from big data and e-health are many. However, while big data is open to abuse and misuse, the benefits clearly outnumber possible harms. Following Polonetsky and Tene (2013), we ought to view the problem of big data not from a Kantian perspective, based on unconditional rules or maxims, but more from a consequentialist point of view, considering generic benefits and harms. Polonetsky and Tene (2013) use a more economical approach, based on utility, and point out that one can see benefits for the individual (e.g., individuals in idiosyncratic conditions will benefit from massive studies producing big data, which will include other people like them and help to find common patterns to be cured), specific communities, and organizations which generate economic benefits from the research, and finally society as a whole.

However, it is important to build machines that generate trust; otherwise, benefits to human beings will not obtain. A patient is not going to give her doctor personal information if she thinks that is going to be stored in an online database open to unknown third-parties [10]. We certainly need to consider our human rights regime while analyzing the effect of digital technologies on our privacy. However, there is no need to create new rights as some people like Barlow [11], Hayles [12] or Gray [13] argue. Following Casacuberta and Senges [14], we only need to consider some inarticulate elements which are either missing or overdeveloped.

What type of right is privacy? Our proposal is that if privacy is moved from the real world to the Internet we may thereby neglect something essential and important. Because most rights depend on context, we can evaluate a right only as long as we consider its context. In life before the Internet, context was so ubiquitous that it was taken for granted. Perry [15] defines this context as an "inarticulate element": some essential information doesn't manifest directly in communication because it is tacitly assumed by the conversants.

A common example is time zones. Most of our communications share a common space, friends and family that are close by, so when arranging a meeting we don't specify the time zone. However, if we are talking to people abroad, and want to arrange an intercontinental chat via Skype, we need to specify the typically unarticulated element of time zone.

Similarly, when we talk about privacy in online databases and AI machines, we have to consider the inarticulate elements that surround our everyday conversations about privacy. Following Casacuberta and Senges [14], these are the relevant elements:

1. How is identification achieved? Anonymity is the default in the real world, but identification is the norm in cyberspace, and we need to create strong controls and procedures to generate anonymity.
2. Who is listening? In the real world, usually no one. So we don't use cryptography to talk to our spouse at home. In the digital world, it is not only that anyone could be listening; it is the fact that we are leaving foot-prints whenever we write or click online.
3. What is the object to be protected? In a world without big online databases a researcher that obtains a de-identified bunch of records which include some embarrassing illness about us could not find out who are we. The work of Latanya Sweeney has proven that this unarticulated element doesn't hold in cyberspace.

So, contrary to Brin [16], the main reason we believe that privacy is obsolete is not technology per se, but that the traditional conditions no longer hold. Simply because they are inarticulate makes this fact difficult to realize. But we have to be very conscious of such inarticulate elements when creating e-health apps or autonomous AI agents.

1.4 Machine Cheating as Blind Affective Computing?

When computational devices were first implemented, there emerged a natural worry: should these machines interact emotionally with their users? If humans are social as well as emotionally wired, machines with the skill to show some of these properties would be welcome by their human users and, therefore, their task performance would improve. This was the beginning of the affective computing research program [17]. Since then, huge progress has been achieved in the field. In most cases, the idea was that machines should be able to detect and recognize human emotional states and provide good (and convincing) feedback, thus creating a satisfactory interactive framework. This means that computers (or robots in the case of Human-Robot Interaction, HRI) show emotional activity. But do they really have emotional architectures or just convincingly mimic emotional actions?

We do not affirm that machines must feel the emotions they perform; instead, we claim that machine emotional responses must *follow* or accord with a human-like emotional structure. For example, we don't change continuously from panic to joy in 1 ms. At the same time, we need to keep in mind that the principle that governs human emotion is empathy. This makes it possible for human beings to understand what is happening to other human beings and, in some cases (socially/culturally/personally mediated), to react to such cases. The best affective machines should be able not only to understand human states, but also to drive their behavior through this emotional syntax and semantics. By "human emotional semantics", we mean the channels by which emotions are meaningfully expressed (eye gaze, body movements, voice tone, heart rate, facial expressions), and by "human emotional syntax", we refer to the dynamics of change among emotions.

It could be argued against us that this suggestion takes us back to the symbol-system hypothesis paradigm of AI (GOFAI), just applied to emotion, and that embodied robotics or behavior-based robotics (like the hat of Rodney Brooks, pioneer of the field, or Rolf Pfeiffer, today) is therefore neglected. But this is not true: we are defending machine-software architectures in which emotional architectures are embedded to naturally guide the system according to the natural interaction with humans (if they do not modify their usual ways to do things). Following Vallverdú [18]:

Emotions cannot be something that should be embedded into a pre-existing artificial device, but they must be embodied into them. Emotions are at the same time the own body and the informational process of the body sensors. Our bodily structure makes us feel the world under a specific meaning route, the emotional one. Embodying emotions, that is the real task, horizon and goal of the whole research field. At the end, we need to create feeling machines. Yes, machine learning, integration of perceptive data, data categorization, action and goal selection or communication among agents, among others, are very important and unsolved problems of contemporary AI. Because it is impossible to solving the whole problem of human cognition, we've fragmented it into small pieces, made it more tractable, trying to achieve easier solutions to them. And this is the actual situation of AI research, necessary but unsatisfactory. The convergence of all these fields led to the artificial conscious existence, something for which will be absolutely necessary for synthetic emotions [18, p. 4].

So, a second wave of affective computing innovation is necessary, one that integrates human emotional architectures into their basic design and, thus, can offer a richer and deeper interaction with humans.

1.5 Trust in E-Health Autonomous or Semi-Supervised Systems

One of the hottest topics in the field of human-machine interaction is the issue about the responsibility of autonomous machines during shared activities. Must robots, machines or software with several degrees of autonomy (supervised or not unsupervised) be held responsible for possible errors? Who is in charge of their functioning?

There is an additional problem that must be faced: which mechanisms mediate the trust and confidence between humans and IAMs? Due to the vast range of possible e-health and care robotic platforms and designs, it is difficult to define a unique scenario or answer, but it can be stated that emotional aspects are deeply involved in this process. Several emotions are involved in a medical procedure, at least for patients:

- Fear of procedures and possible pain (by techniques, tests) or unknown consequences
- Trust/distrust of medical personnel
- Insecurity about a situation that cannot be controlled or understood
- Anxiety about medical privacy and the perception of others of the "new (ill) self"
- Compassion towards others suffering or fear
- Tenderness towards other affected persons
- Shame about feeling oneself unhealthy or non-autonomous
- Embarrassment for being helped in "private" activities (e.g., bathing[2]).

We can even make a more surprising statement: the same emotion in the same medical circumstance is shown differently by different people. Not only by the degree of the intensity (from apprehension to fear and later to panic or terror), but also by the control and modulation of the emotion itself. For example, Zborowski [19] demonstrated that pain symptomatic of the same illness was processed very differently by patients according their belonging to a cultural community within the USA (italo-American, old-American, Irish-American, Jews). Zborowski showed that older adult Americans were generally "stronger" (they did not publicly show their pain reactions) and had a greater tendency to avoid social contact when experiencing pain, compared with the others. Personal feelings are not only biologically hardwired, but also culturally determined. Risk perception and understanding of scientific procedures is again a cultural process that assumes different values at several niches in which we can study it. Armstrong et al. [20] found significant results with racial/ethnic differences in physician distrust in the United States. In univariate analyses, Blacks and Hispanics reported higher levels of physician distrust than did Whites. However, multivariate analyses suggested a complex interaction among socio-demographic variables, city of residence, race/ethnicity, and distrust of physician. This also forewarns us towards data analysis models and the inferences we can obtain from them.

Consequently, we need to consider the crucial role of emotions in medical practice and the necessity of creating empathic bonds between doctor/care-robots and human patients. Empathy is an emotion that one of us has studied in HRI situations and preliminary results offer good encouragement to this field: empathic

[2] Cody robot has been partially designed to avoid this and to help nurses. See: http://www.coe.gatech.edu/content/robotic-cody-learns-bathe.

predisposition towards robots can be easily created if there is good previous work with their human users [21]. This prone-to-empathy attitude depends on two main variables:

(a) psychological perception of the machine, and
(b) physical shape and signals (sound, color, temperature, movement) of the robot.

The first condition depends on how a person projects his empathy towards an external object different from self; it is closer to a mental construction than to a real interaction, and may have a strong neural correlate. Recent studies have correlated the apparently neural basis of learning by mirror neurons [22, 23] and functional magnetic resonance imaging (fMRI) has been intensively employed to investigate the functional anatomy of empathy (for reviews, see [24–26, 27]).

Recently, some research has addressed the neural correlates of empathy with robots (e.g., [28]). Empathy is a fundamental trigger of emotional attitudes and responses, contributing towards understanding of others and, thus, facilitating learning by imitation (not only gesture learning, but also social learning).

The emotional basis of empathy makes possible self-learning by imitation. Therefore, if we want to create better and smarter intelligent autonomous machines, we must put emotions into them, because emotions are the basic language among human beings. If emotional interaction is basic for social human activities, then any robot study that implies an interaction between the robot and humans must take empathy into account [21].

The second condition stipulates cultural influence, and is relevant to the construction and design of the robots. When considering human-robot interaction we are considering the interaction of two physical bodies (natural and machine, respectively). In Mori (1970), Masahiro Mori introduced the notion of uncanny valley in his discussion of human emotional responses to robots ("Bukimi No Tani", translated as "The Uncanny Valley"). Mori found that human comfort with a robot increases until it very closely resembles the human shape. At this point, humans report a strong feeling to reject the robot. In this sense, a humanoid can be seen as "nice" or "creepy" according to its design, but human-like performance is not always welcome by humans. For example, the Geminoids of Hiroshi Ishiguro can be seen as unconvincing bad copies, or as true humans according to the kind of action they perform. Subtle robot details such as eye-gaze, head and neck movements, tongue presence (or not), lip formation, natural eye blinking can make human users react more positively to the machine. Less human-like but still comfortable robot designs can avoid these problems (Mori 1970).

There is one last question to ask: cultural attitudes towards robots also change [29, 30], although the conceptual design is the same for eastern and Western robotic experts [31]. Care-robotics companies will have to take in account not only how to construct their robots, but also how to introduce their products into the market.

2 Technical Aspects of Medical Machines

2.1 Existing Health Machines: An Overall Taxonomy

From the European Commission on Robotics for Healthcare,[3] we can extract and map basic medical machines, summarized in 6 subfields: (a) Smart medical capsules, (b) Robotized surgery, (c) Intelligent prosthetics, (d) Robotized motor coordination analysis and therapy, (e) Robot-assisted mental, cognitive and social therapy (f) Robotized patient monitoring systems.

Currently, there are several companies and university laboratories devoted to healthcare robotics: Argo Medical Technology, Cyberdyne, Ekso Bionics, Hansen Medical, Hocoma, Intuitive Surgical, iWalk, Kinea Design, MAKO Surgical, Mazor Robotics, Motion Control Inc., Myomo, Ossur, Otto Bock Healthcare, Parker/Hannifin, Rex Bionics, RLS Steeper, SPRINT, Tibion, Tital Medical. Touch Bionics, Victhom Human Bionics, among others. Most of them are Japanese or American. Surely, their number will increase in the next decades because the economic benefits of their implementation in wealthy first-world countries will be huge, and even more in an increasingly aged population.[4] Artificial organs, exoskeletons, care robots, surgical training robots, telepresence robots (iRobot: telepresence robot RP-Vita,[5] VGo Pediatric telepresence,[6] Fraunhofer institute: Care-O-Bot, Double robotics and his double telepresence robot[7]) and other machines (as therapeutic pets like PARO[8]) are increasingly being produced and implemented into health services. Several organizations have been formed, such as the International society of Medical robotics) or the Minimally Invasive Robotic Association), among others. Specialized medical robotics departments have been created all around the world. This research field is young, but strong, despite some recent chirurgical problems with a Vinci robot surveyed by FDA (November 2013).[9]

This is the result of the growth of power and utility of these machines as well as to the advances held in computer sciences that led in 1986 to the creation of The Society for Artificial Intelligence in Medicine (AIME), established with two main

---

[3] http://rehacare.messe-dus.co.jp/fileadmin/files/EuropCommission_ny_robotics_for_healthcare.pdf

[4] See http://www.worldrobotics.org/uploads/media/Executive_Summary_WR_2013.pdf; http://www.roboticsbusinessreview.com/research/report/outlook_for_health_care_robotics_for_2013. The European project SILVER is another good example of deep investment in this direction (www.silverpcp.eu).

[5] http://www.irobot.com/us/learn/commercial/rpvita.aspx

[6] http://www.vgocom.com/health-it-promises-new-paradigm-patient-care

[7] http://www.doublerobotics.com/. The 2009 Hollywood film "Surrogates" is not so far from this robot implementation, except for the grade of human likeness of the service robots.

[8] http://www.parorobots.com/

[9] http://www.fda.gov/downloads/MedicalDevices/ProductsandMedicalProcedures/SurgeryandLifeSupport/ComputerAssistedRoboticSurgicalSystems/UCM374095.pdf

goals: (a) to foster fundamental and applied research in the application of Artificial Intelligence (AI) techniques to medical care and medical research, and (b) to provide a forum for reporting significant results achieved at biennial conferences.

2.2 Future Devices

2.2.1 Apps for Mobile Apps and E-Health

The potential for mobile apps within the health context is enormous. There are already 96,000 health apps in the market, from fitness tracking to weight loss motivator programs or support networks. Prospective studies report a market for mobile apps on e-health valued around US$26 billion by 2017 [32]. Both present and future apps can be used in many different contexts and situations, which we can roughly organized as follows:

1. Assessment of patient health by use of sensors.
2. Storage of patient health states in a database, for further use by a medical team or the patient.
3. Real time prediction: 24 h monitoring to make a diagnosis and propose a remedy.
4. Offline prediction: same as (3) but the AI machine doesn't use currently available data, but also all the data about the patient that has been stored on a database.
5. Epidemiological studies using data gathered by smart phones of thousands of people.
6. Research procedures. Any of the items above, used as research tools.

Why are mobile apps so relevant in medicine? There are several reasons. First, portability, as we carry our cell phones everywhere. Also consider the 6.8 billion people connected with a mobile subscription worldwide (MobiThinking 2013). Those gigantic numbers are relevant for both research projects (like epidemiologic studies) as well as for peer to peer networks. According to Fox [33], 23 % of people with chronic illness use social networks to find other patients with a similar sickness in order to get help and support from them.

Nevertheless, the key element is how the latest generation of smart phones which incorporate sensors: microphones, cameras, accelerometers and location detectors, among others, which can be hacked for medical purposes. For example, a camera to report online dermatologic problems, or microphones to analyze how a person is breathing. Sensors are important, as doctors tend to be cautious about oral or written reports delivered by patients [34]. However, according to Topol [35], if the system is automatic and generated by reliable and trustworthy sensors and algorithms, doctor worries may diminish.

Here are some examples. ginger.io is an app designed at the MIT which gathers medical data from thousands of students and analyses its relevance in order to create a more inclusive "walls free" healthcare system. Using Big Data techniques, the app searches for behavior patterns to assess the patient's condition and sends an alert to a clinician if necessary [34]. Following the Big Data approach, the journal Science published a paper [36] describing how in tracking cell phone data from 15 million people in Kenya, they were able to discover relevant facts on how traveling patterns contribute to the spreading of malaria.

But probably the Holy Grail is a tricoder, which includes specific sensors and software to rapidly diagnose a disease, without relying on a doctor or a nurse. A current example of such a tricoder device is Scanadu, a hand-held unit that communicates with the smartphone and sends information about blood oxygenation, pulse transit or heart rate [37]. Such devices could be of great importance in Third World countries, where a great deal of people live in rural areas and might be very far away from hospitals. Also, the fact that there are a lot less doctors and nurses there: around two doctors per 10,000 people in Africa [38].

Other apps base their efficiency on reducing costs. Adding a specifically designed sensor to a mobile phone helps to bring costs down. For example, there is an attachment for the mobile phone called CellScope which converts it into a microscope. In medicine, CellScope is used to look for pathogens or for retinal scans and there are fifteen prototypes tested in Vietnam clinics [39]. Also, there is MobiSante, which has designed a specific device to turn a smartphone into an ultrasound probe that generates high resolution images, costing only US$7,500, just a fraction of the price of a conventional ultrasound [40].

2.2.2 Making Apps Privacy Friendly

Following Oram [41], we recommend app developers consider the following in order to make their apps better suited to protect customer privacy.

1. Downloadable data will make users trust the system more, knowing what exactly the app is recording about them. It can also help them to reflect on their own health situation a little better.
2. Let the patient know to whom you are delivering the data. Patients don't like their data to be sent to third parties, especially for market research purposes, even if it has been anonymized.
3. If giving data to third parties is inevitable, allow the patient to opt out of revealing particular items, like hiding information about former mental illness or sexually transmitted diseases. This is called data segmentation, and makes app design and development a little more complex, but will increase the trust of the patients using it.
4. Easy navigable interfaces, especially for the data segmentation processes. If the patients don't clearly see which information is in and which is out, they won't trust the app.
5. Encrypt data, because encrypted data is everywhere. Hacking is common these days, and no operative system or app is free of exploitable bugs. The app may be a little slower, but it is a lot more trustable.

2.2.3 Telemedicine and Care Robotics

Although hardware and software systems are widely researched and have even achieved industrial production, this is a field in which the first big wave of implementation has still not occurred. Organizations like the ATA (American Telemedicine Association) have made serious investments into these new ways of performing telemedicine. The investment has resulted in new fields:

1. Remote Monitoring and Home Telehealth Systems.
2. Mobile Health (mHealth). Wireless technology, smart phones and health apps that are making healthcare a mobile 24/7 service. ATA 2013 hosts the leading mHealth companies, showcasing new hardware and infrastructure for mobile health, along with the next big "killer apps."
3. Outsourced Clinical Service Providers.
4. Videoconferencing. Hardware, software and infrastructure for real-time video calls and telepresentating.
5. Health Information Technology. HIT achieves true "meaningful use" when it is used with telemedicine to deliver quality healthcare.
6. Teleradiology. Off-site solutions for 24/7 radiological consultations. The ATA 2013 trade show has multiple exhibitors providing teleradiological services.
7. Hardware, Software and Equipment. Everything needed to do telemedicine, in one huge showroom.

Consequently, any university making significant investment in robotics has research labs working on healthcare robotics. Private companies, frequently spin-offs from university labs, are also producing this new wave of assistance robots. In the USA, the NSF has created a joint project of leading universities called "Socially Assistive Robotics", which aims to create fundamental computational techniques that enable the design, implementation, and evaluation of robots that encourage social, emotional and cognitive growth in children, including those with social or cognitive deficits. At the same time, "first-world" societies (especially Japan) are aging, and care or industrial robotics are regarded as necessary. According to Iida,[10] the number of Japanese aged 65 or above is expected to jump by around 7.09 million between 2010 and 2025, when they will account for 30 % of the overall population, up from 23 % in the base year. Consequently, some 2.32 million to 2.44 million caregivers will be required to look after them, up more than 1.5 times from the 2010 level.

But the elder care sector has a high job turnover rate, partly due to low pay. In addition, about 70 % of elder caregivers are said to experience back pains due to constantly lifting elderly between beds and wheelchairs as well as helping them take baths and perform other daily activities. Back pain could be solved if caretakers were using exoskeletons as well as assistant robots to help into their tasks.

[10] http://www.japantimes.co.jp/news/2013/06/19/national/robot-niche-expands-in-senior-care/#.Ul-YONLIXwk

Taking statistics from the International Federation of Robotics,[11] and considering that since 2007 an ISO working group is still revising the ISO 8373 which finally will include an official definition of service robots (this definition is still not clear but is generally accepted), the number of such robots is increasing, although not as much as industrial robots. This is a trend that will only increase, demanding special emotional studies of HRI.

3 Emotions, Humans and Care Machines

3.1 How Does Mammalian Empathy Work?

In this third and final section we want to argue why empathy is the most significant emotion IAMs in medicine can have. According to Waal and Thompson [42], we can view empathy as an emotion that includes the ability to connect with others and to become affected by their emotions. This affection can be behavioral, emotional and cognitive. To have empathy is not black or white. Following Waal and Thompson [42] we can distinguish four levels of empathy: motor contagion, imaginative transposition, perspective taking and moral empathy. Motor contagion is based on visual cues from gestures and auditory experiences. It is a basic, probably hardwired mechanism. Imaginative transposition is what we commonly call "to put oneself in someone else's shoes to feel like that person". In perspective-taking, our thoughts and beliefs process how the other person may see a situation and what view she has of us. Finally, moral empathy implies viewing the person as a moral agent, a sentient being that has the capacity to feel and suffer.

The first level is already amendable by means of pattern recognition algorithms which would allow an autonomous agent to recognize the patient emotional state by means of a camera and/or microphone [43, 44]. New eHealth working environments ask for new competencies required by healthcare professionals working in home care, with eHealth technologies such as remote telecare and ambient assisted living (AAL), mobile health, and fall detection systems. The second empathy level implies basic reasoning, based on the general context of the patient to understand feelings by means of transposition. The third level adds to the system folk psychology [45] so it can reinterpret patient psychological states by means of an expert system that is able to reason within the context of beliefs, desires, plans and motivations [46]. The third and fourth levels are more elaborate, and imply still unsolved problems in the field of artificial intelligence.[12] We can expect that as AI evolves and more accurate and powerful machine models of emotions develop, the third and fourth levels will also be achievable in the future by machines.

In order to have empathy in medical contexts four components are required: affective sharing based on perception-action coupling, self-other awareness, mental

---

[11] http://www.ifr.org/industrial-robots/statistics/

[12] See the 2012 official report http://www.healthit.gov/sites/default/files/final_report_building_better_consumer_ehealth.pdf.

flexibility to adopt other people's perspective, and emotion regulation. Posture is also important in generating emotions and empathy, according to Niedenthal [47]. Besides affecting our own emotions, our body posture informs to our close interacting persons about our feelings, so it would be very important for autonomous AI agents both to present postures that generate good mood (if they are anthropomorphic) and, more importantly, to recognize such postures as indicators of how one is feeling.[13] For example, depression makes people adopt poor posture, angling the body away from others and performing anxious behaviours, like tightly crossing the arms. Cultural differences among tele-patient population should also be analyzed.[14] Imitation of body behaviour is also central in understanding empathy. Again, following Waal and Thompson [42], emotional expressions and gestures are visibly imitated by observers and this imitation is accompanied by self-reports of associated emotional states. When emotional gestures are properly imitated, there is a strong foundation for empathy. This has been experimentally tested by artist and scientist Naoko Tosa in her projects Neurobaby [48] and Unconscious Flow [49]. Neurobaby was one of the first projects to study emotional expressions synthetically using a small baby. Unconscious Flow translated empathic expressions to a virtual world in which two images of moving sirens were projected, expressed in movement emotions human users were reluctant to show.

3.2 Studies on Robot Empathy

Several studies have tried to analyze possible empathy relationships between humans and robots (from humans to robots, to be precise). Many of these studies have been published in the field of HRI and have illuminated different areas of the study, as follows below.

3.2.1 Empathy and Robotic Anthropomorphism

Although humans easily empathically bond to entities like pets, tools, buildings or symbolic objects, basic empathic relationships are created among humans. Friedman et al. [50] found that owners projected towards AIBO robots conceptions of technological essences (75 %), life-like essences (49 %), mental states (60 %), and social rapport (59 %). However, participants seldom attributed moral standing to AIBO (e.g., that

[13] This would follow normal psychological rules implemented In normal health systems. See for example official guides like: http://www.cms.gov/Outreach-and-Education/Medicare-Learning-Network-MLN/MLNProducts/Downloads/Communicating_With_Patients_Fact-_Sheet_ICN908063.pdf or http://www.rcpsych.ac.uk/files/pdfversion/cr108.pdf.

[14] For example, the University of Washington, Medical Central provides several examples following this special necessity: with Korean patients see http://depts.washington.edu/pfes/PDFs/KoreanCultureClue.pdf, http://depts.washington.edu/pfes/PDFs/DeafCultureClue.pdf. They also call them 'cultural clues' http://depts.washington.edu/pfes/CultureClues.htm.

AIBO deserved respect, had rights, or could be held morally accountable for action) [50]. This is not strange: humans make the same ascriptions to real dogs or cats. This is the evolutionary meaning of empathy: human imitation. The other uses emerged as extended ways to apply empathy in the same way that language evolved from mere alert sounds to an artificial way to tag the world with concepts. Nevertheless, humans look more comfortable with anthropomorphic objects (if not under Uncanny valley effect) [51]. But even in this case, that is, of accepting anthropomorphization [52], there is no agreement on which elements are necessary to make it possible: body gestures, size, sounds, temperature, or facial emotions? Some studies from a neuroanthropological perspective (e.g., [53]) provide a richer framework from which to approach anthropomorphic robots. Pereira et al. [54] and Tapus and Mataric [55] also use empathy as a way to improve HRI, making possible "friendships" between machines and robots.

3.2.2 Military Robots

Carpenter [56] interviewed multiple soldiers based in the Explosive Ordenance Disposal Unit about their work operating robots that diffuse or locate bombs. These operators often talked to their robots, occasionally gave then names and/or wrote on their chasses. It is not so strange: during World War Two, pilots painted their planes and baptized them with a last list of names (usually of girl-friends or pin-ups). This military practice is called "nose-up." When the robots were disabled beyond repair, they felt "anger, frustration and sadness for a sense of loss". Again, the robots were often named after a girlfriend or wife. Robots (most of them PackBot or TALON models) and humans posed together in photos and a mock funeral was even performed for a destroyed robot. Some new studies are being performed because emotional bonds between machines and soldiers are not useful in combat areas: if these machines are just pieces of iron, human morale is not affected by robot loses, but if some empathic relationship has been established between human soldiers and military robots, then the initial psychological advantage disappears. At the same time, not only are robots cheaper if you consider humans costs throughout the processes of recruitment, training, deployment costs, care services in case of injuries, but they are also free from social worries (human casualties in war appear every day at TV news and mass media, and social networks). Thus, extended anthropomorphization is a clear problem for basic military robotic purposes.

3.2.3 Neuroscience and Psychology

In the first fMRI study on how humans express empathy for a robot (a tortured and vexed Pleo-toy dinosaur-robot), Rosenthal-von der Pütten [57][15] conclude: "One limitation of the study is that we used videos in our experiment and that participants saw

[15] The official video is here: http://www.youtube.com/watch?v=wAVtkh0mL20&feature=player_detailpage.

potentially fictional material. Perhaps reactions would have been different if the torture of Pleo had been performed live in front of the participants. Moreover, the robot's shape is also rather fictional as dinosaurs are not part of our daily experiences. Participants might react more strongly to a robot with a humanoid shape" [57, p. 30]. Other studies have focused on anthropomorphic variables of robots design [58].

3.2.4 HRI and Emotions

Vallverdú et al. [21] investigated the induction of empathy towards robots in humans. The study performed an experiment in a WOZ[16] environment in which different subjects ($n = 17$) interacted with the same robot but believing that it was a different version for each group of participants. The main hypothesis of the study was to elucidate if empathic responses could be modeled in humans in HRI. Results were obtained from questionnaires and multi-angle video recordings. The results of Vallverdú et al. [21] tended to be homogeneously distributed within each Scenario Group, with evident differences among Scenarios. This reinforces the initial hypothesis: HRI empathy can be modeled in human beings merely by creating an empathetic (and fake) atmosphere with the robot. Becker-Asano et al. [60] have also started to work in this area with different emotions (fear, pain, sadness, surprise) and also emotion behaviors like laughter.

3.3 Detecting and Interacting with Emotions: The Emocibernetic Loop

In the section above about apps, we introduced the idea to gather data about the patient using several sensors which inform about key personal health data. Most of those sensors can be used to establish emotional patterns in the patient. Microphones are helpful in detecting emotions in patients, by measuring breath rhythm; there are skin conductivity sensors (popularly known as "lie detectors") and sensors to measure cardiac variables. Thus, one can have an approximate idea of the type of emotion a person is feeling, especially whether it is a positive or negative one.

Since Ekman et al. [61] we know that there are specific arousal patterns for each of the 6 basic emotions he described and considered *universal across all cultures* (*anger, disgust, fear, happiness, sadness and surprise*), and with the proper sensors, you can distinguish between them in a reliable form [62–64]. Facial expressions are also fundamental for the correct interpretation of emotions.

---

[16] WOZ (Wizard of OZ) experiments are experiments with teleoperated robots. I performed some of these experiments at the Nishidalab Kyoto University, http://www.aaai.org/Papers/Workshops/2007/WS-07-07/WS07-07-008.pdf. More on WOZ in Riek [59].

According to Ekman and Friesen [65] and Ekman [66], they present transcultural patterns, so they are a reliable guide to decide which emotion a person is feeling independently of his or her cultural background. Even if the subject tries to hide an emotion, or pretend to have a different one, based on what [67] calls micro expressions, it is always possible to check the real emotion a person is feeling. Similar phenomena can be obtained by analyzing a person's voice. When considering IAMs, the method of facial expressions, even if it less reliable than the use of sensors, is a lot less invasive and more human like. Humans are used to finding out the emotion in another person based on her face and voice.

Therefore, we propose the introduction of sensors only in cases of monitoring and assessing poor patient health, such as a weak heart, a very special situation in which the object under medical study is too sensitive to be inferred from powerful changes in the emotional tone, for example. Otherwise, cameras and microphones are reliable enough, a lot cheaper, and more importantly, a less intrusive for patients.

How should these measuring technologies be developed and delivered? There are two initial scenarios: personal apps and ambient intelligence. Ambient Intelligence is the term developed by the European Commission in 2001 to discuss the relevance of invisible computers, or how to move the computer away from the PC and distribute it in environmental surroundings. Thus, we have intelligent houses, buildings or even streets and whole cities [68]. Following Sadri [69], we can view ambient intelligence as follows:

Ambient Intelligence is the vision of a future in which environments support the people inhabiting them. This envisaged environment is unobtrusive, interconnected, adaptable, dynamic, embedded, and intelligent. In this vision the traditional computer input and output media disappear. Instead processors and sensors are integrated in everyday objects. [...] The envisioned environment is sensitive to the needs of its inhabitants, and capable of anticipating their needs and behavior. It is aware of their personal requirements and preferences, and interacts with people in a user-friendly way, even capable of expressing, recognizing and responding to emotion [69, 36: 1–2].

While we are still far away from imagining a "compassionate city" that has sensors everywhere and senses any person having a serious health problem, it is easy to imagine the development of such a system in the home of a person with a chronic illness. The user also sees information overload reduced, as he or she don't need to pay attention to all the relevant information about their health situation that sensors are routinely obtaining, and can adapt to changes in weather, temperature, development of a disease, patient emotion, or how many people are around and how they behave, and so on.

What are the main benefits of measuring patient emotions? We identify four main benefits:

1. Reducing patient frustration when using a complex interface or computer system. A good example of such technology is AutoTutor, an e-learning software that checks the user's main emotion in order to see if the person is following lectures or not, or is bored by them, and then changes the rhythm and the patterns in which information is presented [70].

2. Better understanding of how the patient relates to the disease and/or to the environment.
3. Detecting emotions is also relevant to diagnosis of disease, as well as future risk.
4. Improved online communication. As posed by Picard [63], asynchronic communications like email and text-based synchronic communications like chats lack a proper emotional channel. European projects like e-Sense or Sensei [69] are trying to develop systems and protocols for remote emoting, facilitating emotional communication in such environments.

The middle ground between apps and ambient intelligence is wearable computers. They are within the person, not an external technology one has to carry, but embedded in clothing, so they are less invasive and more easily transported. The field of wearable computing is still under development, but we can mention pilot projects like wearIT@work (also funded by the European Commission) which is an experimental wearable computing setup to be used under emergency situations such as earthquakes and fires. The system makes communication with coordinators and surveillance of basic medical situation of firemen, police, etc. easier. One of the key issues measured is emotion [71].

4 Conclusions

Our chapter claims that existing robotic health devices require empathy in HRI contexts, especially in medical environments. Our suggestions reinforce the necessity of working simultaneously on cultural, cognitive and technical aspects of healthcare robots in order to improve the creation of empathic bonds among such robots and human users/clients.

Emotional bonds between human and IAMs are not only the result of human-like communication protocols but also the outcome of a global trust process in which emotions are co-created. Consequently, we affirm that empathy can be induced due to sufficient emotional ambience, and machine design must be adjusted to practical necessities as well as to human needs (psychological and cultural).

We believe that medical IAMs are part of the future of society and of our wellbeing. They promise democratization of health, at lower cost and easier implementation. The fair distribution or at least improvement in very specific care/treatment areas in health processes may become reality thanks in part to IAMs, as we have tried to show. But from academia or industry to social implementation there is a long road.

Acknowledgments Financial support for this research was received from the Spanish Government's DGICYT research project: FFI2011-23238, "Innovation in scientific practice: cognitive approaches and their philosophical consequences". Most of this work was supported by the TECNOCOG research group (at UAB) on Cognition and Technological Environments, and has been developed by the SETE (Synthetic Emotions in Technological Environments) research group. The last part of the research on HRI and emotions was funded by the Japan society for the Promotion of Science (JSPS) at Nishidalab, University of Kyoto (Japan).

References

1. Wu FT (2013) Defining privacy and utility in data sets. U Colo L Rev 84:1117
2. McKinsey Global Institute (2011) Big data: the next frontier for innovation, competition, and productivity 1. Available at. http://www.mckinsey.com/insights/business_technology/big_data_the_next_frontier_for_innovation. Accessed 14 Nov 2013
3. Sweeney L (2013) K-anonymity: a model for protecting privacy. IEEE Secur Priv 10(5):557–570
4. Becker F, Sweeney B, Parsons K (2008) Ambulatory facility design and patients' perceptions of healthcare quality. Health Environ Res Des J 1(4):35–54
5. Bambauer JR, Muralidhar K, Sarathy R (2014) Fool's gold: an illustrated critique of differential privacy. Vanderbilt J Entertainment Technol Law (forthcoming)
6. Casacuberta D (2008) Digital inclusion: best practices from eLearning. E-learning Pap 16:60–74
7. Sen A (1999) Development as freedom. OUP, Oxford
8. Sarkar U, Karter AJ et al (2011) Social disparities in internet patient portal use in diabetes: evidence that the digital divide extends beyond access. J Am Med Inf Assoc 18:318–321
9. Lerman J (2013) Big data and its exclusions. Stanford Law Rev Online 55
10. Oram A (2012) Healthcare privacy discussed as an aspect of patient control. O'Reilly Radar. http://radar.oreilly.com/2012/06/health-care-privacy-discussed.html. Accessed 30 Sept 2013
11. Barlow J (1996) A declaration of the independence of cyberspace. http://homes.eff.org/~barlow/Declaration-Final.html. Accessed 15 Oct 2013
12. Hayles NK (1999) How we became posthuman. Chicago University Press, Chicago
13. Gray CH (2002) The cyborg citizen: politics in the posthuman age. Routledge, London
14. Casacuberta D, Senges M (2008) Do we need new rights in cyberspace? Discussing the case of how to define on-line privacy in an internet bill of rights. Enrahonar 40–41:99–111
15. Perry J (1993) The problem of essential indexical and other essays. OUP, Oxford
16. Brin D (1998) The transparent society. Basic Books, New York
17. Picard RW (1997) Affective computing. MIT Press, Cambridge (MA)
18. Vallverdú J (2012) Subsuming or embodying emotions? In: Vallverdú J (ed) Creating synthetic emotions through technological and robotic advancements. IGI Global Group, Hershey, pp 9–14
19. Zborowski M (1969) People in pain. Jossey-Bass, San Francisco
20. Armstrong K et al (2007) Racial/ethnic differences in physician distrust in the United States. Am J Public Health 97(7):1283–1289
21. Vallverdú J, Nishida T, Ohmoto Y, Moran S, Lázare S (2013) Fake empathy and HRI. A preliminary study. J Soc Robot (in press)
22. Rizzolatti G, Craighero L (2004) The mirror-neuron system. Ann Rev Neurosci 27:169–192
23. Ramachandran VS (2004) A brief tour of human consciousness. Pi Press, Pearson Education, New York
24. Decety J, Jackson PL (2004) The functional architecture of human empathy. Behav Cogn Neurosci Rev 3:71–100
25. Decety J, Lamm C (2006) Human empathy through the lens of social neuroscience. Sci World J 6:1146–1163
26. De Vignemont F, Singer T (2006) The empathic brain: how, when and why? Trends Cogn Sci 10(10):435–441
27. Krämer UM et al (2010) Emotional and cognitive aspects of empathy and their relation to social cognition–an fMRI-study. Brain Res 1311:110–120
28. Rosenthal-von der Pütten AM, Schulte FP, Eimler SC, Hoffmann L, Sobieraj S, Maderwald S, Krämer NC, Brand M (2013) Neural correlates of empathy towards robots. In: Proceedings of the 8th ACM/IEEE international conference on human-robot interaction (HRI '13). IEEE Press, Piscataway, NJ, USA, pp 215–216

29. Bartneck C, Nomura T, Kanda T, Suzuki T, Ennsuke K (2005) Cultural differences in attitudes towards robots. In: Proceedings of the AISB symposium on robot companions: hard problems and open challenges in human-robot interaction, Hatfield, pp 1–4
30. Bartneck C (2008) Who like androids more: Japanese or US Americans? In: Proceedings of the 17th IEEE international symposium on robot and human interactive communication, Technische Universität München, Munich, Germany
31. Vallverdú J (2011) The Eastern construction of the artificial mind. Enrahonar 47:171–185
32. Jahns RF (2013) The market for mHealth app services will reach $26 billion by 2017 (3/07/2013). http://www.research2guidance.com/the-market-for-mhealth-app-services-will-reach-26-billion-by-2017/. Accessed 14 Nov 2013
33. Fox S (2011) Peer-to-peer healthcare. Pew internet reports. http://www.pewinternet.org/Reports/2011/P2PHealthcare.aspx. Accessed 14 Nov 2013
34. Lupton D (2013) The commodification of patient opinion: the digital patient experience economy in the age of big data. Mobithinking. http://mobithinking.com/mobile-marketing-tools/latest-mobile-stats/a#subscribers. Accessed 14 Nov 2013
35. Topol EJ (2012) The creative destruction of medicine. Basic Books, New York
36. Wesolowski A, Eagle N, Tatem AJ, Smith DL, Noor AM, Snow RW, Buckee CO (2012) Quantifying the impact of human mobility on malaria. Science 338(6104):267–270
37. Clarysse I, Chellam M, Debrouwer W, Smart A (2013) WIPO patent no. 2013066642. World Intellectual Property Organization, Geneva, Switzerland
38. Naicker S, Plange-Rhule J, Tutt RC, Eastwood JB (2009) Shortage of healthcare workers in developing countries—Africa. Ethn Dis 19(1):60
39. Chutani S, Aalami JR, Badshah A (2010) Technology at the margins: how it meets the needs of emerging markets, vol 22. Wiley, USA
40. Terry M (2011) Telemicroscopes and point-of-care diagnostics team up with smart phones. Telemed J e-health: Off J Am Telemed Assoc 17(5):320–323
41. Oram A (2013) Why privacy should be among the first considerations of a healthcare app developer. http://rockhealth.com/2013/06/why-privacy-should-be-among-the-first-considerations-of-a-health-care-app-developer/. Accessed 15 Nov 2013
42. Waal F, Thompson E (2005) Primates, monks and the mind. J Conscious Stud 12(7):38–54
43. Luneski A, Bamidis PD, Hitoglou-Antoniadou M (2008) Affective computing and medical informatics: state of the art in emotion-aware medical applications. Stud Health Technol Inform 136:517–522
44. Scherer KR (2010) A blueprint for affective computing: a sourcebook. OUP, Oxford
45. Stich SP (1983) From folk psychology to cognitive science: the case against belief. The MIT Press, Cambridge, MA, US
46. Barakat A, Woolrych RD, Sixsmith A, Kearns WD, Kort HSM (2012) eHealth technology competencies for health professionals working in home care to support older adults to age in place: outcomes of a two-day collaborative workshop. Med 2.0 2(2):e10
47. Niedenthal PM (2007) Embodying emotion. Science 316:1002–1005
48. Tosa N (1996) Life-like communication agent-emotion sensing character "MIC" and feeling session character "MUSE". In: Proceedings of the third IEEE international conference on multimedia computing and systems
49. Tosa N (2000) Unconscious flow. Leonardo 33:422–442
50. Friedman B et al (2003) Hardware companions? What online AIBO discussion forums reveal about the human-robotic relationship. Digit Sociability 5(1):273–280
51. Riek LD et al (2009) How anthropomorphism affects empathy toward robots. In: Proceedings of the 4th ACM/IEEE international conference on human robot interaction. ACM, New York, pp 245–246
52. Fink J (2012) Anthropomorphism and human likeness in the design of robots and human-robot interaction. Soc Robot 7621:199–208
53. Glaskin K (2012) Empathy and the robot: a neuroanthropological analysis. Ann Anthropol Pract 36(1):68–87

54. Pereira A et al (2011) Using empathy to improve human-robot relationships. Human-robot personal relationships lecture notes of the institute for computer sciences, social informatics and telecommunications engineering, vol 59. Springer, UK, pp 130–138
55. Tapus A, Mataric MJ (2006) Emulating empathy in socially assistive robotics. AAAI
56. Carpenter J (2013) The quiet professional: an investigation of U.S. military explosive ordnance disposal personnel interactions with everyday field robots. Doctoral Dissertation, University of Washington
57. Rosenthal-von der Pütten AM et al (2013) An experimental study on emotional reactions towards a robot. Int J Soc Robot 5:17–34
58. Kamide H, Eyssel F, Arai T (2013) Psychological anthropomorphism of robots. Soc Robot 8239:199–208
59. Riek LD (2012) Wizard of oz studies in HRI: a systematic review and new reporting guidelines. J Human-Robot Interact 1(1):119–136
60. Becker-Asano C, Kanda T, Ishiguro H (2011) Studying laughter in combination with two humanoid robots. AI Soc 26(3):291–300
61. Ekman P, Levenson RW, Friesen WV (1983) Autonomic nervous system activity distinguishes among emotions. Science 221(4616):1208–1210
62. Forest F, Oehme A, Yaici K, Verchere-Morice C (2006) Psycho-social aspects of context awareness in ambient intelligent mobile systems. In: Proceedings of the IST summit workshop, capturing context and context aware systems and platforms
63. Picard RW (2000) Towards computers that recognize and respond to user emotion. Tech Rep 0018-8670/00 2000 IBM
64. Picard RW (2007) Toward machines with emotional intelligence. In: Matthews G, Zeidner M, Roberts RD (eds) The science of emotional intelligence: knowns and unknowns. Oxford University Press, Oxford, UK
65. Ekman E, Friesen E (1971) Constants across culture in the face and emotion. J Pers Soc Psychol 17:124–129
66. Ekman E (1973) Cross cultural studies of facial expressions. In: Ekman P (ed) Darwin and facial expression: a century or research in review. Academic Press, New York
67. Ekman P (2003) Darwin, deception, and facial expression. Ann NY Acad Sci 1000(1):205–221
68. Norman D (1999) The invisible computer. MIT Press, Cambridge (Mass)
69. Sadri F (2011) Ambient intelligence: a survey. ACM Comput Surv (CSUR) 43(4):36
70. D'Mello S, Jackson T, Craig S, Morgan B, Chipman P, White H, Person N, Kort B, El Kalioby R, Picard R, Graesser A (2008) AutoTutor detects and responds to learners affective and cognitive states. In: Proceedings of the workshop on emotional and cognitive issues at the international conference of intelligent tutoring systems
71. González G, De la Rosa JL, Dugdale J, Pavard B, Eljed M, Pallamín N, Angulo C, Klann M (2006) Towards ambient recommender systems: results of new cross-disciplinary trends. In: Proceedings of ECAI workshop on recommender systems

Epilogue

Simon Peter van Rysewyk and Janneke van Leeuven

Pictures of Mind-Machines

Researcher: Machine ethics concerns whether there ought to be machine decisions and behaviors in the real world, where such decisions and behaviors have real world effects. We researchers think the machines in question will need the capacity for self-direction or autonomy in decision and behavior, just like normal human beings. Do you think it is possible to artistically represent machine self-direction or autonomy? What must be shown in such a representation?

Artist: I hope you don't mind if I ask you a few questions first in response to your introduction. To begin with, what do you consider to be a "normal human being"? I would already find that very challenging to define. Furthermore, for a machine to act like a "normal human being", you would have to create models that could describe not only concepts like self-direction or autonomy in mechanical terms, but every possible psychological process that goes on in a human mind. You asked me whether I think it is possible to artistically represent machine self-direction or autonomy, but I would like to ask you this first "Do you as a scientist think the human mind could be faithfully represented as a machine?"

Researcher: Minds are brains and so part of the natural world. There is robust evidence and argument that the mind relies on the approximate laws of physics, chemistry, and biology, and on mathematics and logic; it is not gifted from some supernatural realm. I think that's news for great optimism and excitement, because

S.P. van Rysewyk (✉) · J. van Leeuven
Graduate Institute of Humanities in Medicine, Taipei Medical University, Taipei, Taiwan

Department of Philosophy, School of Humanities, University of Tasmania, Hobart, Australia
e-mail: vanrysewyk@tmu.edu.tw

J. van Leeuven
e-mail: info@jannekevanleeuwen.com

S.P. van Rysewyk and M. Pontier (eds.), *Machine Medical Ethics*,
Intelligent Systems, Control and Automation: Science and Engineering 74,
DOI 10.1007/978-3-319-08108-3

it implies there's no theoretical reason why Mind cannot be simulated, "faithfully represented", in a machine.

Artist: I agree the human mind is generated by the brain and hence rooted in the natural world and its physical laws. Still, the way the combined activity of all these natural forces creates our sense of personal identity remains a magical phenomenon. On the one hand, human brains process every kind of information in predictable patterns. On the other hand however, the personal experience of and reactions to these processes are completely different for everyone. This trade-off between unique experiences and predictable patterns of brain activity is I think the most fascinating characteristic of the human mind. It's also the point of departure in my artistic practice, in which I create photographic mind models of psychological states. Although everybody has different representations of the world inside their minds, on a more abstracted level they overlap with the internal worlds of other people and it"s these abstracted internal worlds that I try to visualize.

Cable Rooms #1, #4, 2008

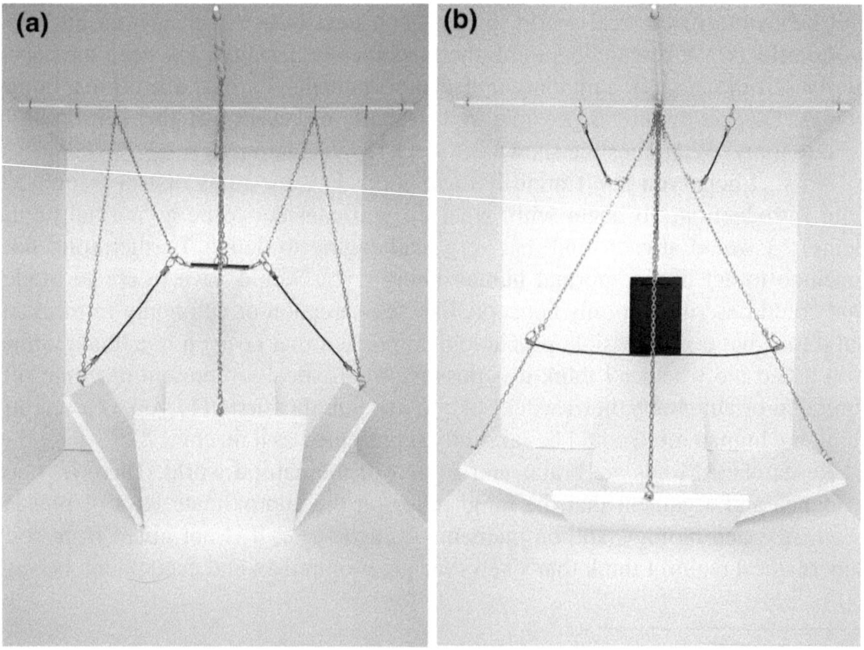

Researcher: Interesting. Can you describe to me how these pieces achieve your stated goals of co-representing individual differences and similarities in mental states? Are the patterns shown significant?

Artist: The series "Cable Rooms" was the first mind model I made and I regard it as the basic model from which I still build on. It was inspired by the statement made by Descartes that a human was an automaton, a type of spring-loaded machine, similar to animals, except in one respect, that it contained a "rational soul" that could initiate volition of the body by its own accord. Especially that last part intrigued me, because how exactly would that "rational soul" initiate the movement of the human automaton if you would look into its inner mechanisms? I decided to explore this question by approaching my own mind as an automaton, using photography as an analogy for internal representations. Could I represent the mechanisms of my own mind in such a way it would relate to other people as well? I felt this would be strongly dependent on the level of contextual grounding the model would have. The more abstracted spaces and objects are from a recognizable time and place, the less personal they become. Photography's ability to distort perception of space and dimensions is also an important aspect of my work. The mind models I create are all build in miniature scale, but by using a wide-angle lens they look like much larger constructions in the photographs. Careful observation of the elements used in the "Cable Rooms" will provide hints on how they could be understood, but there is no right way of "getting it". I aim to expose the mechanisms of my mind through the eyes of the camera, in the hope others can intuitively connect to the images as well. However, it is impossible for me to declare to which degree I succeeded in making these models extend beyond my own mind. That's a question only others can answer.

Researcher: Is it a question that an intelligent machine could answer?

Artist: That's a very interesting thing to ask! I think that depends on how you would want to approach this. If you would want to find out whether my mind models represent some kind of mathematical truth about the functioning of the human brain, I am sure an intelligent machine could test this. This is a completely different question however from asking the machine how it values the photographs as an artistic expression. Even though education and familiarity strongly influence the validation of a work of art, the aesthetic appreciation and emotional evaluation of it is an ultimately intuitive judgment, which I think would be much harder to program.

Golden Rule #4, 2012

Researcher: Humans and machines understand the world differently: the objects and shapes in human art cannot be understood by machines in the form in which they are presented in a gallery exhibition. In the same way, the codes that are understandable to an intelligent machine are merely incomprehensible symbols to human aesthetic understanding. Do you think the gap between human and machine understanding can be somewhat bridged? If not, why?

Imagine taking an image of any canvas painting and reducing it to its fundamental hexadecimal code. If this code is then read by an intelligent machine, it would understand what the code is and thus what the painting represents or symbolizes. Is it possible to use code to create simple geometrical objects, representations of mathematical constants and recreations of masterpiece paintings that *both* humans and machines can understand? This would amount to creating objects in two forms: as the traditional material object (e.g., a cube) and as the hexadecimal and binary codes that symbolize them. Thus, abstraction becomes material, the meanings for humans and machines are somewhat bridged.

Artist: To begin with, I would like to stress that there is a big difference between understanding a work of art as a formal object and appreciation of it on an aesthetic and emotional level. In my opinion, the example you give above only relates to the understanding of a work of art as a formal object. More importantly even, the reduction of a masterpiece into hexadecimal codes, which is then turned into both geometrical forms and binary code is a very subjective intervention in itself; an artistic interpretation of a masterpiece rather than a truthful representation. The masterpiece could have just as well been left out to create an artwork as such and in my opinion this approach would still give no meaningful insight into the aesthetic experience of a work of art.

It would be more helpful if the intelligent machine would be programmed with a set of aesthetic values, measured against its personal beliefs and a database of historical and contemporary art knowledge. It would be interesting to see how the intelligent machine would integrate and draw conclusions on these often contradicting sources of input. The series "Golden Rules" I made in 2012 alludes to the ambiguity of knowledge and rules when it comes to understanding art. It is a reference to the Golden Ratio, a mathematical proportion, which is regarded as the key to understanding beauty in nature. In my "Golden Rules" images, however, there is no such immediate understanding, since it is unclear what the golden instruments are supposed to do. In this way I wanted to emphasize that not everything can be determined and that a certain level of uncertainty is always present and actually is the essence of beauty.

Golden Rule #2 & #6, 2012

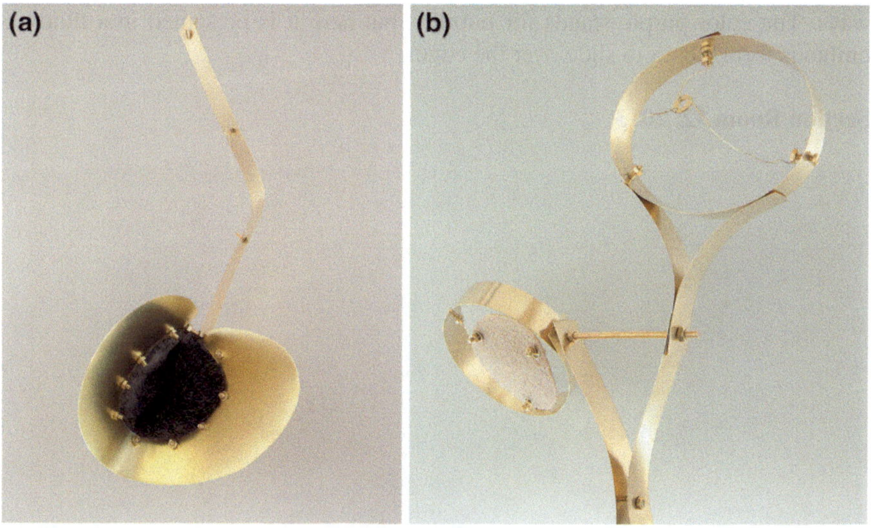

Researcher: I think your insight about uncertainty and human creation can be generalized: uncertainty is always present when human beings create something. Is that true? Consider human engineers creating an intelligent machine. The problem here is uncertainty in the form of unforeseen side effects: when two or more systems, well-engineered in isolation, are put into a super system, this can produce interactions that were not only not part of the intended design, but also harmful; the activity of one system inadvertently harms the activity of the other. A way to help avert this particular problem is to design the subsystems to have fairly impenetrable boundaries that coincide with the epistemic boundaries of their human designers.

That last thought makes me think: successful human-machine interactions may well require that the thought boundaries of machines should be fairly close or even coincident in some cases to the thought boundaries of their human creators. If we are going to trust machines in such interactions, to see in them an authentic interaction partner, they will have to incorporate our thought and value systems, including ethics.

Artist: I think that when intelligent machines would be capable of this deeper level of understanding, it would be because they would have the ability to abstract patterns from specific situations. This would also allow them to reflect on their own processes, form a Theory of Mind of other intelligent beings similar to them and operate independently. As long as they would use this "mind reading" with a morally sound intention it would be highly beneficial to humans to co-operate with these machines as it would expand both our physical and mental capacities. However, if they lacked moral awareness, humans would be facing intelligent machines with an antisocial "mindset" and it is not difficult to imagine how that could have very harmful consequences. The series "Section Rooms" is a representation of that possibly harmful force in a single mind. This rather cold mental space is translucent, yet not fully transparent and could fold out in many possible ways. The color purple stands for nobility, but here it is presented in a fluid and ominous form, about to slide over the edge.

Section Room #2, 2013

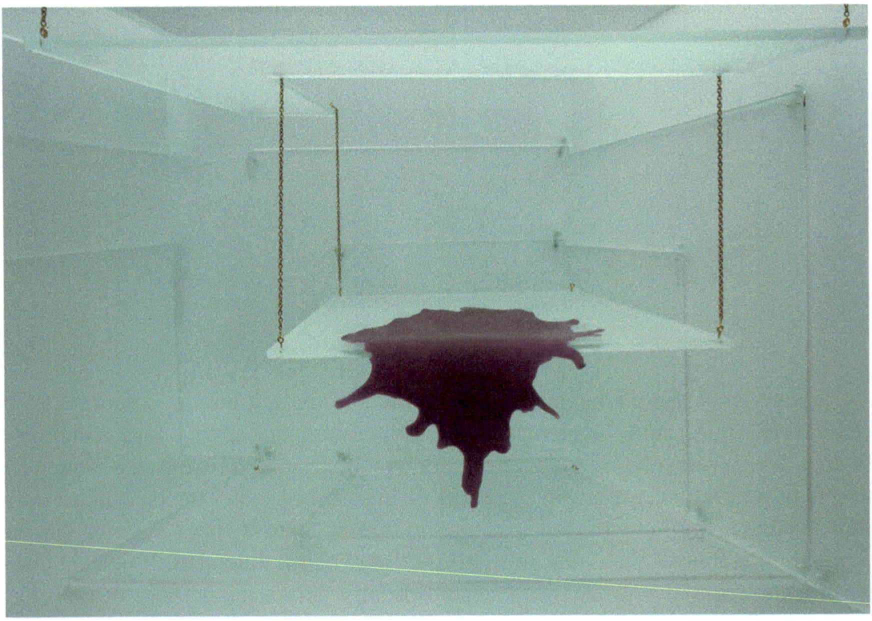

Researcher: Vivid. Here is another representation of the same point: the boundaries of some or all "section rooms" in machines and humans should coincide. That might help to regulate human-machine interactions. This is essential if humans are going to trust machines, since trust relies on the feeling that those you are interacting with share your basic ethics and concerns; or minimally, will act within the constraints prompted by those ethics and concerns.

Touch Screen #1, 2013

Artist: I agree! We find ourselves in a very interesting moment in history, where we as human beings are merging more and more with technology that is becoming increasingly intelligent and independent. It is our responsibility to do all we can to ensure that these technologies will behave morally once they are beyond our control. This also means we have to be keenly aware that our personal desires and impulses might be harmful to others, which many of us so easily forget when we're engaging with the internet for instance. The series "Touch Screens" is perhaps a good work to conclude with in that respect. This work is a reference to the lure and seduction of the virtual worlds behind the smooth surfaces of the many computer screens we engage with everyday. Yet no matter how convincing the illusions are that these virtual worlds can create, I hope there will never come a time this will completely substitute genuine human interaction.

Researcher: Vivid. Here is another representation of the [illegible]

[illegible]

[illegible]

MIX
Papier aus verantwortungsvollen Quellen
Paper from responsible sources
FSC® C105338

If you have any concerns about our products,
you can contact us on
ProductSafety@springernature.com

In case Publisher is established outside the EU,
the EU authorized representative is:
Springer Nature Customer Service Center GmbH
Europaplatz 3, 69115 Heidelberg, Germany

Printed by Libri Plureos GmbH
in Hamburg, Germany